AF581657

Wearable Sensors in the Evaluation of Gait and Balance in Neurological Disorders

Wearable Sensors in the Evaluation of Gait and Balance in Neurological Disorders

Editors

Antonio Suppa
Fernanda Irrera
Joan Cabestany

MDPI • Basel • Beijing • Wuhan • Barcelona • Belgrade • Manchester • Tokyo • Cluj • Tianjin

Editors

Antonio Suppa
Department of Human
Neurosciences, Sapienza
University of Rome
Italy

Fernanda Irrera
Department of Information
Engineering, Electronics and
Telecommunications, Sapienza
University of Rome
Italy

Joan Cabestany
Technical Research Centre for
Dependency Care and
Autonomous Living (CETpD),
Universitat Politècnica de
Catalunya
Spain

Editorial Office
MDPI
St. Alban-Anlage 66
4052 Basel, Switzerland

This is a reprint of articles from the Special Issue published online in the open access journal *Sensors* (ISSN 1424-8220) (available at: https://www.mdpi.com/journal/sensors/special_issues/wearable_sensors_neurological_disorders).

For citation purposes, cite each article independently as indicated on the article page online and as indicated below:

LastName, A.A.; LastName, B.B.; LastName, C.C. Article Title. *Journal Name* **Year**, *Article Number*, Page Range.

ISBN 978-3-03943-144-1 (Hbk)
ISBN 978-3-03943-145-8 (PDF)

Contents

About the Editors

Antonio Suppa, MD, PhD, is working at the Department of Human Neurosciences, Sapienza University of Rome, Italy. He is currently Delegate of the Italian Society of Clinical Neurophysiology (SINC) and Scientific Advisor of the Sapienza information-based Technology InnovaTion Center for Health (STITCH). His research activity mainly focuses on the pathophysiology of motor symptoms in Parkinson's disease and other movement disorders. His current clinical activity also concerns the objective diagnosis of motor symptoms in patients with movement disorders, by means of advanced wireless and wearable technologies.

Fernanda Irrera joined the Sapienza University of Rome in 1989, where she is now a Full Professor of Electronics and a Scientific Advisor of the Sapienza information-based Technology InnovaTion Center for Health (STITCH). Since 2012 she coordinates the IEEE—Electron Device Society—Italy Chapter. Her main research activities are stand-alone and embedded integrated sensors for health and image applications, reliability in CMOS nanoelectronics, and non-volatile memories.

Joan Cabestany, PhD, Telecommunication Engineer, is currently working at the Electronic Engineering Department, Universitat Politècnica de Catalunya (UPC), Spain. His main activity is related to the research and development of systems for human gait and movement disorder monitoring, mainly related to Parkinson's disease. He is also a co-founder of the Sense4Care Company distributing related medical solutions in the market.

Preface to "Wearable Sensors in the Evaluation of Gait and Balance in Neurological Disorders"

The aging population and the increased prevalence of neurological diseases have raised the issue of gait and balance disorders as a major public concern worldwide. Indeed, gait and balance disorders are responsible for harmful consequences, such as falls, frequently leading to hospitalization and even death. The high healthcare and economic burden of gait and balance disorders on society, therefore, require new diagnostic and therapeutic strategies to promptly address this issue.

Advances in wearable technologies have offered innovative solutions to objectively assess different biological parameters, including motor behaviors, thus providing new opportunities in the management of health-related issues. Accordingly, over recent years, researchers have increasingly devoted greater efforts to assessing gait and balance through wearable sensors in healthy subjects and patients affected by neurological disorders. The use of wearable sensors has multiple appealing prospects, including applications in telemedicine and telerehabilitation in neurological patients with gait and balance disorders.

This book is a printed edition of the Special Issue "Wearable Sensors in the Evaluation of Gait and Balance in Neurological Disorders". It collects sixteen original research articles that provide the most up-to-date information about the objective evaluation of gait and balance disorders by means of wearable sensors in patients with various neurological diseases, such as Parkinson's disease, multiple sclerosis, stroke, traumatic brain injury, and cerebellar ataxia. Overall, this book offers a detailed overview of the most recent achievements in the field and encourages the development of new wearable solutions to address gait and balance disorders in patients with neurological diseases.

Antonio Suppa, Fernanda Irrera, Joan Cabestany
Editors

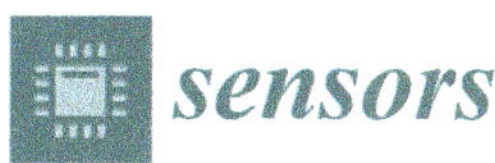

Article

Indoor Trajectory Reconstruction of Walking, Jogging, and Running Activities Based on a Foot-Mounted Inertial Pedestrian Dead-Reckoning System

Jesus D. Ceron [1], Christine F. Martindale [2], Diego M. López [1,*], Felix Kluge [2] and Bjoern M. Eskofier [2,*]

[1] Telematics Engineering Research Group, Telematics Department, Universidad Del Cauca (Unicauca), Popayán 190002, Colombia; jesusceron@unicauca.edu.co

[2] Machine Learning and Data Analytics Lab, Computer Science Department, Friedrich-Alexander University Erlangen-Nürnberg (FAU), 91052 Erlangen, Germany; christine.f.martindale@fau.de (C.F.M.); felix.kluge@fau.de (F.K.)

* Correspondence: dmlopez@unicauca.edu.co (D.M.L.); bjoern.eskofier@fau.de (B.M.E.)

Received: 27 September 2019; Accepted: 1 November 2019; Published: 24 January 2020

Abstract: The evaluation of trajectory reconstruction of the human body obtained by foot-mounted Inertial Pedestrian Dead-Reckoning (IPDR) methods has usually been carried out in controlled environments, with very few participants and limited to walking. In this study, a pipeline for trajectory reconstruction using a foot-mounted IPDR system is proposed and evaluated in two large datasets containing activities that involve walking, jogging, and running, as well as movements such as side and backward strides, sitting, and standing. First, stride segmentation is addressed using a multi-subsequence Dynamic Time Warping method. Then, detection of Toe-Off and Mid-Stance is performed by using two new algorithms. Finally, stride length and orientation estimation are performed using a Zero Velocity Update algorithm empowered by a complementary Kalman filter. As a result, the Toe-Off detection algorithm reached an F-score between 90% and 100% for activities that do not involve stopping, and between 71% and 78% otherwise. Resulting return position errors were in the range of 0.5% to 8.8% for non-stopping activities and 8.8% to 27.4% otherwise. The proposed pipeline is able to reconstruct indoor trajectories of people performing activities that involve walking, jogging, running, side and backward walking, sitting, and standing.

Keywords: trajectory reconstruction; stride segmentation; dynamic time warping; pedestrian dead-reckoning

1. Introduction

Indoor positioning systems (IPS) enable the provision of several location-based services such as home monitoring, rehabilitation, navigation for blind and visual impaired people, and finding and rescuing people/firefighters in emergencies. IPSs can be divided into two approaches: infrastructure-based and infrastructure-free [1,2]. Infrastructure-based IPS require the deployment of devices in the indoor environment to calculate the position of the person. Among the technologies used by this type of IPS are Wi-Fi [3], radio frequency identification (RFID) [4], Bluetooth [5], ultra-wide band (UWB) [6], infrared [7], and video cameras [4]. Infrastructure-free IPS do not need the deployment of devices and mainly use dead-reckoning algorithms. Those systems are called inertial pedestrian dead-reckoning (IPDR) because they use body movement information measured by inertial measurement units (IMU) to estimate a person's position changes based on a previously estimated or known position [2]. The sum of these changes of position allows the reconstruction of the person's trajectory [2]. An IMU usually consists of a triaxial accelerometer and gyroscope. Although some IMUs also incorporate a triaxial magnetometer, alterations of the magnetic field indoors make it unreliable for indoor positioning [8].

The advantages of IPDR systems over infrastructure-based systems are generally lower cost, data privacy, and ease of deployment. However, IPDR systems without correction suffer from severe drift, as person displacement is often calculated by integrating acceleration data from the accelerometer twice and integrating the rotational angle from the gyroscope. In consequence, intrinsic errors and IMU noise are raised to the third power, making a person's trajectory reconstruction by direct integration without correction impractical [9–11].

The literature review done in this study is aimed at foot-mounted IMU IPDR systems that only use the accelerometer and/or gyroscope. Foot-mounted IPDRs, together with a zero velocity update (ZUPT) algorithm, have been the most widely and successful method used to mitigate the drift in trajectory reconstruction [9]. We use only the accelerometer and gyroscope because in indoor environments, different sources might produce alterations in the magnetic field that make the magnetometer readings unreliable for trajectory reconstruction [8]. Most of the foot-mounted IPDR systems that only use accelerometer and gyroscope data are based on trajectory reconstruction during normal walking. Natural movements like avoiding obstacles, sitting, swinging legs, stopping, or performing activities like jumping, jogging, or running have rarely been considered [9,10]. In consequence, the literature review is focused on the foot-mounted IPDR systems that have reconstructed the trajectory of walking, jogging, and/or running activities. Thus, only six studies met the inclusion criteria and are part of the literature review. The foot-mounted IPDR systems are usually evaluated in closed-loop trajectories by measuring the return position error (RPE). The RPE indicates the distance between the final position of the person obtained by the system and the actual physical final position of the person at the end of the trial [8].

Threshold-based and machine learning-based foot-mounted IPDR approaches have been proposed to deal with walking and running activities [12–16]. Li et al. [12,13] proposed a threshold-based stance-phase detector that consists of one footstep detector and two zero velocity detectors, one for walking and another for running. The evaluation of the system was done with one pedestrian who followed two closed-loop trajectories while walking and running. For the square-shape path (195.7 m), the RPE was 0.24% for walking and 0.42% for running. For the eight-shape path (292.1 m), the RPE was 0.2% for walking and 1.01% for running. An adaptive zero-velocity detector that selects an optimal threshold for zero-velocity detection depending on the movement (walking or running) of the person was proposed by Wagstaff et al. [15]. This system was evaluated by five people who walked and ran a distance of 130 m in an "L" shaped path. The RPE reported were 1% for walking and 3.24% for running.

Considering that zero-velocity detection using machine learning-based IPDR systems is free of threshold-tuning, Wagstaff et al. proposed a method for zero-velocity detection by using a long short-term memory neural network (LSTM) [16]. Five people walked and ran a 220-m "L" shaped path. The RPE in walking was 0.49% and running 0.93%. Similarly, Ren et al. proposed a zero-velocity detection algorithm based on HMM [14]. The system was evaluated by one person in an oval-shaped sports field of 422 m. The RPE when walking and running was 0.6% and 1.61%, respectively.

The described works have obtained very high precision in the trajectory reconstruction of walking and running. However, the systems were evaluated with very few participants, and the evaluated trajectories involved continuous walking and running activities. Currently, trajectory reconstruction methods in realistic scenarios—with several people, and considering walking, jogging, and running strides—are still missing.

Physical activity classification and gait event detection are key components of the trajectory reconstruction process using IPDR. Machine learning has played an important role in both topics. In [17] it is shown how different machine learning-based algorithms are able to classify different physical activities, including standing, sitting, walking, and running. Gait event detection has been performed by using several machine learning algorithms such as deep learning [18], hidden Markov models (HMM) [19,20] and neural networks [21,22].

The aim of the present work was to propose a pipeline for trajectory reconstruction using a foot-mounted IPDR system able to reconstruct the trajectories of activities that involve walking, jogging, and running strides as well as natural movements like stopping, standing, sitting, and side-walking.

This paper contributes to foot-mounted IPDR systems by (1) comprehensively evaluating the trajectory reconstruction of activities that involve walking, jogging, and running strides including the discrimination of natural activities such as stopping, sitting, and side-walking; and (2) evaluating two algorithms for Toe-off and Mid-Stance detection during walking, jogging, and running strides adapted from the ones proposed by Barth et al. [23].

The proposed pipeline is able to recognize walking, jogging and running strides and detect the Toe-off and Mid-Stance events in each of them. With this information, a foot-mounted IPDR system is able to reconstruct the person's trajectory regardless of their gait speed. This allows the development of new ambient assisted living applications in which indoor tracking is a ground technology as well as the development of new applications for indoor sports.

2. Datasets

2.1. Unicauca Dataset

The objective of the Unicauca dataset was to evaluate the trajectory reconstruction of walking, jogging, and running in similar settings as the state-of-the-art methods, which are usually evaluated in close-loop trajectories and the activities performed by the participants include continuous walking, jogging, or running. This dataset was collected at the University of Cauca, Popayán, Colombia. Ten participants (mean age: 30 ± 3 years) walked, jogged, and ran a closed-loop P-shaped path of approximately 150 m (Figure 1) with an IMU attached to the lateral side of the left shoe with a Velcro strap (Figure 2).

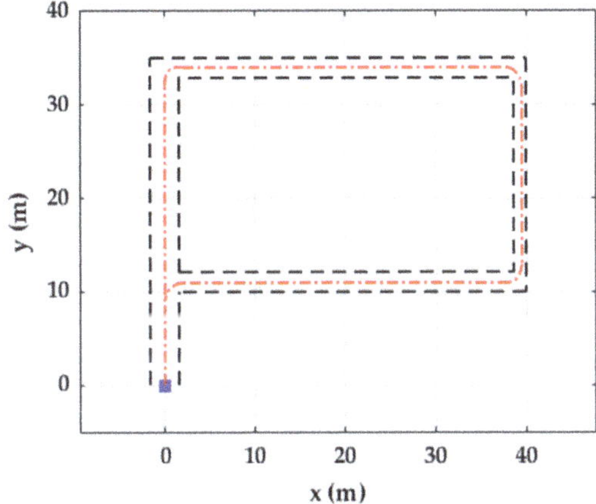

Figure 1. Illustration of the path used for walking, jogging and running in the Unicauca dataset. It is a "P" shaped path. The dotted red line represents the trajectory followed by one person, dotted black lines show outer edges (walls) of the path, and the blue square shows the start and end point of the trajectory.

The IMU was a Shimmer3 GSR+ (Shimmer Sensing, Dublin, Ireland). Acceleration (range: ±16 g) and angular velocity (range: ±2000 dps) data were collected at a frequency of 200 Hz. Accelerometer calibration consisted in leaving the sensor still for a few seconds lying on each of its 6 sides on a flat surface. For gyroscope calibration, the sensor is rotated around the three axes. At the beginning of each trial, the participant was asked to remain standing without moving the IMU for at least 10 s for gyroscope bias calculation.

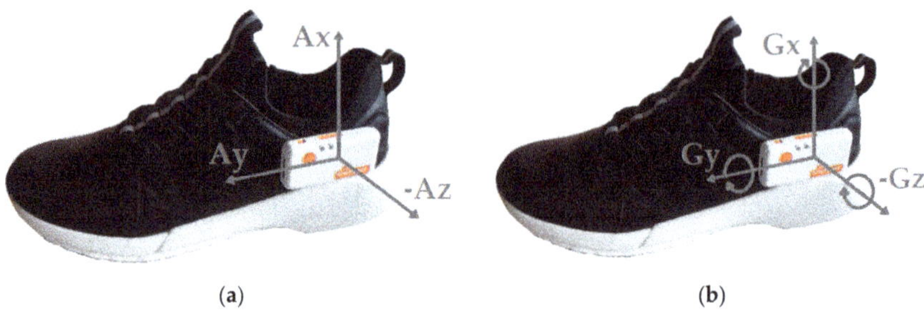

(a) (b)

Figure 2. IMU sensor placement and axis alignment. (**a**) Accelerometer. (**b**) Gyroscope.

2.2. FAU Dataset

The FAU dataset is based on a previous study evaluating a method for smart labeling of cyclic activities [24] and is publicly available at www.activitynet.org. The dataset provides gait data in a relatively natural setting, and its protocol consisted in the execution of 12 different task-driven activities performed in random order for each participant. It includes data from 80 healthy participants with a mean age of 27 ± 6 years. Data were collected from 56 participants at the Friedrich-Alexander University Erlangen-Nürnberg (Germany) and from 24 participants at the University of Ljubljana (Slovenia). In this study, data collected at Slovenia from 20 of the 24 participants (mean age of 28 years) was used as training dataset [25] and data collected in Germany from the 56 participants were used as evaluation dataset. Only the data collected from the IMU worn on the left foot was used for trajectory reconstruction of ten activities (Table 1). Sensor placement and axis alignment are the same used in the Unicauca dataset (Figure 2). The acceleration (range: ±8 g) and angular velocity (range: ±2000 dps) were collected at a frequency of 200 Hz. The on-ground and off-ground phases of each stride are labeled. The accelerometer was calibrated using six static positions and the gyroscope was calibrated using a complete rotation about each of the three axes. Data were acquired in an indoor environment which including chairs and tables (Figure 3). Jogging was described to the participants as "if one would jog for exercise in the evening" and running as "if one is late for a bus". These instructions were the same used in the Unicauca dataset.

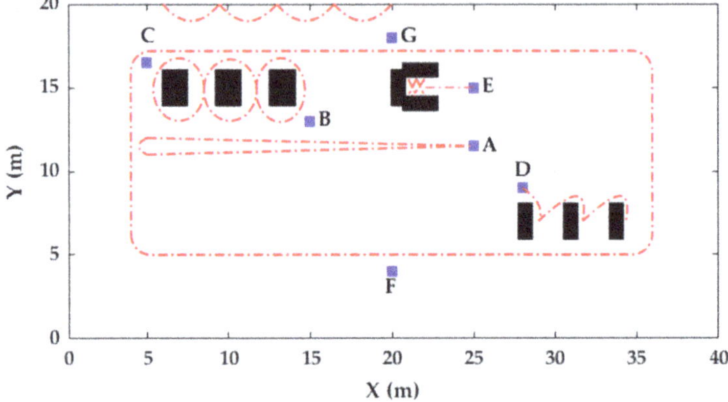

Figure 3. Map of the indoor environment used for collecting the FAU dataset. Blue squares represent chairs that denote start/end positions of activities. Black rectangles represent tables, and dotted red lines represent the possible trajectories followed by participants in each activity.

Table 1. Activity descriptions and abbreviations, shown with their relevant start and end points as labeled in Figure 3 as well as approximated distances.

Activity	Description	Start/End Position	Approximated Distance (m)
W-Slalom	Walk slalom through 3 tables	B→B	31
W-Posters	Sign name on 5 posters on the wall	C→G	21
W-Tables	Perform task at 3 different tables while sitting	D→D	20
W-Cards	Perform task on a table while standing	E→E	6
W, J, R-20	Walk, jog, run 2 times 20 m	A→A	40
W, J, R-Circuit	Walk, jog and run half a circuit each	F, G→G,F	43

3. Methods

A trajectory reconstruction pipeline was carried out separately for each activity of both datasets (Figure 4). This pipeline is based on previous work by Hannink et al. [26]. A type of activity classification step was included. Toe-Off and Mid-Stance algorithms were modified in order to deal with non-walking strides as well as a complementary filter added for stride length and orientation estimation.

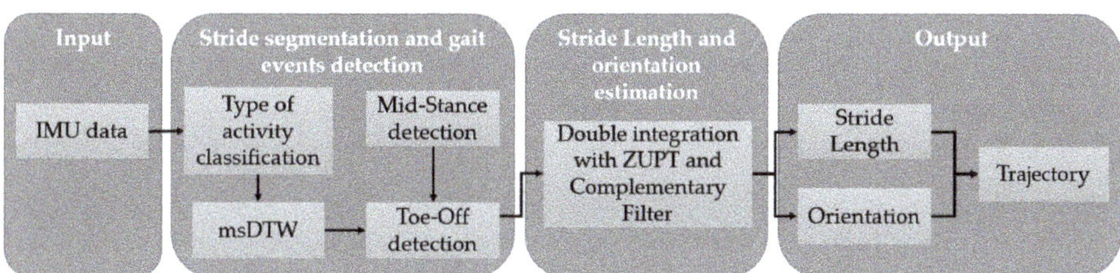

Figure 4. Pipeline for trajectory reconstruction for each activity.

3.1. Stride Segmentation

As shown by Zrenner et al., a threshold-based stride segmentation and a double integration with the ZUPT algorithm performed better than other approaches based on stride time, foot acceleration, and deep learning for calculating stride length in running using a foot-mounted IMU [27]. Thus, multi-dimensional subsequence dynamic time warping (msDTW) and a double integration with ZUPT were used as the stride segmentation and stride length and orientation estimation methods, respectively, in this study [23].

msDTW is used to find a subsequence of continuous signal sequences similar to a given reference pattern. In the context of stride segmentation, that pattern consists of a template of one stride. The stride start was set to the negative peak before the swing phase and stride end to the negative peak at the end of the stance phase (Figure 5a), according to the definition of stride given in [20]. Using that template, msDTW looks for similarities in a movement sequence. msDTW has been shown to be a robust method to segment strides from healthy, geriatric, and Parkinson's patients using foot-mounted IMUs [28].

3.1.1. Template Generation

A MatLab script was developed for template generation. It included two steps: interpolation and averaging. Interpolation consisted of taking each stride and interpolating it to a fixed duration of 200 samples. After interpolation, the template was obtained by averaging, sample by sample, all the strides. The templates for walking, jogging, and running were built using the 8724, 1688, and 1360 walking, jogging, and running strides, respectively, of the training dataset. Unlike other studies, which used only straight strides for building templates [23,28,29], the three templates were built with all the strides of the activities. Thus, both straight and non-straight strides were included in the templates.

The swing-phase starts when the foot leaves the ground (Toe-Off) and ends when the heel strikes the ground (Heel Strike). The portion of the gyroscope z-signal after Heel Strike (HS) describes the stance-phase. A Mid-Stance (MS) event is defined as the part of the stance-phase when the signal energy is zero [30].

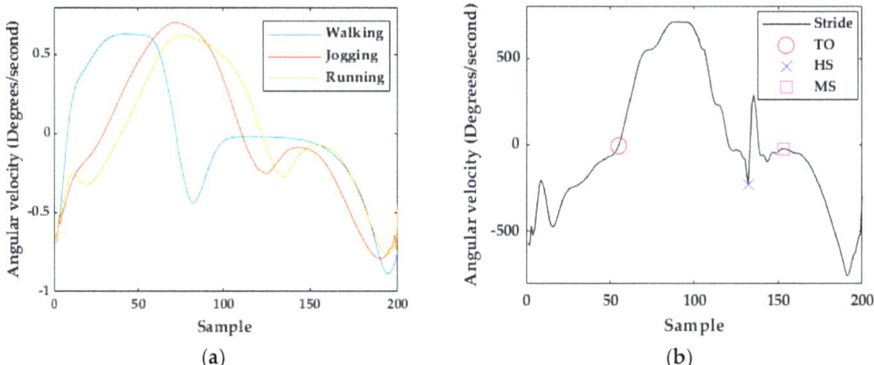

Figure 5. (**a**) Walking, jogging, and running templates (gyroscope z-axis). (**b**) Running stride example (gyroscope z-axis).

3.1.2. Classification of Walking, Jogging, and Running Activities

In order to automatically select the walking, jogging, or running template that will be used in the stride segmentation process, the machine learning algorithms included in the Matlab Classification Learner app were trained using the activities of the training dataset. A window size of 200 samples (1 s of data) and an overlap of 100 samples were used for feature extraction. The features extracted were velocity (by integrating accelerometer readings), angular velocity (by integrating gyroscope readings) and energy of accelerometer and gyroscope axes. The most frequent value in the result was chosen as the final classification. The evaluation was performed using ten-fold cross-validation. As a result, the highest accuracy (98.1%) was achieved by the SVM classifier with a polynomial kernel function of third-order.

3.1.3. Multi-Subsequence Dynamic Time Warping Implementation

The output of the stride segmentation based on msDTW is a set of segments [31]. Each segment describes a possible stride. One issue using these resulting segments for trajectory reconstruction is that often the end of a segment does not coincide with the start of the next segment even for consecutive strides (Figure 6a). The solution to this issue is based on the Toe-Off (TO) detection, which is described in the next section. Using the templates (Figure 5a), the first event detected in each stride is TO. For this reason, TO was defined as the beginning of a stride. For consecutive strides, the end of the stride corresponds with the beginning of the next stride (next TO), resulting in a stride segmentation without "holes" (Figure 6b).

The precision and sensitivity of the stride segmentation using msDTW can be tuned using a threshold. The threshold needed to detect a stride indicates the similarity between that stride and the template used, that is, a large threshold indicates a large difference between the template and the segmented stride [23]. Therefore, with a very small threshold, the number of false negatives strides would increase, and a very large threshold would generate false positives strides. Thresholds from 0 to 100 in steps of 5 were tested on the training dataset. As a result, it was found that a fixed threshold of 65 maximizes the F-score of the stride segmentation in walking, jogging, and running activities (Figure 7).

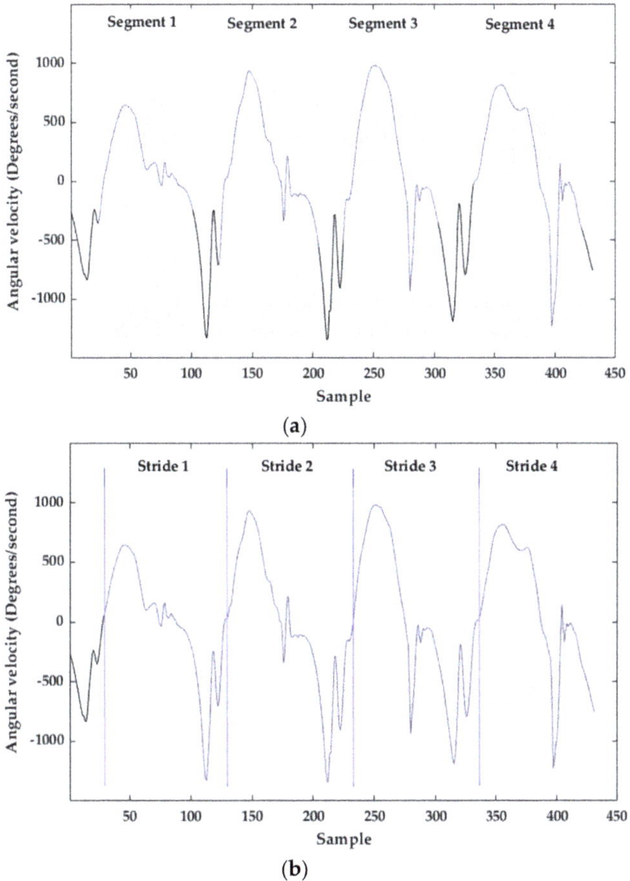

Figure 6. (**a**) Result of stride segmentation with msDTW. (**b**) Final stride segmentation with TO detection. Blue vertical lines depict TOs. Light blue rectangles are segmented strides.

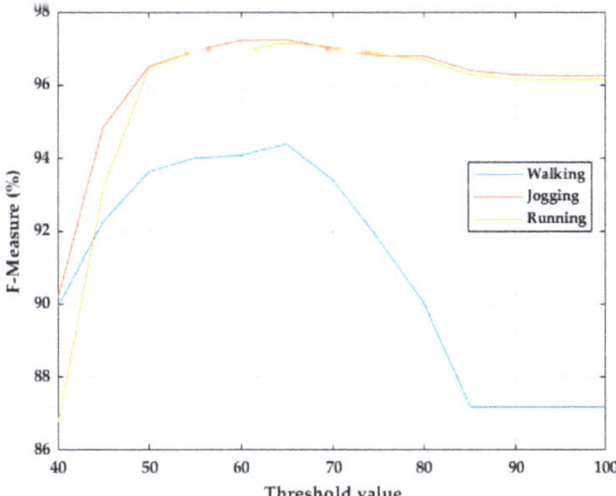

Figure 7. Threshold choice for stride segmentation of walking, jogging, and running strides using msDTW.

3.2. Toe-Off and Mid-Stance Detection

The previous algorithms for TO and MS detection [31] were modified in order to improve detection accuracy in jogging and running. These modifications are described in this section. Both previous and proposed algorithms use the signal of the gyroscope *z*-axis for TO and MS detection.

3.2.1. To Detection

At TO, the gyroscope *z*-axis describes a zero-crossing because of the ankle joint changes from plantar flexion to a dorsal extension position in the sagittal plane [23]. The algorithm included in [31] for TO detection consists of detecting the first zero-crossing in the gyroscope *z*-axis. Due to the abrupt movements in jogging and running strides, in a few cases, a peak located at the beginning of the stride causes a zero crossing. This would lead to a wrong TO detection (red circle in Figure 8). Consequently, the adapted algorithm for TO detection (Algorithm 1) find the maximum peak of the signal and then find the nearest zero crossing before it (blue circle in Figure 8). After the detection of all the TOs that belong to the activity, all the portions corresponding from TO to TO are considered as strides (Figure 6b). Considering that the stride time of walking strides is around one second [24], if one TO to TO portion is greater than 2 s (400 samples), only the signal until 1.5 s was taken into account. This often happens because the participant is standing still or sitting.

Algorithm 1: Toe-off (TO) detection algorithm.

```
1: xMP ← getMaximumPeak(stride)
2: xZC ← getZeroCrossings(stride(1 : xMP))
3: TO ← getNearestZCtoMP(xZC, xMP)
```

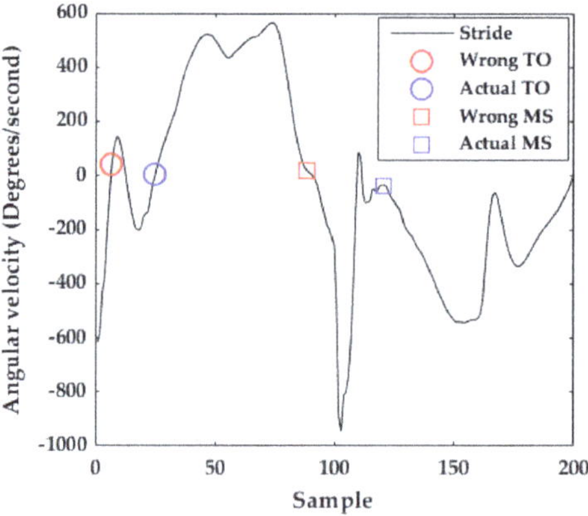

Figure 8. Example of TO and MS detection. The red circle and square show a wrong TO and MS detection, respectively, using the previous TO detection algorithm. The blue circle and square show an adequate TO and MS detection, respectively, using the proposed algorithms.

3.2.2. Mid-Stance Detection

At Mid-Stance (MS) we define that the foot is entirely stationary on the ground [23,28] and its velocity is zero. The gyroscope *z*-signal is minimal at that moment. As the speed of movement increases from walking to running, the stance-phase time decreases (Figure 5a) making MS detection more difficult [10]. The previous algorithm for MS detection in walking strides consists of calculating the middle of the window with the lowest energy in the full stride's gyroscope *z*-signal [23,28,31]. For jogging and running strides, the MS is often confused with other parts of the signal like the valley just before the HS or the peak before the next TO (red square in Figure 8).

The adaptation of the MS detection algorithm (Algorithm 2) consisted of (1) taking only the stride portion from HS to 80% of the stride—this portion was chosen taking into account that the stance-phase of walking strides is approximately the last 60% of the stride and for jogging and running strides it is approximately the last 40% of the stride [25]; (2) calculating the middle of the window with the lowest energy within that portion—to this end, a window size of 20 samples (100 ms) and a window overlap of 10 samples (Blue square in Figure 8) are used.

Algorithm 2: Mid-Stance (MS) detection algorithm.

```
windowSize ← 20
overlap ← 10
stride ← interpolateStrideTo200Samples(stride)
xMP ← getMaximumPeak(stride)
stride ← stride(xMP : 160)
xHS ← getMinimumPeak(stride)
stride ← stride(xHS : end)
MS getMinimumEnergy(stride, windowSize, overlap)
```

3.2.3. Stride Length and Orientation Estimation

The biggest challenge to adequately estimate stride length using IMU data is the significant bias derived from the use of IMUs, which leads to large drifts after the double-integration process. For that reason, the ZUPT method was used. Zero-velocity detection was done by evaluating a threshold on the magnitude of the gyroscope rate of turn of each measurement. If the measurement is less than a threshold of 0.6 dps, that measurement is considered as a zero-velocity measurement. It has been proved that this simple approach works properly in walking strides [11,30]. However, this approach does not work correctly in jogging and running strides due to the abrupt signal variations. The solution to this problem is the use of the MS detected previously. Taking into account that the average stance-phase time in running strides is around 100 ms (20 samples), it was empirically found that taking 5 samples to each side of the MS (which corresponds to 50 ms with the sampling frequency used) leads to better zero-velocity detection in jogging and running strides.

After zero-velocity detection, a complementary Kalman filter (CF) was used in order to model the error in velocity and position estimates using the ZUPTs as measurements (see Appendix A for details). When zero-velocity is detected, but the estimated velocity is different to zero, the CF adjusts the velocity and the corresponding displacement. The CF used in this work is based on the proposed work by Fischer et al. [11]. Three main parameters have to be set up for CF initialization: accelerometer and gyroscope noise (σ_a and σ_w) and the ZUPT detection noise (σ_v). Accelerometer and gyroscope noise were set to equal value in both datasets ($\sigma_a = 0.01\ m/s^2$ and $\sigma_w = 0.01\ rad/s$). ZUPT detection noise depends on the velocity of the participant. That parameter was established by evaluating from $\sigma_v = 0.001\ m/s$ to $\sigma_v = 0.05\ m/s$ in steps of 0.001 m/s for each trajectory performed. The σ_v chosen was the one that produced the least error in the final distance evaluated. The stride length and orientation estimation are obtained using the position increments in each MS event. Stride length, where ∇P_k is the position increment from stride k-1 to stride k, is calculated as follows:

$$SL_k = \sqrt{\nabla P_k(x)^2 + \nabla P_k(y)^2}, \tag{1}$$

4. Results

4.1. Unicauca Dataset

4.1.1. Classification of the Type of Activity

The accuracy in the activity classification was 90%. There were only three misclassifications: two running activities were classified as jogging activities and one jogging activity was classified as a running activity (Figure 9).

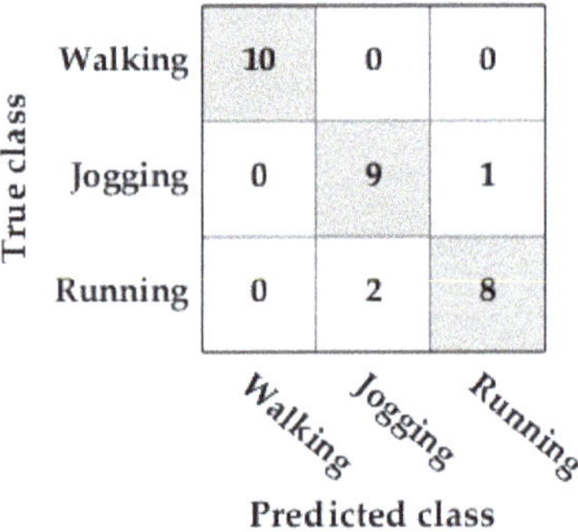

Figure 9. Confusion matrix of the classification of the type of activity in the Unicauca dataset.

4.1.2. Toe-Off and Mid-Stance Detection

In this dataset, TO and MS were manually labeled. A TO/MS is considered as a true positive (TP) if it is located within 15% of the total number of samples of the stride to the right and left of the TO/MS ground truth. A false positive (FP) occurs when a TO/MS is detected outside this range. A false negative (FN) indicates that a TO/MS for a stride was not detected. Having in mind that 40% and 60% of the stride corresponds to the stance-phase of walking and running strides, respectively [25], the TO detection performance was evaluated in the training dataset using error ranges from 5% to 21% of the total stride in steps of 3% (Figure 10). As a result, 15% was chosen as an acceptable error range for TP calculation.

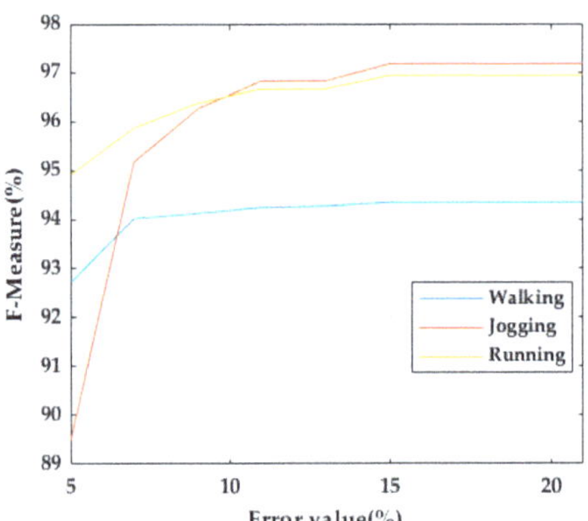

Figure 10. TO performance evaluation using error ranges from 5% to 21% in steps of 3%.

Results of the evaluation of the TO and MS detection using the previous and proposed algorithms are shown in Tables 2 and 3, respectively.

Table 2. Averaged results of TO and MS detection for the 10 participants in the Unicauca dataset using the previous TO and MS detection algorithms.

	Toe-Off					Mid-Stance				
Activity	**TO GT**	**TP**	**FP**	**FN**	**F-Score (%)**	**MS GT**	**TP**	**FP**	**FN**	**F-Score (%)**
Walking	105.5	105.4	0.1	0.1	99.9	104.5	104.4	0.1	0.1	99.9
Jogging	75.4	39.4	37.1	36.2	51.5	74.4	41.2	34.2	33.3	54.9
Running	59.6	21.7	37.5	37.1	36.4	58.6	25.8	31.5	30.7	45.3

TO GT: ground truth TO rate. MS GT: ground truth MS rate. TP: true-positive rate. FP: false-positive rate. FN: false-negative rate.

Table 3. Averaged results of TO and MS detection for the 10 participants in the Unicauca dataset using the proposed TO and MS detection algorithms.

	Toe-Off					Mid-Stance				
Activity	**TO GT**	**TP**	**FP**	**FN**	**F-Score (%)**	**MS GT**	**TP**	**FP**	**FN**	**F-Score (%)**
Walking	105.5	105.5	0	0	100	104.5	104.5	0	0	100
Jogging	75.4	75.2	0.1	0.2	99.8	74.4	74.4	0	0	100
Running	59.6	59.3	0.3	0.2	99.7	58.6	58.5	0.1	0.1	99.8

TO GT: ground truth TO rate. MS GT: ground truth MS rate. TP: true-positive rate. FP: false-positive rate. FN: false-negative rate.

A perfect F-score was obtained for TO and MS detection in walking strides. Very few mistakes occurred for jogging and running, but the F-score remains high.

4.1.3. Trajectory Reconstruction

Two evaluation measures were used. (1) Return position error (RPE): the distance between the coordinates of the actual final point of the activity and the coordinates of the participant's final stride of the corresponding activity. (2) Strides out of trajectory (SOT): All strides of the reconstructed trajectory should be within the boundaries of the corridors represented by black dotted lines (Figure 11). Otherwise, those strides will be counted as out of trajectory.

Higher velocity corresponds to more SOT and RPE. Although, on average, 5.7 % of the strides are out of trajectory in the running trial, the RPE remains less than 1.0% (Table 4). Trajectories of the three trials are mostly within the boundaries (Figure 11).

Table 4. Average results of trajectory reconstruction for each type of activity performed by the 10 participants using the previous and the proposed TO and MS detection algorithms.

Activity	SOT				RPE			
	[31]		New A		[31]		New A	
	#	%	#	%	meters	%	meters	%
Walking	1.7	1.6	1.7	1.6	0.8	0.5	0.8	0.5
Jogging	6.6	8.6	2.9	3.8	2.2	1.4	1.4	0.9
Running	5.3	9.2	3.3	5.7	2.6	1.6	1.4	0.9

SOT: strides out of trajectory, RPE: return position error, [31]: previous TO and MS detection algorithms, New A: proposed TO and MS detection algorithms.

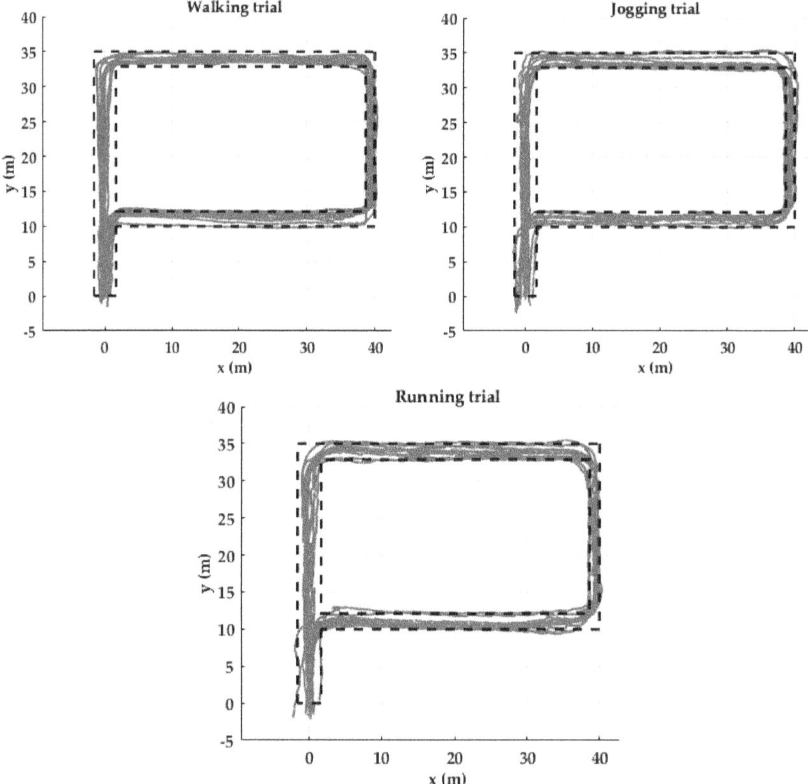

Figure 11. Trajectory reconstruction for the ten participants of the Unicauca dataset in a P shaped path. Black dotted lines show outer edges (walls) of the possible path. Gray lines are the trajectories reconstructed of the ten participants by using the proposed pipeline.

4.2. FAU Dataset

4.2.1. Classification of the Type of Activity

The accuracy obtained by the SVM classifier was 93%. Most of the misclassifications occurred when classifying between running and jogging (Figure 12).

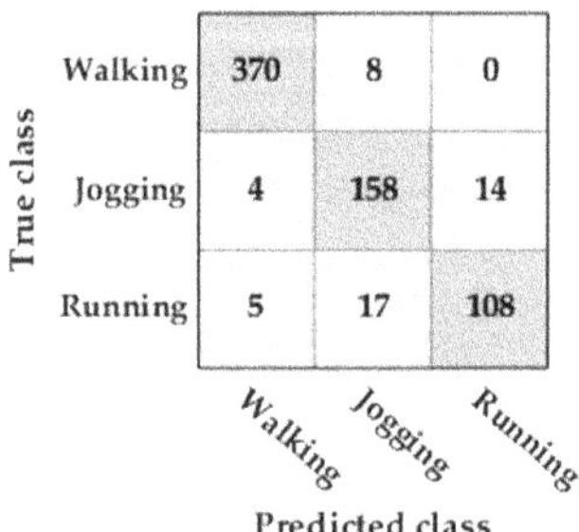

True class \ Predicted class	Walking	Jogging	Running
Walking	370	8	0
Jogging	4	158	14
Running	5	17	108

Figure 12. Confusion matrix of the classification of the type of activity classification in the FAU dataset activities.

4.2.2. Toe-Off Detection

The last sample of the on-ground phase of each stride was used as ground truth for the evaluation of the TO detection algorithm (Table 5). The same criteria used in the Unicauca dataset for TP, FP, and FN calculations were used. The evaluation was carried out on the data collected from the 56 participants at the Friedrich-Alexander University Erlangen-Nürnberg (Germany) of FAU dataset.

Table 5. Average results of TO detection for each type of activity performed by the 56 participants in the FAU dataset using the previous and the proposed TO detection algorithms.

Activity	TO	TP		FP		FN		F-Score (%)	
		[31]	New A.	[31]	New A.	[31]	New A.	[31]	New A.
W-Slalom	21.5	21.2	21.2	0.8	0.8	0.4	0.3	97	97
W-Posters	13.0	10.6	10.6	3.6	3.5	2.3	2.3	77	78
W-Tables	11.9	9.5	9.5	3.7	3.7	2.4	2.4	75	75
W-Cards	4.33	3.7	3.7	1.6	1.6	2.4	0.6	71	71
W-20	28.4	28.2	28.2	1.3	1.0	0.4	0.2	99	98
J-20	22.3	13.7	21.6	9.7	1.1	8.6	0.7	56	96
R-20	18.4	8.1	17.0	12.3	2.1	10.6	1.3	48	90
W-Circuit	28.2	27.6	27.7	0.7	0.7	0.4	0.3	98	98
J-Circuit	21.9	11.8	21.3	10.8	0.7	10	0.5	49	97
R-Circuit	17.7	7.6	17.3	10.7	0.8	10.2	0.4	40	96

TO: toe-off rate, TP: true positives rate, FP: false positives rate, FN: false negatives rate, [31]: previous TO and MS detection algorithms, New A: proposed TO and MS detection algorithms.

4.2.3. Body Trajectory Reconstruction

For RPE estimation in FAU dataset (Table 6), it is important to note that the start/end activity positions were defined by chairs in the indoor environment. For that reason, the actual positions where the participants started and finished the activities were not precisely the same as the chairs' positions since participants began each activity near the corresponding chair and did not necessarily return to the exact point where they started the activity. Based on the videos of the data collection, participants started and finished the activities within a radius of 1.5 m around the chairs. Light blue and gray rectangles in Figures 13 and 14, respectively, indicate the path where all the strides related to a certain activity should take place. If a stride is out of this path, it is considered as a Stride Out of Trajectory (SOT). A SOT can be caused by the accumulative error of stride lengths and angle calculation of previous strides. These zones were defined taking into account the coordinates of the chairs and tables and the boundaries of the indoor environment.

Table 6. Averaged results of trajectory reconstruction of activities performed by the 56 participants in the FAU dataset using the previous and the proposed TO and MS detection algorithms.

Activity	Activity distance	SOT				RPE			
		[31]		New A.		[31]		New A.	
	meters	#	%	#	%	meters	%	meters	%
W-Slalom	31	1.1	5.2	1.1	5.2	1.7	5.5	1.7	5.5
W-Posters	21	1.0	8.0	1.0	8.0	1.9	9.0	1.8	8.8
W-Tables	20	3.1	25.9	3.1	25.9	2.8	14.1	2.8	14.1
W-Cards	6	1.3	30.5	1.3	30.5	1.6	27.4	1.6	27.4
W-20	40	3.7	13.1	3.7	13.1	1.7	4.2	1.7	4.2
J-20	40	7.5	34.2	3.9	17.9	5.5	14.2	2.0	5.1
R-20	40	4.1	22.5	3.1	17.0	5.2	13.9	2.5	6.0
W-Circuit	43	4.1	14.4	4.1	14.4	2.9	6.7	3.0	6.7
J-Circuit	43	6.4	30.2	4.5	20.4	14.5	33.7	3.6	8.8
R-Circuit	43	5.1	29.9	3.9	22.4	16.2	37.7	3.7	8.7

SOT: strides out of trajectory, RPE: return position error, [31]: previous TO and MS detection algorithms, New A: proposed TO and MS detection algorithms.

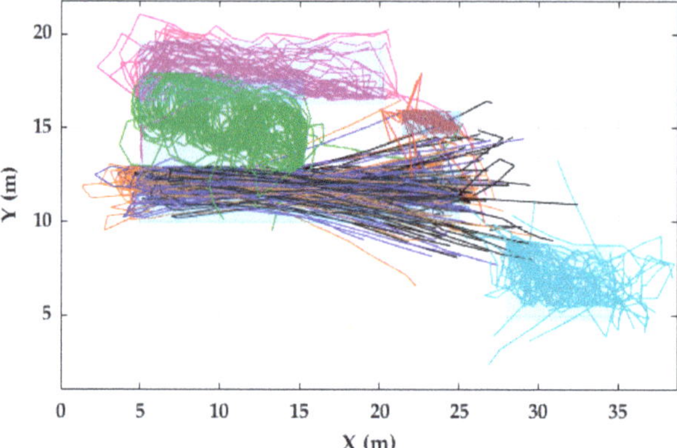

Figure 13. Trajectory reconstruction of non-circuit activities for all 56 participants of the FAU dataset. Black, blue and orange lines denote R-20, J-20, and W-20, respectively. Red, green, violet and light green lines represent W-Cards, W-Slalom, W-Posters, and W-Tables, respectively. Gray rectangles represent zones where all the strides related to certain activity should take place.

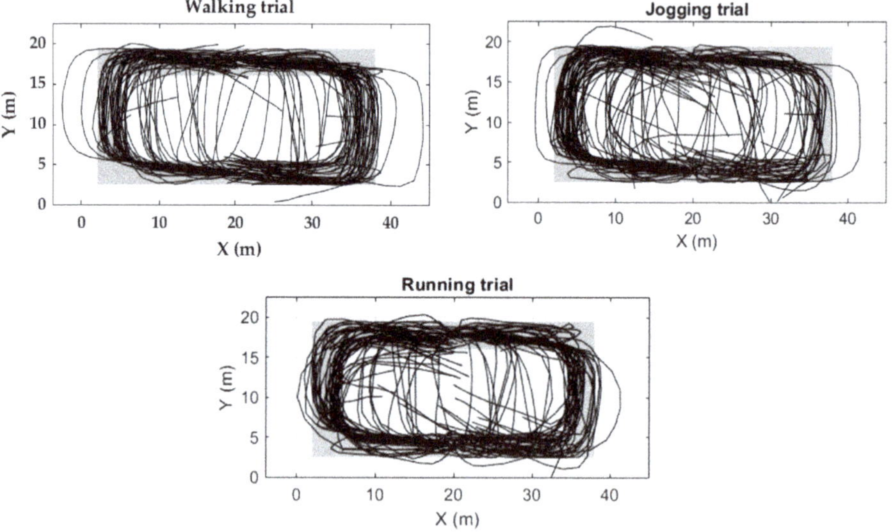

Figure 14. Trajectory reconstruction of circuit activities for all 56 participants of the FAU dataset. Black lines denote the trajectory follows by the participants. Gray zones represent the zone where all the strides should take place.

Most of the trajectories were inside the zones (Figures 13 and 14). The trajectory reconstruction of activities W-20, J-20, and R-20 describes two straight trajectories, joined by a 180-degree turn. The trajectory reconstruction of W-Slalom allows sight of the area where the tables are located. The W-Posters activity includes non-straight strides, which are well described in the trajectory obtained. Regarding the circuit activities, although most of the strides are inside the activity zones, some trajectories lead towards the outer part of the activity zone. Others lead towards the internal part of the circuit (Figure 13).

5. Discussion

We have proposed a pipeline for indoor trajectory reconstruction of walking, jogging, and running activities. The proposed pipeline was evaluated with two datasets. The results showed that it is able to reconstruct a person's trajectory regardless of their gait speed.

5.1. Classification of the Type of Activity

It was found that the classification model obtained with the SVM algorithm is able to classify the three types of activities performed: walking, jogging, and running. The classification between jogging and running is the one in which the classifier made more mistakes. This is possibly due to the jogging and running speeds of some participants being similar. The use of personal models to avoid this problem could be promising.

5.2. TO and MS Detection

Previous studies focused on the reconstruction of the trajectory during walking and running and do not show results of segmentation or detection of strides [18–22]. The two datasets used in this study allow TO evaluation. In the case of MS detection, ground truth information was not available in the FAU dataset. Therefore, it was not possible to evaluate MS detection in that dataset. However, a high F-score was obtained in the detection of MS in the Unicauca dataset.

While the F-score obtained for the proposed TO and MS detection algorithms is similar to that obtained for the previous algorithms for walking activities, the F-score achieved for the proposed TO and MS detection algorithms outperformed that achieved for the previous algorithms for all jogging and running activities. That suggests that the proposed algorithms can detect those gait events in walking, jogging, and running strides. The number of false positives (FP) was always higher than the number of false negatives (FN). This could indicate that the threshold used for stride segmentation with msDTW might have been overestimated, since stride segmentation using a large threshold implies that there is a large difference between the template used and the segmented strides, leading to the detection of FP strides. However, it was checked that by reducing that threshold, the number of FN increased, causing a decrease in the F-score. Threshold-free methods based on machine learning techniques such as those used by Ren [20] and Wagstaff [22] would make the stride segmentation process straightforward by avoiding setting any threshold.

The lowest F-scores are obtained for three walking activities: W-Posters, W-Tables, and W-Cards, which might be due to the fact that those activities involve non-stride movements such as stopping, sitting, lateral and backward steps. This could be because the signal generated for those foot movements is different from the walking/running templates. This could be accounted for by using templates generated by those specific movements, as previously demonstrated in [29], where specific templates were generated for each specific activity such as ascending and descending stairs. Unfortunately, the wide range of possible natural foot movements makes this alternative hard to implement. A hierarchical hidden Markov model (hHMM) approach has proved to be a robust method for stride segmentation of walking activities that include non-stride movements in Parkinson's patients [14] and for stride segmentation of jogging activities [15]. Furthermore, hHMM is a threshold-free approach, therefore it should be explored in order to improve the results obtained for the walking activities that include non-stride movements such as W-Posters, W-Tables, and W-Cards, as well as for stride segmentation of jogging and running activities.

5.3. Trajectory Reconstruction

Usually, the foot-mounted IPDR systems have been evaluated in closed-loop trajectories and by measuring the Return Position Error (RPE) [18–22]. The purpose of the Unicauca dataset was, therefore, to provide a starting point to allow a fair comparison with the state-of-the-art papers.

Sometimes the RPE is small, although the reconstructed trajectory does not fit the actual trajectory performed by the person. That is why we proposed the number of strides out of the trajectory as an additional evaluation metric. The RPEs obtained with the pipeline proposed in this paper for the three trials collected in the Unicauca dataset are less than 1%. The results obtained by the works described in the literature review section are also lower than 1%.

As a result of the better detection of TO and MS obtained by using the algorithms proposed in this study, there is also a better trajectory reconstruction since there were fewer strides out of trajectory (SOT) and shorter RPE for jogging and running activities. This demonstrates two things. The first is the importance of performing a correct detection of TO and MS for trajectory reconstruction. The second is that if the complementary filter does not have precise data to perform the ZVUs, it is not capable of modeling errors in speed on its own, even if its parameters were tuned. It has also been demonstrated that by properly detecting TO and MS, the complementary filter is capable of modeling errors in walking, jogging, and running strides.

RPE obtained for trajectories in the FAU dataset are higher than for the Unicauca dataset. It is important to highlight two limitations that the FAU dataset has for trajectory reconstruction. Firstly, the position of the participants at the beginning and end of the activities is not exactly the same. When analyzing the videos of the FAU dataset collection, it was concluded that these positions vary approximately in a radius of one and a half meters, taking as reference the chairs that indicated the start and end of the activities. Therefore, the RPEs calculated have an error of ±1.5 m. This fact should be taken into account for the preparation of the protocol for the collection of a future dataset. Secondly, it was not possible to subtract the gyroscope bias in all activities performed in the FAU dataset, because the activities were performed continuously. A prerequisite for bias computation is that the person stands still for a few seconds for the calculation of the mean of the gyroscope readings and then subtracting it from the entire movement sequence.

The number of strides out of trajectory is directly related to the RPE obtained; the more strides out of the acceptable path range, the higher the RPE. When observing the trajectory reconstruction of the activities W-20, J-20, R-20, and W-Circuit, J-Circuit, R-Circuit, it appears that the difficulty in trajectory reconstruction increases with stride velocity (from walking to jogging and running). This also occurred in the five papers described in the literature review section [18–22]. In those papers, the evaluation was performed with very few people. From our study, we can confirm that there is still a gap in trajectory reconstruction using foot-mounted IPDR systems of jogging/running activities regarding the trajectory reconstruction of walking activities.

The RPE of the trajectory reconstruction of W-Cards, W-Tables, and W-Posters activities are particularly high, due to the bad detection of TOs. These activities should be treated with special care in future works since they describe movements of daily living activities that happen frequently.

The trajectories obtained have a very well-defined shape and could be used for mapping an indoor environment.

One important recommendation for future work in the field of trajectory reconstruction using IPDR systems is that the datasets collected for evaluation are labeled at activity and stride/step levels, as the FAU dataset used in this paper. Additionally, the participants of the data collection process must start and end precisely at the indicated coordinates.

6. Conclusions

In this paper, we have proposed and evaluated a pipeline for trajectory reconstruction of walking, jogging, and running activities using a foot-mounted inertial pedestrian dead-reckoning system. The dynamic time warping method was adapted within this paper to segment walking, jogging, and running strides. Stride length and orientation estimation were performed using a zero velocity update algorithm adapted for walking, jogging, and running strides and empowered by a complementary Kalman filter.

The presented results showed that the proposed pipeline provides good trajectory estimations during walking, jogging, and running. TO detection algorithm reached an F-score between 92% and 100% for activities that do not involve stopping, and between 67% and 70% otherwise. Resulting return distance errors were in the range of 0.51% to 8.67% for non-stopping activities and 8.79% to 27.36% otherwise.

To the best of the authors' knowledge, this is the most comprehensive evaluation of a foot-mounted IPDR system regarding the type and number of activities and quantity of people included in the datasets and can serve as a baseline for the comparison of future systems. Future work will be focused on using hidden Markov models in order to improve stride segmentation and fusing symbolic location from an RSSI signal to update the indoor localization when possible.

Author Contributions: Conceptualization, J.D.C., C.F.M., D.M.L., F.K. and B.M.E.; Formal analysis, J.D.C. and D.M.L.; Resources, C.F.M., F.K. and B.M.E.; Software, J.D.C.; Writing—original draft, J.D.C. and D.M.L.; Writing—review & editing, C.F.M., F.K. and B.M.E. All authors have read and agreed to the published version of the manuscript.

Funding: This research was funded by Departamento Administrativo de Ciencia, Tecnología e Innovación (COLCIENCIAS), (Call 727-2015).

Acknowledgments: Jesus Ceron gratefully acknowledges the support of the Departamento Administrativo de Ciencia, Tecnología e Innovación (COLCIENCIAS) within the national doctoral grants, call 727-2015. Bjoern Eskofier gratefully acknowledges the support of the German Research Foundation (DFG) within the framework of the Heisenberg professorship programme (grant number ES 434/8-1).

Conflicts of Interest: The authors declare no conflict of interest. The funders had no role in the design of the study; in the collection, analyses, or interpretation of data; in the writing of the manuscript, or in the decision to publish the results.

Appendix A. Complementary Filter

The initialization of the Complementary Filter (CF) implies to establish a series of matrices. First, the state of the CF includes the errors in orientation, position, and velocity. (A1) shows the state in an array representation. Each array element is a 1×3 array containing the errors in the three-axis.

$$E = \left[E_o \; E_p \; E_v\right] \tag{A1}$$

The error covariance matrix accumulates the error in orientation, position, and velocity produced in each sample k:

$$P_k = [0_{9x9}] \tag{A2}$$

The state transition function is a matrix that is multiplied with the previous state to get the next state, as shown in (A7). 'S' is the Skew-symmetric cross-product operator matrix formed from the n-frame accelerations and is the time step equals to 0.005 s, which results from dividing 1 s between the IMU data collection frequency (200Hz).

$$F_k = \begin{matrix} I_{3X3} & 0_{3x3} & 0_{3x3} \\ 0_{3x3} & I_{3X3} & I_{3X3}\Delta t \\ -S\Delta t & 0_{3x3} & I_{3X3} \end{matrix} \tag{A3}$$

The process noise covariance matrix is calculated for each sample by multiplying the accelerometer and gyroscope noise by:

$$Q_k = \left[\left(\sigma_{w_x} \; \sigma_{w_y} \; \sigma_{w_z} \; 0 \; 0 \; 0 \; \sigma_{a_x} \; \sigma_{a_y} \; \sigma_{a_z}\right)\Delta t\right] \tag{A4}$$

The uncertainty in velocity during each ZUPT is represented using the measurement noise covariance matrix (A5). It is a diagonal matrix because no correlation in velocity is supposed to exist between axes.

$$R = \begin{matrix} \sigma_{v_x}^2 & 0 & 0 \\ 0 & \sigma_{v_y}^2 & 0 \\ 0 & 0 & \sigma_{v_z}^2 \end{matrix} \tag{A5}$$

The measurement function matrix is used to move from the state variables space to the measurement variables states. In this implementation, the measurements are the ZUPTs that is when velocity is supposed to be zero. That way, the measurement function has to contain an identity matrix in the position of the velocity error state as follows:

$$H_k = [(0_{3x3}\ 0_{3x3}\ I_{3x3})] \tag{A6}$$

Before running the CF, the gyroscope bias has to be removed. Gyroscope bias is obtained by calculating the mean of the gyroscope readings while IMU is not moving just before the beginning of the activity. The resulting value is subtracted to all gyroscope signals.

After gyroscope bias subtraction, the CF is executed. It has two phases: Prediction and update. In the prediction phase, the error covariance matrix (P_k) is propagated using (A7):

$$P_k = F_k P_{k-1} F_k^T + Q_k \tag{A7}$$

Only when a sample k is a ZUPT, the Update phase comes into play. In this case, the Kalman gain is calculated with (A8), and with that gain, the error is obtained using (A9).

$$K_k = P_k H^T (H P_k H_T + R)^{-1} \tag{A8}$$

$$E = \left[E_o\ E_p\ E_v\right] = K_k V_k \tag{A9}$$

Finally, the velocity and position estimates are corrected as well as P_k:

$$V_k = V_k - E_v \tag{A10}$$

$$Pos_k = Pos_k - E_p \tag{A11}$$

$$P_k = (I_{9x9} - K_k H) P_k \tag{A12}$$

References

1. Zheng, L.; Zhou, W.; Tang, W.; Zheng, X.; Peng, A.; Zheng, H. A 3D indoor positioning system based on low-cost MEMS sensors. *Simul. Model. Pract. Theory* **2016**, *65*, 45–56. [CrossRef]
2. Susanti, R.M.; Adhinugraha, K.M.; Alamri, S.; Barolli, L.; Taniar, D. Indoor Trajectory Reconstruction Using Mobile Devices. In Proceedings of the IEEE 32nd International Conference on Advanced Information Networking and Applications (AINA), Krakow, Poland, 16–18 May 2018.
3. Alarifi, A.; Al-Salman, A.; Alsaleh, M.; Alnafessah, A.; Al-Hadhrami, S.; Al-Ammar, M.A.; Al-Khalifa, H.S. Ultra Wideband Indoor Positioning Technologies: Analysis and Recent Advances. *Sensors* **2016**, *16*, 707. [CrossRef] [PubMed]
4. Leong, C.Y.; Perumal, T.; Peng, K.W.; Yaakob, R. Enabling Indoor Localization with Internet of Things (IoT). In Proceedings of the IEEE 7th Global Conference on Consumer Electronics (GCCE), Nara, Japan, 9–12 October 2018; pp. 571–573.
5. Correa, A.; Barcelo, M.; Morell, A.; Vicario, J.L. A Review of Pedestrian Indoor Positioning Systems for Mass Market Applications. *Sensors* **2017**, *17*, 1927. [CrossRef] [PubMed]
6. Farid, Z.; Nordin, R.; Ismail, M. Recent Advances in Wireless Indoor Localization Techniques and System. *J. Comput. Netw. Commun.* **2013**, *2013*, 185138. [CrossRef]

7. Mainetti, L.; Patrono, L.; Sergi, I. A survey on indoor positioning systems. In Proceedings of the 22nd International Conference on Software, Telecommunications and Computer Networks (SoftCOM), Split, Croatia, 17–19 September 2014; pp. 111–120.
8. Muhammad, M.N.; Salcic, Z.; Wang, K.I.-K. Indoor Pedestrian Tracking Using Consumer-Grade Inertial Sensors with PZTD Heading Correction. *IEEE Sens. J.* **2018**, *18*, 5164–5172. [CrossRef]
9. Harle, R. A Survey of Indoor Inertial Positioning Systems for Pedestrians. *IEEE Commun. Surv. Tutor.* **2013**, *15*, 1281–1293. [CrossRef]
10. Wu, Y.; Zhu, H.B.; Du, Q.X.; Tang, S.M. A Survey of the Research Status of Pedestrian Dead Reckoning Systems Based on Inertial Sensors. *Int. J. Autom. Comput.* **2019**, *16*, 65–83. [CrossRef]
11. Fischer, C.; Sukumar, P.T.; Hazas, M. Tutorial: Implementing a pedestrian tracker using inertial sensors. *IEEE Pervasive Comput.* **2012**, *12*, 17–27. [CrossRef]
12. Li, Y.; Wang, J.J. A robust pedestrian navigation algorithm with low cost IMU. In Proceedings of the International Conference on Indoor Positioning and Indoor Navigation (IPIN), Sydney, Australia, 13–15 November 2012; pp. 1–7.
13. Li, Y.; Wang, J.J. A Pedestrian Navigation System Based on Low Cost IMU. *J. Navig.* **2014**, *67*, 929–949. [CrossRef]
14. Ren, M.; Pan, K.; Liu, Y.; Guo, H.; Zhang, X.; Wang, P. A Novel Pedestrian Navigation Algorithm for a Foot-Mounted Inertial-Sensor-Based System. *Sensors* **2016**, *16*, 139. [CrossRef] [PubMed]
15. Wagstaff, B.; Peretroukhin, V.; Kelly, J. Improving foot-mounted inertial navigation through real-time motion classification. In Proceedings of the International Conference on Indoor Positioning and Indoor Navigation (IPIN), Sapporo, Japan, 18–21 September 2017; pp. 1–8.
16. Wagstaff, B.; Kelly, J. LSTM-Based Zero-Velocity Detection for Robust Inertial Navigation. In Proceedings of the International Conference on Indoor Positioning and Indoor Navigation (IPIN), Nantes, France, 24–27 September 2018; pp. 1–8.
17. Mannini, A.; Sabatini, A.M. Machine Learning Methods for Classifying Human Physical Activity from On-Body Accelerometers. *Sensors* **2010**, *10*, 1154–1175. [CrossRef] [PubMed]
18. Hannink, J.; Kautz, T.; Pasluosta, C.F.; Gasmann, K.G.; Klucken, J.; Eskofier, B.M. Sensor-Based Gait Parameter Extraction with Deep Convolutional Neural Networks. *IEEE J. Biomed. Health Inform.* **2016**, *21*, 85–93. [CrossRef] [PubMed]
19. Ghassemi, N.H.; Hannink, J.; Martindale, C.F.; Gaßner, H.; Muller, M.; Klucken, J.; Eskofier, B.M. Segmentation of Gait Sequences in Sensor-Based Movement Analysis: A Comparison of Methods in Parkinson's Disease. *Sensors* **2018**, *18*, 145. [CrossRef] [PubMed]
20. Mannini, A.; Sabatini, A.M. Gait phase detection and discrimination between walking–jogging activities using hidden Markov models applied to foot motion data from a gyroscope. *Gait Posture* **2012**, *36*, 657–661. [CrossRef] [PubMed]
21. Stetter, B.J.; Ringhof, S.; Krafft, F.C.; Sell, S.; Stein, T. Estimation of Knee Joint Forces in Sport Movements Using Wearable Sensors and Machine Learning. *Sensors* **2019**, *19*, 3690. [CrossRef] [PubMed]
22. Wouda, F.J.; Giuberti, M.; Bellusci, G.; Maartens, E.; Reenalda, J.; Van Beijnum, B.-J.F.; Veltink, P.H. Estimation of Vertical Ground Reaction Forces and Sagittal Knee Kinematics During Running Using Three Inertial Sensors. *Front. Physiol.* **2018**, *9*, 218. [CrossRef] [PubMed]
23. Barth, J.; Oberndorfer, C.; Kugler, P.; Schuldhaus, D.; Winkler, J.; Klucken, J.; Eskofier, B.; Barth, J. Subsequence dynamic time warping as a method for robust step segmentation using gyroscope signals of daily life activities. In Proceedings of the 35th Annual International Conference of the IEEE Engineering in Medicine and Biology Society (EMBC), Osaka, Japan, 3–7 July 2013; Volume 2013, pp. 6744–6747.
24. Martindale, C.F.; Sprager, S.; Eskofier, B.M. Hidden Markov Model-Based Smart Annotation for Benchmark Cyclic Activity Recognition Database Using Wearables. *Sensors* **2019**, *19*, 1820. [CrossRef] [PubMed]
25. Martindale, C.F.; Roth, N.; Hannink, J.; Sprager, S.; Eskofier, B.M. Smart Annotation Tool for Multi-sensor Gait-based Daily Activity Data. In Proceedings of the IEEE International Conference on Pervasive Computing and Communications Workshops (PerCom Workshops), Athens, Greece, 19–23 March 2018; Institute of Electrical and Electronics Engineers (IEEE): Piscataway, NJ, USA, 2018; pp. 549–554.
26. Hannink, J.; Ollenschläger, M.; Kluge, F.; Roth, N.; Klucken, J.; Eskofier, B.M. Benchmarking Foot Trajectory Estimation Methods for Mobile Gait Analysis. *Sensors* **2017**, *17*, 1940. [CrossRef] [PubMed]

27. Zrenner, M.; Gradl, S.; Jensen, U.; Ullrich, M.; Eskofier, B.M. Comparison of Different Algorithms for Calculating Velocity and Stride Length in Running Using Inertial Measurement Units. *Sensors* **2018**, *18*, 4194. [CrossRef] [PubMed]
28. Barth, J.; Oberndorfer, C.; Pasluosta, C.; Schülein, S.; Gassner, H.; Reinfelder, S.; Kugler, P.; Schuldhaus, D.; Winkler, J.; Klucken, J.; et al. Stride Segmentation during Free Walk Movements Using Multi-Dimensional Subsequence Dynamic Time Warping on Inertial Sensor Data. *Sensors* **2015**, *15*, 6419–6440. [CrossRef] [PubMed]
29. Leutheuser, H.; Doelfel, S.; Schuldhaus, D.; Reinfelder, S.; Eskofier, B.M. Performance Comparison of Two Step Segmentation Algorithms Using Different Step Activities. In Proceedings of the 11th International Conference on Wearable and Implantable Body Sensor Networks, Zurich, Switzerland, 16–19 June 2014; pp. 143–148.
30. Skog, I.; Nilsson, J.-O.; Händel, P. Evaluation of zero-velocity detectors for foot-mounted inertial navigation systems. In Proceedings of the International Conference on Indoor Positioning and Indoor Navigation, Zurich, Switzerland, 15–17 September 2010; pp. 1–6.
31. Hannink, J. *Mobile Gait Analysis: From Prototype towards Clinical Grade Wearable*; FAU University Press: Erlangen, Germany, 2019.

Technical Note

The Coefficient of Variation of Step Time Can Overestimate Gait Abnormality: Test-Retest Reliability of Gait-Related Parameters Obtained with a Tri-Axial Accelerometer in Healthy Subjects

Shunrou Fujiwara [1,2,*], Shinpei Sato [1], Atsushi Sugawara [1], Yasumasa Nishikawa [1], Takahiro Koji [1], Yukihide Nishimura [3] and Kuniaki Ogasawara [1]

1 Department of Neurosurgery, Iwate Medical University, 1-1-1 Idaidori, Yahaba, Iwate 028-3695, Japan; ssato.neuro@gmail.com (S.S.); asuga@iwate-med.ac.jp (A.S.); yasnishi777@yahoo.co.jp (Y.N.); thkoji@hotmail.com (T.K.); kuogasa@iwate-med.ac.jp (K.O.)

2 Institute for Open and Transdisciplinary Research Initiative, Osaka University, 3-1 Yamadaoka, Suita, Osaka 565-0871, Japan

3 Department of Rehabilitation Medicine, Iwate Medical University, 1-1-1 Idaidori, Yahaba, Iwate 028-3695, Japan; ynishi@iwate-med.ac.jp

* Correspondence: shunfuji@iwate-med.ac.jp; Tel.: +81-19-651-5111

Received: 25 October 2019; Accepted: 16 January 2020; Published: 21 January 2020

Abstract: The aim of this study was to investigate whether variation in gait-related parameters among healthy participants could help detect gait abnormalities. In total, 36 participants (21 men, 15 women; mean age, 35.7 ± 9.9 years) performed a 10-m walk six times while wearing a tri-axial accelerometer fixed at the L3 level. A second walk was performed ≥1 month after the first (mean interval, 49.6 ± 7.6 days). From each 10-m data set, the following nine gait-related parameters were automatically calculated: assessment time, number of steps, stride time, cadence, ground force reaction, step time, coefficient of variation (CV) of step time, velocity, and step length. Six repeated measurement values were averaged for each gait parameter. In addition, for each gait parameter, the difference between the first and second assessments was statistically examined, and the intraclass correlation coefficient (ICC) was calculated with the level of significance set at $p < 0.05$. Only the CV of step time showed a significant difference between the first and second assessments ($p = 0.0188$). The CV of step time also showed the lowest ICC, at <0.50 (0.425), among all parameters. Test–retest results of gait assessment using a tri-axial accelerometer showed sufficient reproducibility in terms of the clinical evaluation of all parameters except the CV of step time.

Keywords: gait assessment; tri-axial accelerometer; CV; healthy subjects; test-retest

---

1. Introduction

Walking is naturally performed by humans without any deficits, and it is one of the indexes that show normality of motor and/or cognitive function in healthy subjects and abnormality in patients [1–6]. In patients with neurodegenerative diseases such as Parkinson's or Huntington's disease, freezing of gait or reduction of gait performance is often observed [7–9]. For assessing the gait in post-stroke patients during the follow-up period, a special portable stride analyzer was used in a previous work [2]. This device consisted of special insoles with compression sensors in shoes, and the insole was connected to a mobile data collection box worn on the belt. The other study to assess the gait in patients with subcortical capsular encephalopathy also used a similar device [10]. On the other hand, an electronic walkway that detects the spatial and temporal characteristics of footfalls during gait was used in the assessment of walking behavior in multiple sclerosis [11,12]. Then, a camera-based device was also used for gait assessments in patients with spinal cord injury in addition to multiple sclerosis [1,13]. Various techniques for quantitative gait assessment have been proposed in previous works; however, gait performance is difficult to assess, requiring the use of special or large machines and facilities in clinical scene and/or multicenter trials [1,10–15]. In practice, qualitative assessments have often been performed without any devices, but with the use of clinical scores [16,17], and no standard device for quantitative gait assessment has been established.

The portable devices for gait assessment, particularly a tri-axial accelerometer [18–21], that have been developed recently exhibit an accuracy equal to that of large devices such as treadmills in the assessment of gait performance and are used in practical and clinical studies [18–24]. Tri-axial accelerometers have the advantage of easy attachment to the body or foot of subjects with a belt and the estimation of various gait parameters such as movements of the body trunk, step time, and ground reaction force from the acceleration wave dataset in the three axial directions during walking. On the other hand, the reproducibility of the accelerometer in healthy subjects remains unclear. In the present study, we investigated whether gait-related parameters obtained by a tri-axial accelerometer are reliable in terms of reproducibility by performing test-retest gait measurement in healthy subjects.

2. Materials and Methods

2.1. Subjects

All subjects participated in this study between July 2017 and October 2017 after providing written informed consent and primary medical check interviewing history of disorders, age and height by authors (S.S. and Y.N.). The inclusion criteria were as follows: >20 but <60 years of age; no history of brain-related disorders, including surgical operation, irradiation, stroke, infection, remarkable atrophy, or demyelinating disease; no history of hypertension, diabetes mellitus, atrial fibrillation, pulmonary disease, leukoaraiosis, no musculoskeletal deficits or the other diseases showing gait abnormalities without neurological deficits [24–27]. In the first stage, each subject performed a 10-m walk six times with a tri-axial accelerometer (MG-M1110-HW, LSI Medience, Tokyo, Japan) fixed at the L3 level of the subject by a nylon belt (Figure 1).

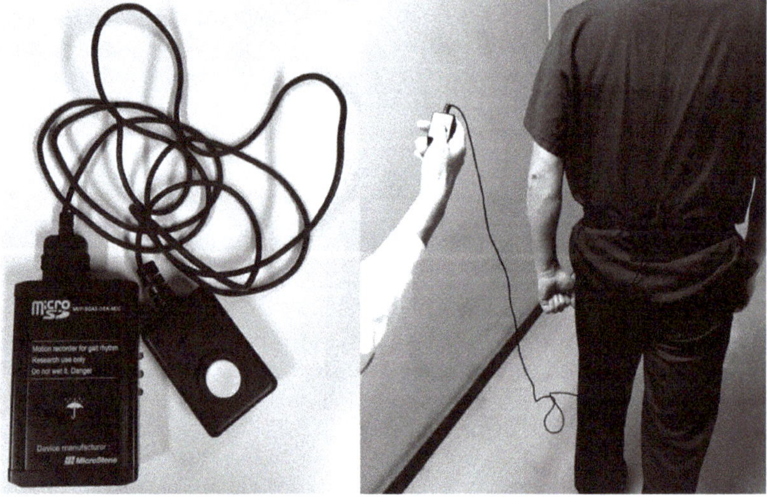

Figure 1. Tri-axial accelerometer (MG-M1110) with a switch cable for marking points of a 10-m walking interval for the dataset (**left**) and the accelerometer fixed at the L3 level by a nylon belt in a subject (**right**).

The device can measure tri-axial (vertical, anteroposterior, and mediolateral) acceleration by detecting limb and trunk movements at a sampling rate of 100 Hz during step-in and kick-off motions. The tests were performed on a 30-m straight walkway in our hospital. All subjects were instructed by an author (S.F.) for walking at each usual pace and they walked 16 m, including 3 m before the starting point and 3 m after the end point, as intervals to obtain the 10-m walking dataset in Table 1. To mark the 10-m segment of the dataset, an operator (S.F.) pushed a button connected to the accelerometer with a cable at both the start and end points while following the subject. The second test was also performed with the same subjects under the same conditions (the tri-axial accelerometer, an operator, and the walkway) within a 3-month period at least 1 month after the first evaluation. The statistical comparison of two datasets, each showing 10% standard deviation relative to each mean value and 10% mean difference to an averaged value of the two mean values, requires the examination of more than 28 subjects with alpha and beta levels less than 0.01 [28], indicating Type I error and Type II error, respectively. Thus, we determined that the present study required more than 28 subjects in order to compare the first and second tests.

2.2. Data Analysis

From each tri-axial acceleration wave dataset of six repetitions of the 10-m walk measurement, the following nine gait parameters were calculated using commercial software (LSI Medience): assessment time (s); number of steps (step); stride time (s; time from initial contact of one foot to subsequent contact of the same foot); cadence (step/m); ground floor reaction ($\times 9.8$ m/s^2); step time (s; time from initial contact of one foot to initial contact of the other foot); coefficient of variation (standard deviation/mean $\times$ 100) of step time (%; CV); velocity (m/min); and step length (cm). To calculate these gait parameters, pre-processing, called "step extraction", which marks each wave indicating a step in the tri-axial acceleration wave dataset (Figure 2), was needed.

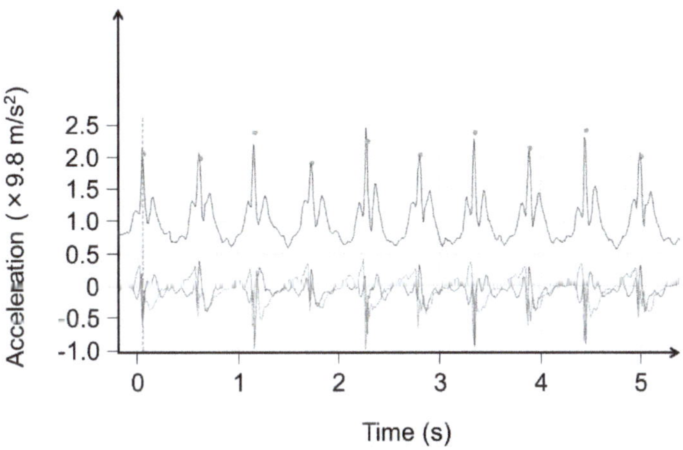

Figure 2. Wave dataset obtained using a tri-axial accelerometer during walking.

By a pre-processing "step extraction", markers were automatically placed on top of each wave, indicating one step.

During pre-processing with the software, the number of steps can be automatically estimated from each 10-m walking wave dataset. If the number of steps was clearly low (e.g., less than half of the number calculated using the other 10-m walking wave datasets in the same subject) by missing the wave peak due to the limitation of the peak detection algorithm, the subject was excluded from the analysis in this study without the manual error correction for the same conditions and protocols for all wave dataset. Finally, the six values from each gait parameter obtained from the six measurements were averaged, and the averaged value was defined as the representative value of the parameter in a subject. The protocol of this study was reviewed and approved by the institutional ethics committee.

2.3. Statistical Analysis

The differences in each gait parameter between the first and second assessments were examined using the Wilcoxon signed-rank test. To validate the reproducibility in two gait assessments with a tri-axial accelerometer, the intraclass correlation coefficient (ICC) was also calculated in each gait parameter. Grading of the ICC was defined as follows: excellent, ICC ≥ 0.9; good, 0.7 ≤ ICC < 0.9; moderate, 0.5 ≤ ICC < 0.7; and poor, ICC < 0.50. Subsequently, a Bland-Altman plot was performed to confirm the tendency of the relationship between the two measurements in each parameter. Furthermore, the correlation between height and each gait parameter at the 1st test was statistically examined with Spearman's correlation coefficient for confirming the effect from the individual factor. All statistical analyses were performed on MedCalc ver. 17.9.7 (MedCalc Software bvba, Ostend, Belgium) with a significance level of $p < 0.05$.

3. Results

Forty-four subjects were included in this study. Two of the 44 subjects could not perform the second test within 3 months. The other 42 subjects performed the 10-m walk at both stages; however, some step waves of the six measurements in six subjects could not be appropriately extracted for each 10-m walk wave, indicating that step extraction errors during pre-processing for gait analysis occurred because of the deterioration of the waveform and the limitation of the wave extraction algorithm. Finally, 36 subjects (Figure 3) (21 men and 15 women; mean age, 35.7 ± 9.9 years; range, 22–58 years) were able to complete both the first and second analyses (mean interval, 49.6 ± 7.6 days; range, 40–65 days).

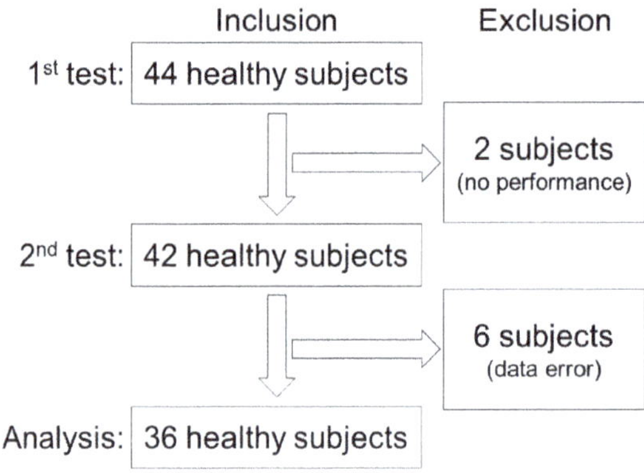

Figure 3. Flowchart for including the subjects.

Only the CV showed a significant difference between the first and second measurements (median CV: first, 2.16; second, 2.50; $p = 0.0188$), while the other parameters showed no significant differences (Table 1). Among all nine gait parameters, stride time (ICC: 0.803), step time (0.788), and cadence (0.784) showed good correlation with a high ICC of ≥0.70. The number of steps (0.685), step length (0.663), ground reaction force (0.615), velocity (0.598), and assessment time (0.565) showed a moderate ICC between 0.50 and 0.70. The ICC of the CV indicated poor correlation and was the lowest value, being <0.50 (0.425) in all the parameters (Table 1). The Bland-Altman plot in the CV showed a negative trend, with the mean of the first and second assessments being larger, while those in the other parameters showed no trends. The plot of the parameters shows good ICCs in Figure 4a–c, and the CV shows poor ICC in Figure 4d.

Table 1. Median and intra-correlation coefficient for each parameter at and between first and second 10-m walks in healthy subjects (n = 36).

	1st	95% CI	2nd	95% CI	*p* Value *	ICC	95% CI
Stride time [sec]	1.02	1.01–1.05	1.03	0.99–1.05	0.689	0.803	0.647–0.894
Cadence [step/min]	119	115–120	117	114–121	0.765	0.784	0.616–0.884
Step time [sec]	0.505	0.500–0.523	0.515	0.500–0.523	0.697	0.788	0.624–0.886
Number of steps [step]	13.8	13.5–14.2	13.9	13.5–14.2	0.765	0.685	0.462–0.827
Step length [cm]	72.3	69.7–73.8	72.0	70.7–73.6	0.981	0.663	0.429–0.813
Ground reaction force [×9.8 m/s^2]	0.360	0.330–0.383	0.355	0.327–0.373	0.980	0.615	0.361–0.784
Velocity [m/min]	85.3	82.1–87.1	84.3	82.3–86.3	0.753	0.598	0.339–0.773
Assessment time [s]	7.04	6.89–7.33	7.13	6.96–7.30	0.831	0.565	0.293–0.752
Coefficient of variance [%]	2.16	1.98–2.57	2.50	2.15–2.95	0.0188	0.425	0.129–0.655

ICC: intra-correlation coefficient; CI: confidential interval. * examined using Wilcoxon signed-rank test.

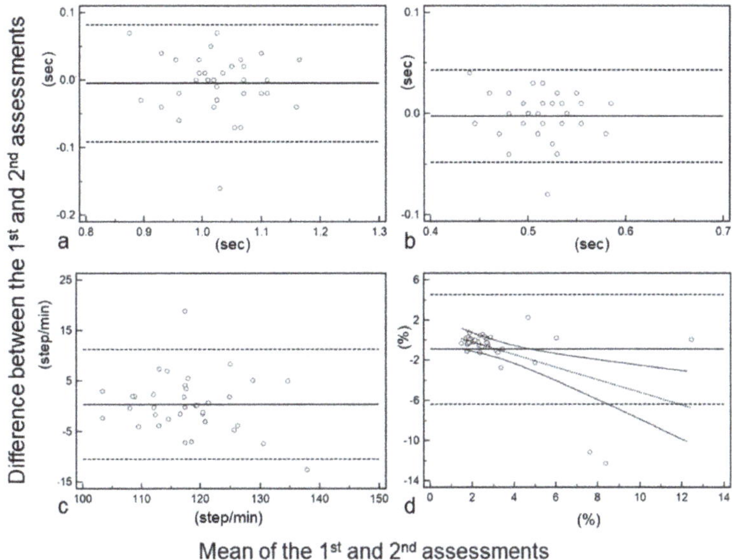

Figure 4. Bland-Altman plots of the gait parameters, showing good intraclass correlation coefficient (ICC; (**a**) stride time, (**b**) step time, (**c**) cadence) and poor ICC ((**d**) coefficient of variation).

Height significantly correlated with number of steps (ρ, *p*-value and the 95% confidential interval: −0.358, 0.0323, −0.614 to −0.0328), stride time (0.468, 0.0040, 0.164 to 0.690), cadence (−0.467, 0.0041, −0.690 to −0.163), step time (0.483, 0.0028, 0.184 to 0.701) or step length (0.356, 0.0331, 0.0311 to 0.613). On the other hand, no significant correlation was observed between height and assessment time (ρ, *p* value and the 95% confidential interval: 0.0312, 0.8567, −0.300 to 0.356), ground force reaction (−0.139, 0.4174, −0.447 to 0.198), CV (0.0464, 0.7883, −0.287 to 0.369), or velocity (−0.0340, 0.8441, −0.359 to 0.298).

4. Discussion

Portable devices for quantitative gait analysis have been developed, and these devices have the advantage in clinical scenarios of needing only a walkway, rather than any large space or facility [18,19,21–23,29–31]. In the present study, we validated the reproducibility of a tri-axial accelerometer used in gait assessment by performing test-retest measurements within a 3-month period 1 month after the first evaluation. All gait parameters except the CV showed adequate reproducibility for practical clinical use, with no significant differences and with practical ICCs between test-retest measurements. The present findings suggest that the tri-axial accelerometer can sufficiently evaluate gait by using just the device and an operator, and without large machines and experts to analyze the dataset.

Three gait parameters (stride time, step time, and cadence) showed good ICC in the present study. In a previous work, the parameters showed no significant differences between controls (patients showing transient ischemic stroke/asymptomatic carotid artery stenosis) and patients without symptoms 1 month after stroke but significantly changed in patients with symptoms after stroke as compared with controls [2]. These three parameters may be more robust than the other gait parameters because they only change in cases that show severe deficits. Furthermore, the device we used may have more sensitivity in such parameters than previous devices; thus, it may also indicate a remarkable robustness in these three parameters.

By contrast, the parameters that show moderate ICC (number of steps, step length, ground reaction force, velocity, and assessment time) may change depending on the condition and/or intention of the subjects participating in the gait measurement. We presume that velocity and assessment time can change more easily than the other parameters because the same pace is difficult for subjects to retain between the first and second measurements, even if an operator carefully performs the measurements with healthy subjects under the same conditions. Thus, when we use these parameters for clinical research, it may be insufficient for identifying the gait abnormalities in the pathological groups to use the standard cutoff values at the 95% confidential intervals ($p = 0.05$) from the healthy groups because the sensitivity to detect the abnormalities may be low with the cutoff values, especially for assessing improvement of ambulation by velocity [32] or assessment time [12,17], as described in a previous study. We have to pay attention to the use of such gait parameters.

A previous report indicated that the CV was significantly larger in patients with Parkinson's or Huntington's disease (HD) than in controls, and that the CV was the best predictor of HD [9]. On the other hand, the averaged value of each gait parameter during gait assessment showed no significant difference between the pathological and control groups in the previous study. The CV showed a significant difference in the present study and a poor ICC compared to that of the healthy subject group. This result indicates that the CV in the previous work may potentially underestimate the gait performance in the pathological group. With the high accuracy of the accelerometer used in the present study, we could identify the significant variation in the CV observed even in the healthy subjects. Therefore, the averaged value of each gait parameter during gait assessment may show the significance in the pathological group if the tri-axial accelerometer that shows high sensitivity is used.

The present study has some limitations. First, the sample size is not so large. Second, the gait dataset may include errors due to a slight slipping of the nylon belt from the L3 level of the subject. Third, the 10-m distance was marked by manually pushing a button connected to the accelerometer with a cable in the present study. Only one operator performed the measurements for gait assessment; therefore, interrater reliability remains unclear. In future studies, a comparison of the tri-axial accelerometer as used in the present study with an accelerometer based on an infrared system (as proposed in a previous study [31]) tell us the difference in the accuracy between manual and automatic procedures.

5. Conclusions

In the test-retest gait assessment using a tri-axial accelerometer, a reproducibility sufficient for clinical research was observed in all parameters except the CV. The present results suggest careful evaluation of the CV because it may potentially overestimate gait disturbance in the pathological group owing to the comparably low reproducibility. Portable accelerometers can assess gait performance noninvasively and with practical accuracy without the need for other huge machines. In future works, the devices may be used for long-term gait assessment over a few days in the elderly and in patients with neurodegenerative and/or spinal disease because subjects only need to attach the device to their back with a belt.

Author Contributions: Conceptualization, S.F. and K.O.; Methodology, S.F., Y.N. (Yasumasa Nishikawa), K.O.; Software, S.F.; Validation, S.F., S.S. and K.O.; Formal Analysis, S.F. and S.S.; Investigation, S.F., S.S. and K.O.; Resources, Y.N. (Yasumasa Nishikawa) and K.O.; Data Curation, S.F., S.S., and K.O.; Writing—Original Draft Preparation, S.F.; Writing—Review & Editing, A.S., T.K., Y.N. (Yukihide Nishimura) and K.O.; Visualization, S.F.; Supervision, Y.N. (Yukihide Nishimura) and K.O.; Project Administration, S.F. and K.O.; Funding Acquisition, A.S., S.F. and K.O. All authors have read and agreed to the published version of the manuscript.

Funding: This research received no external funding.

Acknowledgments: This study was partially supported by Grant-in-Aid for Scientific Research (C) (No. 18K08948, 2018–2021 from the Ministry of Education, Culture, Sports, Science and Technology of Japan.

Conflicts of Interest: The authors declare no conflict of interest.

References

1. Nooijen, C.F.; Ter Hoeve, N.; Field-Fote, E.C. Gait quality is improved by locomotor training in individuals with SCI regardless of training approach. *J. Neuroeng. Rehabil.* **2009**, *6*, 36. [CrossRef] [PubMed]
2. von Schroeder, H.P.; Coutts, R.D.; Lyden, P.D.; Billings, E., Jr.; Nickel, V.L. Gait parameters following stroke: A practical assessment. *J. Rehabil. Res. Dev.* **1995**, *32*, 25–31. [PubMed]
3. Imms, F.J.; Edholm, O.G. The assessment of gait and mobility in the elderly. *Age Ageing* **1979**, *8*, 261–267. [CrossRef] [PubMed]
4. Miyazaki, S.; Iwakura, H. Foot-force measuring device for clinical assessment of pathological gait. *Med. Biol. Eng. Comput.* **1978**, *16*, 429–436. [CrossRef]
5. Grieve, D.W. The assessment of gait. *Physiotherapy* **1969**, *55*, 452–460.
6. Kennedy, P.E. Locomotion and gait: The physiotherapists assessment. *Ir. Nurs. Hosp. World* **1966**, *32*, 11–12.
7. Bohnen, N.I.; Kanel, P.; Zhou, Z.; Koeppe, R.A.; Frey, K.A.; Dauer, W.T.; Albin, R.L.; Muller, M. Cholinergic system changes of falls and freezing of gait in Parkinson disease. *Ann. Neurol.* **2019**. [CrossRef]
8. Bohnen, N.I.; Frey, K.A.; Studenski, S.; Kotagal, V.; Koeppe, R.A.; Constantine, G.M.; Scott, P.J.; Albin, R.L.; Muller, M.L. Extra-nigral pathological conditions are common in Parkinson's disease with freezing of gait: An in vivo positron emission tomography study. *Mov. Disord. Off. J. Mov. Disord. Soc.* **2014**, *29*, 1118–1124. [CrossRef]
9. Hausdorff, J.M.; Cudkowicz, M.E.; Firtion, R.; Wei, J.Y.; Goldberger, A.L. Gait variability and basal ganglia disorders: Stride-to-stride variations of gait cycle timing in Parkinson's disease and Huntington's disease. *Mov. Disord. Off. J. Mov. Disord. Soc.* **1998**, *13*, 428–437. [CrossRef]
10. Bazner, H.; Oster, M.; Daffertshofer, M.; Hennerici, M. Assessment of gait in subcortical vascular encephalopathy by computerized analysis: A cross-sectional and longitudinal study. *J. Neurol.* **2000**, *247*, 841–849. [CrossRef]
11. Wajda, D.A.; Sandroff, B.M.; Pula, J.H.; Motl, R.W.; Sosnoff, J.J. Effects of walking direction and cognitive challenges on gait in persons with multiple sclerosis. *Mult. Scler. Int.* **2013**, *2013*, 859323. [CrossRef] [PubMed]
12. Motl, R.W.; Pilutti, L.; Sandroff, B.M.; Dlugonski, D.; Sosnoff, J.J.; Pula, J.H. Accelerometry as a measure of walking behavior in multiple sclerosis. *Acta Neurol. Scand.* **2013**, *127*, 384–390. [CrossRef]
13. Grobelny, A.; Behrens, J.R.; Mertens, S.; Otte, K.; Mansow-Model, S.; Kruger, T.; Gusho, E.; Bellmann-Strobl, J.; Paul, F.; Brandt, A.U.; et al. Maximum walking speed in multiple sclerosis assessed with visual perceptive computing. *PLoS ONE* **2017**, *12*, e0189281. [CrossRef]
14. Sehle, A.; Mundermann, A.; Starrost, K.; Sailer, S.; Becher, I.; Dettmers, C.; Vieten, M. Objective assessment of motor fatigue in Multiple Sclerosis using kinematic gait analysis: A pilot study. *J. Neuroeng. Rehabil.* **2011**, *8*, 59. [CrossRef] [PubMed]
15. Al-Amri, M.; Al Balushi, H.; Mashabi, A. Intra-rater repeatability of gait parameters in healthy adults during self-paced treadmill-based virtual reality walking. *Comput. Methods Biomech. Biomed. Eng.* **2017**, *20*, 1669–1677. [CrossRef] [PubMed]
16. Sosnoff, J.J.; Socie, M.J.; Sandroff, B.M.; Balantrapu, S.; Suh, Y.; Pula, J.H.; Motl, R.W. Mobility and cognitive correlates of dual task cost of walking in persons with multiple sclerosis. *Disabil. Rehabil.* **2014**, *36*, 205–209. [CrossRef] [PubMed]

17. Phan-Ba, R.; Pace, A.; Calay, P.; Grodent, P.; Douchamps, F.; Hyde, R.; Hotermans, C.; Delvaux, V.; Hansen, I.; Moonen, G.; et al. Comparison of the timed 25-foot and the 100-m walk as performance measures in multiple sclerosis. *Neurorehabil. Neural Repair* **2011**, *25*, 672–679. [CrossRef] [PubMed]
18. Terui, Y.; Suto, E.; Konno, Y.; Kubota, K.; Iwakura, M.; Satou, M.; Nitta, S.; Hasegawa, K.; Satake, M.; Shioya, T. Evaluation of gait symmetry using a tri-axial accelerometer in stroke patients. *NeuroRehabilitation* **2018**, *42*, 173–180. [CrossRef]
19. Supratak, A.; Datta, G.; Gafson, A.R.; Nicholas, R.; Guo, Y.; Matthews, P.M. Remote Monitoring in the Home Validates Clinical Gait Measures for Multiple Sclerosis. *Front. Neurol.* **2018**, *9*, 561. [CrossRef]
20. Donath, L.; Faude, O.; Lichtenstein, E.; Nuesch, C.; Mundermann, A. Validity and reliability of a portable gait analysis system for measuring spatiotemporal gait characteristics: Comparison to an instrumented treadmill. *J. Neuroeng. Rehabil.* **2016**, *13*, 6. [CrossRef]
21. Manor, B.; Yu, W.; Zhu, H.; Harrison, R.; Lo, O.Y.; Lipsitz, L.; Travison, T.; Pascual-Leone, A.; Zhou, J. Smartphone App-Based Assessment of Gait During Normal and Dual-Task Walking: Demonstration of Validity and Reliability. *JMIR Mhealth Uhealth* **2018**, *6*, e36. [CrossRef] [PubMed]
22. Silsupadol, P.; Teja, K.; Lugade, V. Reliability and validity of a smartphone-based assessment of gait parameters across walking speed and smartphone locations: Body, bag, belt, hand, and pocket. *Gait Posture* **2017**, *58*, 516–522. [CrossRef]
23. Ojeda, L.V.; Rebula, J.R.; Kuo, A.D.; Adamczyk, P.G. Influence of contextual task constraints on preferred stride parameters and their variabilities during human walking. *Med. Eng. Phys.* **2015**, *37*, 929–936. [CrossRef] [PubMed]
24. Yentes, J.M.; Rennard, S.I.; Schmid, K.K.; Blanke, D.; Stergiou, N. Patients with Chronic Obstructive Pulmonary Disease Walk with Altered Step Time and Step Width Variability as Compared with Healthy Control Subjects. *Ann. Am. Thorac. Soc.* **2017**, *14*, 858–866. [CrossRef] [PubMed]
25. Liu, W.Y.; Spruit, M.A.; Delbressine, J.M.; Willems, P.J.; Franssen, F.M.E.; Wouters, E.F.M.; Meijer, K. Spatiotemporal gait characteristics in patients with COPD during the Gait Real-time Analysis Interactive Lab-based 6-minute walk test. *PLoS ONE* **2017**, *12*, e0190099. [CrossRef] [PubMed]
26. Heredia-Jimenez, J.; Orantes-Gonzalez, E.; Soto-Hermoso, V.M. Variability of gait, bilateral coordination, and asymmetry in women with fibromyalgia. *Gait Posture* **2016**, *45*, 41–44. [CrossRef]
27. Clermont, C.A.; Barden, J.M. Accelerometer-based determination of gait variability in older adults with knee osteoarthritis. *Gait Posture* **2016**, *50*, 126–130. [CrossRef]
28. Neely, J.G.; Karni, R.J.; Engel, S.H.; Fraley, P.L.; Nussenbaum, B.; Paniello, R.C. Practical guides to understanding sample size and minimal clinically important difference (MCID). *Otolaryngol. Head Neck Surg.* **2007**, *136*, 14–18. [CrossRef]
29. Pau, M.; Caggiari, S.; Mura, A.; Corona, F.; Leban, B.; Coghe, G.; Lorefice, L.; Marrosu, M.G.; Cocco, E. Clinical assessment of gait in individuals with multiple sclerosis using wearable inertial sensors: Comparison with patient-based measure. *Mult. Scler. Relat. Disord.* **2016**, *10*, 187–191. [CrossRef]
30. Mutoh, T.; Mutoh, T.; Takada, M.; Doumura, M.; Ihara, M.; Taki, Y.; Tsubone, H.; Ihara, M. Application of a tri-axial accelerometry-based portable motion recorder for the quantitative assessment of hippotherapy in children and adolescents with cerebral palsy. *J. Phys. Ther. Sci.* **2016**, *28*, 2970–2974. [CrossRef]
31. Hsu, C.Y.; Tsai, Y.S.; Yau, C.S.; Shie, H.H.; Wu, C.M. Test-Retest Reliability of an Automated Infrared-Assisted Trunk Accelerometer-Based Gait Analysis System. *Sensors* **2016**, *16*, 1156. [CrossRef] [PubMed]
32. Schmid, A.; Duncan, P.W.; Studenski, S.; Lai, S.M.; Richards, L.; Perera, S.; Wu, S.S. Improvements in speed-based gait classifications are meaningful. *Stroke J. Cereb. Circ.* **2007**, *38*, 2096–2100. [CrossRef] [PubMed]

Article

Comparing Gait Trials with Greedy Template Matching

Aliénor Vienne-Jumeau [1], Laurent Oudre [1,2,3,*], Albane Moreau [1], Flavien Quijoux [1,4], Pierre-Paul Vidal [1,5] and Damien Ricard [1,6,7]

1 COGNAC-G (UMR 8257), CNRS Service de Santé des Armées University Paris Descartes, 75006 Paris, France
2 L2TI, University Paris 13, 93430 Villetaneuse, France
3 CMLA (UMR 8536), CNRS ENS Paris-Saclay, 94235 Cachan, France
4 ORPEA Group, 92813 Puteaux, France
5 Hangzhou Dianzi University, 310005 Hangzhou, China
6 Service de Neurologie, Hôpital d'Instruction des Armées Percy, Service de Santé des Armées, 92190 Clamart, France
7 Ecole du Val-de-Grâce, Ecole de Santé des Armées, 75005 Paris, France
* Correspondence: laurent.oudre@univ-paris13.fr; Tel.: +33-1-49-40-40-63

Received: 9 May 2019; Accepted: 11 July 2019; Published: 12 July 2019

Abstract: Gait assessment and quantification have received an increased interest in recent years. Embedded technologies and low-cost sensors can be used for the longitudinal follow-up of various populations (neurological diseases, elderly, etc.). However, the comparison of two gait trials remains a tricky question as standard gait features may prove to be insufficient in some cases. This article describes a new algorithm for comparing two gait trials recorded with inertial measurement units (IMUs). This algorithm uses a library of step templates extracted from one trial and attempts to detect similar steps in the second trial through a greedy template matching approach. The output of our method is a similarity index (SId) comprised between 0 and 1 that reflects the similarity between the patterns observed in both trials. Results on healthy and multiple sclerosis subjects show that this new comparison tool can be used for both inter-individual comparison and longitudinal follow-up.

Keywords: inertial measurement units; gait analysis; biomedical signal processing; pattern recognition; step detection; physiological signals

1. Introduction

Gait semiology is of major importance in neurological practice, as abnormalities are associated with high comorbidities. The quantification of gait using inertial measurement units (IMUs) has become a democratic method for the follow-up of subjects with locomotion alterations in healthcare. The use of such embedded technologies has already shown its usefulness in the detection of postural strategies during walking [1], partitioning gait during the stance phase [2] or motor supplementation for switch-activated simulators [3]. However, these clinical applications require the detection of steps within the IMU signals. Spatio-temporal gait parameters can also be extracted for the healthy or disabled and stored in databases that enable a longitudinal follow-up of patients with gait disorders due to ageing [4], orthopedic or rheumatic diseases [4–6] or neurological alterations [4,7–9]. It has proven useful to help clinicians refine the description of individual gait disorder and strengthen their insights into the patients' movements and compensation patterns. This quantification of characteristics related to altered gait using signals from IMUs that are collected inside databases allows inter-individual comparisons to assess the distance of the patients' gait from a control group [10] or intra-individual comparisons for longitudinal follow-up [11,12]. Common gait features most often rely on basic statistics such as averages or standard deviations over a whole exercise [13,14]. On the one hand, they provide useful and interpretable information for the clinician. On the other hand, they have not proven sensitive enough for detecting subtle changes in several pathologies [4,15,16]. Besides, they display high inter-session variability for diseases that present with day-to-day changes [17]. To assure robustness of these parameters, it is usually necessary to increase the number of steps within a trial [18–20]. However, repeated measures or treadmill exercises are incompatible with common clinical practice in patients with limited walking perimeter, which is frequent in neurological practice. In order to obtain a more integrative perspective, some authors resort to global indexes, which are composed of several parameters [21,22]. These scores are promising but careful consideration should be given to their evolution inasmuch as the absence of evolution of a multicomponent score does not necessarily reflect the reproducibility of the gait pattern between two measurements [20]. Indeed, maintaining a steady value of an overall score over time may mask gait adaptations. For instance, gait velocity may be maintained despite a decreased step length if the cadence increases concomitantly. What is more, these parameters heavily rely on accurate step detection, which is problematic in severely altered steps: Some patients may require manual painstaking and time-consuming detection [23]. It is therefore key to evaluate this detection or be exempt from it.

Progressive multiple sclerosis (pMS) is one of the disorders that benefits heavily from the use of IMUs in routine clinical practice to assess indices of the disease evolution [24–26]. IMUs have been applied to reliably monitor patients' health status with regard to their risk of falling [27], their physical abilities [28,29] or the neuromotor strategies used to adapt to their disability [22,30–32]. However, patients suffering from severe pMS may impose high constraints both on measurements, which should be short and controlled to abstract from fatigue and day-to-day variations, and processing, which should adapt to very abnormal patterns and confounders such as false gait events triggered off by loading and unloading of walking aids.

In this study, a new metric to compare two gait trials recorded with IMUs, which we called "similarity index" (SId), is introduced. It aims at overcoming previously mentioned withdrawals of statistical methods in pathologies such as pMS. The SId is an asymmetrical metric that takes as input two gait trials and computes an index, comprised between 0 and 1, that assesses the similarity between them. It is hypothesized that such a metric provides a valid characterization of a change in gait patterns between two measurements, and can therefore be used either for inter-individual comparison or longitudinal follow-up.

First, the SId is compared between pairs of trials of increasing distance. Second, it is evaluated against more conventional features to estimate its capacity to assess changes in gait. Eventually, its ability to indicate the level of confidence of the underlying step detection is appraised.

2. Data, Protocol and Subjects

2.1. Protocol

Two XSens® sensors (Xsens® Technologies, Enschede, the Netherlands)—hereafter XS—were placed on the participant's body (one on the dorsal part of each foot) using Velcro bands. The XS showed high reliability at heel impact for the ankle joint in the sagittal plane (inter-correlation coefficient (ICC) > 0.8) [33]. With a standard error of the mean (SEM) below 3°, between- and within-rater reliability of kinematic variables obtained from XS across joints and planes, its consistency is comparable to or better than that obtained from optoelectronic motion capture systems [34]. The GaitRite® mat—hereafter GR—exhibits strong concurrent validity [35,36] and excellent reliability (with ICC > 0.8) for most temporo-spatial gait parameters in both young and older subjects [37–39]. The GR can be used to assess people with altered gait with good reliability even though walking ability does influence it. The ICCs for the older subjects [37] or patients with neurological diseases [40,41] can be somewhat lower but are still adequate for measuring step parameters of gait in these populations. Based on this data, GR is used in this paper as the gold standard. The data were sampled at 100 Hz for the XS and at 120 Hz for the GR. Both systems were synchronized in time by using the PC clock connected to the XS. Participants performed four walks of 12 m with a U-turn (6 m on the way in and 6 m on the way out): Two at the first visit (M0) and again two at the second visit six months later (M6). The choice of a six-month period between measurement was driven by the fact that patients from the pMS group undergo routine evaluation of their gait every six months. The protocol is schematized in Figure 1.

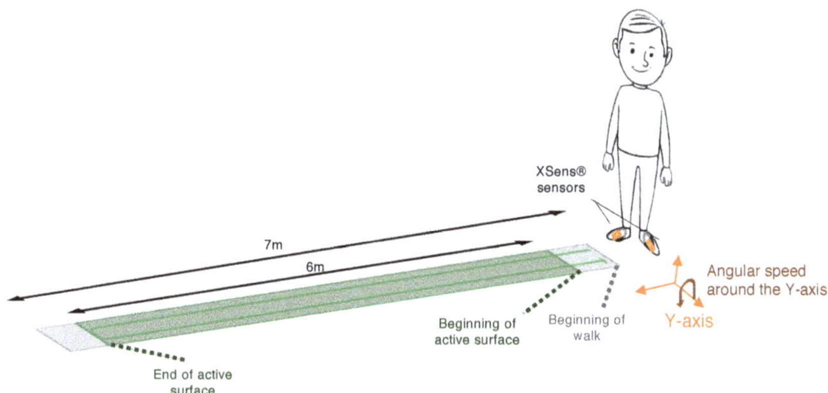

Figure 1. Measurement protocol. The XSens® sensors (XS) inertial measurement units and the GaitRite® mat (GR) are synchronized by using the PC clock connected to the inertial measurement units. The active surface (green) is covered with pressure sensors. The rest of the mat (grey) is inactive and does not detect any pressure from the subject.

2.2. Subjects

Twenty-two patients with progressive multiple sclerosis (pMS) and ten young healthy subjects (HS) were enrolled in this longitudinal study. The characteristics of the subjects are displayed in Table 1. pMS patients were consecutively recruited from the outpatient clinic of Percy Hospital (Clamart, France) between June 2018 and September 2018. The inclusion criteria for participation in this cohort required patients to be at least 18 years old, be diagnosed with primary progressive or secondary progressive multiple sclerosis according to the 2010 International Panel criteria [42], be capable of walking 20 m with U-turn and be free of any other conditions that affect gait. HS participants were recruited from the hospital and research unit staff between June 2018 and September 2018. The inclusion criteria were: No report of falls in the past five years prior to inclusion and no disease that could affect their walk. The sex ratio was comparable between the two groups and no major differences were seen between other anthropometric characteristics. pMS patients were aged 58 (±11) years old and the HS group mean age was 26 (±2) years old. The two groups were not matched for age as one aim of this analysis was to analyze the performance of the algorithm on two opposite groups, one with highly altered steps (pMS) and one with the most normal steps. Severity of the disease was evaluated using the Expanded Disease Status Scale (EDSS) [43], which is a score of 0 to 10, ranging from normal neurological examination (0) to total impotence (9.5) or even death (10). Included participants in the pMS group had an EDSS between 3.0 and 6.5, as disabilities greater than 7.0 impede walking even a few steps. Seven out of the 22 participants had an advanced disease requiring permanent walking aid (cane(s), walker and/or human help). Two patients needed human help to perform the walking test. All subjects provided a written informed consent prior to their inclusion. The study protocol followed the principles of the Declaration of Helsinki and was approved by the Ethics Committee "Protection des Personnes Nord Ouest III" under the ID RCB: 2017-A01538-45.

Table 1. Baseline characteristics of patients with progressive multiple sclerosis (pMS) and healthy subjects (HS). For the age, height, weight, BMI, Multiple Sclerosis Walking Scale (MSWS) and Fatigue Impact Scale (FIS), the mean and the standard deviation (SD) are displayed. For the Expanded Diseases Status Scale (EDSS) and the functional scores (subscores of EDSS), the statistics are reported as median and interval quartile range (IQR).

	pMS (n = 22)	HS (n = 10)
Sex (M/F)	9/13	4/6
Age (years)	58 (11)	26 (1)
Height (m)	1.71 (0.09)	1.72 (0.09)
Weight (kg)	71.2 (16.6)	58.2 (10.9)
BMI (kg/m^2)	24.3 (5.1)	21.0 (3.0)
EDSS	5.0 [3.5–6]	-
EDSS—pyramidal	3.0 [3.0–3.8]	-
EDSS—cerebellar	1.5 [0.0–3.0]	-
EDSS—bulbar	0 [0.0–0.8]	-
EDSS—sensitive	2.0 [1.0–2.0]	-
EDSS—cognitive	1.0 [0.0–2.0]	-
MSWS	65.0 (17.3)	-
FIS	43.4 (24.9)	-
Use of walking aid for the walk test (/total number)	7/22	0/10
Cane (1 or 2)	4	-
Walker	1	-
Human help	1	-
Cane + human help	1	-

3. Method

We now define the similarity index (SId) between two gait trials. Let us consider two gait trials:

- One train trial denoted i_{train} and composed of both GR data and XS data.
- One test trial denoted i_{test}, only composed of XS data.

The aim of the algorithm presented in this section is to compute a similarity index $\mathrm{SId}_{i_{test}|i_{train}}$, comprised between 0 and 1, that will assess the proximity between trials i_{train} and i_{test}. This metric is based upon the following question: How well can the group of steps present in trial i_{train} predict those observed in trial i_{test} ?

The computation of this index is based on three main stages, detailed below and illustrated in Figure 2.

1. The GR and XS data from trial i_{train} are used to build a library of templates $\mathcal{P}_{train}$;
2. This library is used to detect the steps in trial i_{test}, according to a greedy template-based approach inspired by [44];
3. The Pearson coefficients between the detected steps in i_{test} and the patterns in $\mathcal{P}_{train}$ used for their detection are merged to compute the similarity index $\mathrm{SId}_{i_{test}|i_{train}}$.

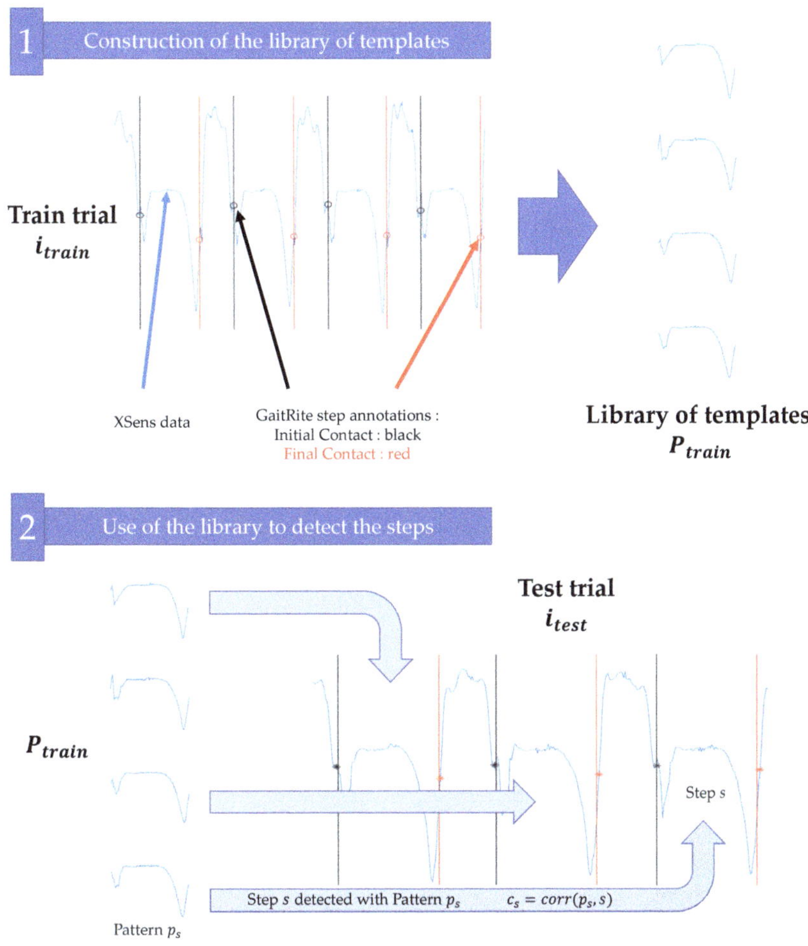

Figure 2. Main stages for the computation of the similarity index (SId). First, the GR and XS data from the trial i_{train} are used to build a library of templates $\mathcal{P}_{train}$. In the second stage, the library is used to detect the steps in the trial i_{test}, according to a greedy template-based approach inspired by [44]. Each detected step s is associated with one template p_s. The correlation coefficients c_s between the steps s and their associated templates p_s are then averaged to obtain the similarity index $\mathrm{SId}_{i_{test}|i_{train}}$.

3.1. Construction of the Library of Templates

Let us consider a train gait trial i_{train} composed of GR and XS data. We first use the GR recordings to extract the exact timings for initial contacts (ICs) and final contacts (FCs). This process is automatically performed thanks to the GR software. Only the steps occurring while the subject is on the active surface of the instrumented mat are used; steps occurring during the U-turn are not considered. Then, we use the XS synchronized data to build the library of templates. We consider for each right/left foot XS sensor the Y-axis angular velocities (swing in the direction of the walk) and construct a library of templates by extracting the steps in the XS signals. More precisely, given a step identified with the GR with the initial contact time t_{IC} and final contact time t_{FC}, we consider the XS signal x_{train} corresponding to the adequate foot and define the pattern $p = x_{train}[t_{IC} : t_{FC}]$. This pattern p can be seen as a signal of length $N_p = t_{FC} - t_{IC} + 1$ that represents the typical angular velocity of a foot during a step. The process is iterated for all the steps and for both feet: Each step identified with the GR forms a different pattern p. All patterns corresponding to the trial i_{train} are stored in a library $\mathcal{P}_{train}$.

3.2. Use of the Library to Detect the Steps

The library $\mathcal{P}_{train}$ is used to detect the steps for the trial i_{test} (which does not necessarily belong to the same subject and/or the same session). To that end, we consider the XS Y-axis angular velocity x_{test} for trial i_{test}. Each pattern $p \in \mathcal{P}_{train}$ is slid along signal x_{test} and for each possible shift we compute the Pearson correlation coefficient. The final result is a matrix $\mathbf{C}$ of size $N_{\mathcal{P}} \times N_{test}$, where $N_{\mathcal{P}}$ is the number of templates in $\mathcal{P}_{train}$ and N_{test} is the number of samples of signal x_{test}, where

$$\forall i_p \in [\![1, N_{\mathcal{P}}]\!],\ \forall i_t \in [\![1, N_{test}]\!] \qquad c(i_p, i_t) = \operatorname{corr}\left(p, x_{test}[i_t : i_t + N_p - 1]\right), \tag{1}$$

and corr(.,.) is the Pearson correlation coefficient.

The matrix $\mathbf{C}$ is then processed with an iterative and greedy detection strategy, described in [44], which detects steps by iteratively selecting the largest Pearson correlation coefficients in the matrix until all of them are lower than a threshold $\lambda = 0.6$. The influence of threshold λ is discussed in [44] and the value 0.6 insures that the algorithm does not consider irrelevant matches. The main idea behind this procedure is that we select the best possible templates in train trial i_{train} to detect the steps in test trial i_{test}.

The output of the algorithm is a list of steps $\mathcal{S}_{i_{test}|i_{train}}$ (steps of trial i_{test} detected with the library of trial i_{train}). For each detected step $s \in \mathcal{S}_{i_{test}|i_{train}}$, we also have access to the template $p_s \in \mathcal{P}_{train}$ that was used for the detection, and to the Pearson correlation coefficient c_s between s and p_s. Those additional outputs, which were not investigated in [44], are actually of interest since:

- Knowledge on p_s allows to characterize the step s. Since we know that p_s was the template in $\mathcal{P}_{train}$ that most resembled step s, any available information on p_s can be used to understand the shape, duration, length, etc., of step s. For instance, several steps detected with the same template are likely to be similar.
- Coefficient c_s informs us of how strongly p_s and s matched. A c_s close to 1 implies that a pattern exists in $\mathcal{P}_{train}$ that is very close to the phenomenon observed in step s; in other words, the confidence in step s is high. If, on the contrary, c_s is small, no templates in $\mathcal{P}_{train}$ exactly fitted the detected step s. This could mean that the locomotion of trial i_{train} is different than the one of i_{test}.

3.3. Similarity Index

In order to use this additional information for the gait characterization, we propose to introduce a new parameter, called SId (similarity index). Given a library of template $\mathcal{P}_{train}$ and a test trial i_{test}, this quantity is defined as

$$\text{SId}_{i_{test}|i_{train}} = \operatorname*{mean}_{s \in \mathcal{S}_{i_{test}|i_{train}}} (c_s). \tag{2}$$

The SId is the mean of the Pearson correlation coefficients computed between detected steps and their respective closest templates. This quantity measures the ability of trial i_{train} to detect the steps in trial i_{test}. It can be interpreted as a similarity index between trials i_{train} and i_{test} (assuming that if both trials were identical the step detection would be easy to perform and would produce large Pearson coefficients). It can also be seen as a confidence index on the detection (if this index is close to 1, it means that all detected steps were very similar to the annotated steps in the library and thus are likely to be well detected).

Note that the SId is not symmetrical, as using steps in trial i_{train} to detect the steps in trial i_{test} might not be the same as using steps in trial i_{test} to detect the steps in trial i_{train}.

3.4. Use of the SId Index in Various Configurations

According to the chosen train and test sets, the SId index can be used either for longitudinal follow-up or inter-individual comparison. In this article, we consider four different configurations referred to as A1–A4.

- A1: Intra-individual intra-session. Two different trials belonging to the same subject and the same session (M0 | M0 or M6 | M6).
- A2: Intra-individual inter-session. Two trials belonging to the same subject but not to the same session (M0 | M6 or M6 | M0).
- A3: Inter-individual intra-group. Two trials belonging to different subjects of the same group (pMS | pMS or HS | HS).
- A4: Inter-individual inter-group. Two trials belonging to two subjects in different groups (pMS | HS or HS | pMS).

In order to investigate the properties of the SId index, we computed all SId between all trials of all subjects and merged the SId values according to these four different configurations, as illustrated in Figure 3.

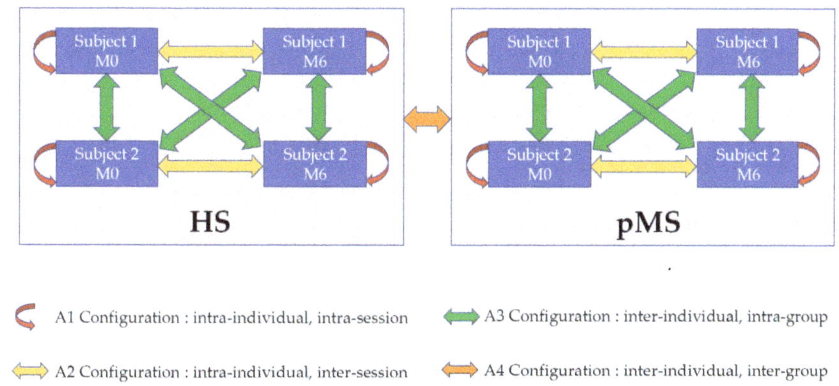

Figure 3. Definitions of the different pairs of extraction/detection trials that are analyzed in the article.

3.5. Conventional Features

In addition to the SId, the following conventional gait parameters were computed:

- Average velocity: Velocity was computed for the way in (or the way out) as the total length of the detection part of the GR divided by the total time of the way in (or the way out). The average velocity was then the average of the velocity of the way in and the velocity of the way out.
- Step length: This parameter was extracted from the GR output.
- Step time: This parameter was computed as the average of the differences between the final contact time and the initial contact times given by the GR.
- Double stance time: This parameter was computed as the average of the differences between the final contact time of one foot and the initial contact times of the contralateral foot given by the GR.
- Variation coefficient of step time: This parameter was computed as the standard deviation of the differences between the final contact time and the initial contact times given by the GR divided by the step time.
- Variation coefficient of double stance time: This parameter was computed as the standard deviation of the differences between the final contact time of one foot and the initial contact times of the contralateral foot given by the GR divided by the step time.

Given two gait trials, we computed the differences between the parameter values and merged these differences with the four different configurations illustrated in Figure 3.

3.6. Link to the Performance of the Step Detection

The similarity index (SId) can be interpreted as a confidence index for the step-detection algorithm. Indeed, a large SId suggests that the patterns present in the train library fit those observed in the test signal and are therefore likely to provide efficient detection. To investigate this question, we computed the correlation between the SId values and some evaluation metrics commonly used for assessing the performances of step-detection algorithms [44]. These metrics are based on the ground truth step annotations provided by the GR.

- **Precision (or positive predictive value).** A detected step is counted as correct if the mean of its start and end times lies inside an annotated step. An annotated step can only be detected one time. If several detected steps correspond to the same annotated step, all but one are considered as false. The precision is the number of correctly detected steps divided by the total number of detected steps.
- **Recall (or sensitivity).** An annotated step is counted as detected if the mean of its start and end times lies inside a detected step. A detected step can only be used to detect one annotated step. If several annotated steps are detected with the same detected step, all but one are considered undetected. The recall is the number of detected annotated steps divided by the total number of annotated steps.
- **F-measure (or F1 score).** The F-measure is the harmonic mean of precision and recall.
- $\Delta Start$**.** For a correctly detected step, this is the difference between the detected start time and the annotated start time.
- ΔEnd**.** For a correctly detected step, this is the difference between the detected end time and the annotated end time.
- $\Delta Duration$**.** For a correctly detected step, this is the difference between the duration of the detected step and the duration of the annotated step.

3.7. Statistics

All parameters were tested for normality using Shapiro-Wilks tests. Parametric tests were applied for normal distributions and non-parametric tests were resorted to when this hypothesis was rejected. Means and standard deviations (SD) were reported, except for ordinal distributions (EDSS) where mean and interquartile range were reported.

3.7.1. Comparisons between Configurations

SId and change in gait conventional features were compared between configurations of pairs of extraction/detection trials using the absolute difference between the mean value in the two groups. For all these non-parametric variables, the Krustkall-Wallis test—a rank-based non-parametric test used to assess more than two independent groups—was used. Rejection of the null-hypothesis was followed by subsequent Wilcoxon tests to test differences in medians. All tests were corrected for multiple comparisons using Bonferroni adjustment. For each group (HS and pMS, respectively), the percentile score of SId from A2 was computed from the distribution of SId from A3. The percentile score of SId from A2 was also computed from the distribution of SId from A4.

3.7.2. Correlations

SId was correlated to performance, accuracy and conventional gait features using Pearson moment product correlation coefficients, which remains a valid method, even in the case of non-normal datasets [45]. Pearson correlation coefficient is interpreted as very high for absolute values between 0.9 and 1.0, high for absolute values between 0.7 and 0.9, moderate for absolute values from 0.5 to 0.7, low for absolute values from 0.3 to 0.5 and negligible for absolute values below 0.3 [46].

Primary data analysis (extraction and detection process) was done using MATLAB® R2019a. Statistical analysis was performed using R v3.5.1. All tests were corrected for multiple comparisons using Bonferroni adjustment.

4. Results

In this section, we investigate the ability of this index to effectively compare two trials. To investigate the potential of SId as a gait biomarker, three different and complementary questions are investigated:

- Comparison of SId based on these four configurations: Comparing SIds computed within the same session (A1), SIds computed from different sessions of the same subject (A2), SIds computed between subjects of the same group (A3) and SIds computed between groups (A4).
- Correlation of SId with more conventional features used to characterize gait (average velocity, step length, step time, double stance time, variation coefficient of step time, variation coefficient of double stance time).
- Correlation of SId with the detection performance of the step-detection algorithm.

4.1. Comparison of SId Based on Configurations

In this experiment, the values of SId are compared between the four configurations: Intra-individual intra-session comparison (A1), intra-individual inter-session comparison (A2), intra-group inter-individudal comparison (A3) and inter-group inter-individudal comparison (A4). Boxplots are displayed in Figure 4. SId shows its highest values for the A1 comparison (HS: 0.99 (0.00), pMS: 0.97 (0.01)) and decreases from A1 to A4, both in the HS and the pMS group (p-value of the Krustkall-Wallis test: <0.0001, p-value of the subsequent Wilcoxon tests: <0.0001 for all paired comparisons). In particular, it shows that trials from a given subject are closer to each other than to trials from another subject both in the HS group (mean difference: 0.046; p-value: <0.0001) and in the pMS group (mean difference: 0.055; p-value: <0.0001). Comparisons of A3 (inter-individual intra-group) and A4 (inter-individual inter-group) show that SIds obtained for intra-group comparison are larger than inter-group ones in the HS group (mean difference: 0.190; p-value: <0.0001) but not in the pMS group (mean difference: 0.070; p-value $= 0.52$).

For a given individual k inside the HS or the pMS groups, SIds for comparison of one trial to another trial are reproduced in Table 2. This table shows that, on average, trials from a given subject are closer to other trials from the same subject than to trials from other subjects. For HS subjects, SId from A2 prediction belongs to the 90th (SD: 14) percentile of the distribution of SId from A3 prediction and is always higher than SId from A4. For pMS subjects, SId from A2 prediction belongs to the 96th (SD: 8) percentile of the distribution of SId from A3 prediction and the 99th (SD: 2) percentile of the distribution of SId from A4.

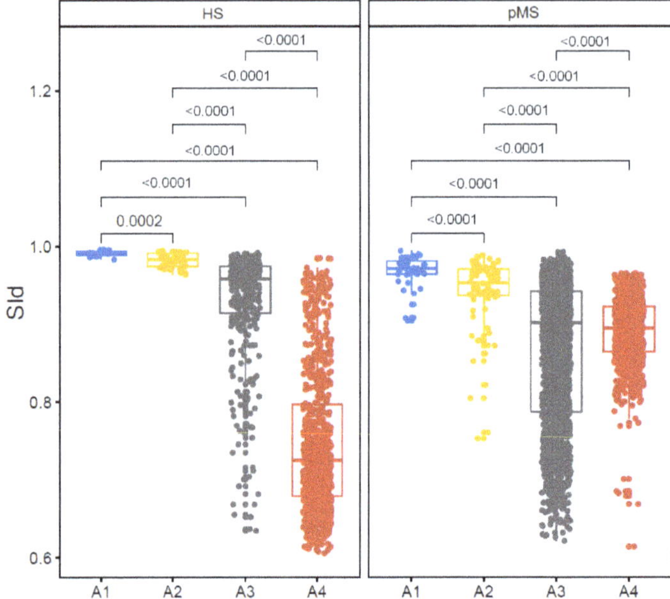

Figure 4. Comparison of SId predictions across configurations: Intra-individual intra-session prediction (A1) vs. intra-individual inter-session prediction (A2) vs. intra-group inter-individual prediction (A3) vs. inter-group inter-individual prediction (A4).

Table 2. Similarity index scores for comparing one gait trial depending on the training trial (intra-individual inter-session, intra-group inter-individual, inter-group inter-individual). Means and standard deviations are displayed for both pMS and HS groups.

	HS		**pMS**	
	Individual k	**Other Individual**	**Individual k**	**Other Individual**
HS (individual k)	0.98 (0.01)	0.93 (0.07)	-	0.75 (0.09)
pMS (individual k)	-	0.89 (0.04)	0.94 (0.05)	0.87 (0.09)

4.2. Correlation of SId with Conventional Features

Comparisons were also carried out for the average walking velocity (Figure 5a), step length (Figure 5b), step time (Figure 5c), double stance time (Figure 5d), coefficient of variation of step time (Figure 5e) and coefficient of variation of double stance time (Figure 5f), which are classical gait features [4]. After controlling for multiple comparisons, difference in average velocity (Figure 5a) and differences in double stance time (Figure 5d) proved significantly higher in the A2 (intra-individual inter-session) comparison as compared to the A1 (intra-individual intra-session) comparison in the HS group (p-values of 0.002 and 0.003, respectively, with a threshold of 0.017) and the pMS group (p-values < 0.001 and < 0.0001, respectively, with a threshold of 0.017). Difference in step length (Figure 5b) was also higher in the A2 comparison as compared to the A1 comparison in the HS group (p-values of 0.007, with a threshold of 0.017). All other comparisons of configurations were highly significant (p-value < 0.0001).

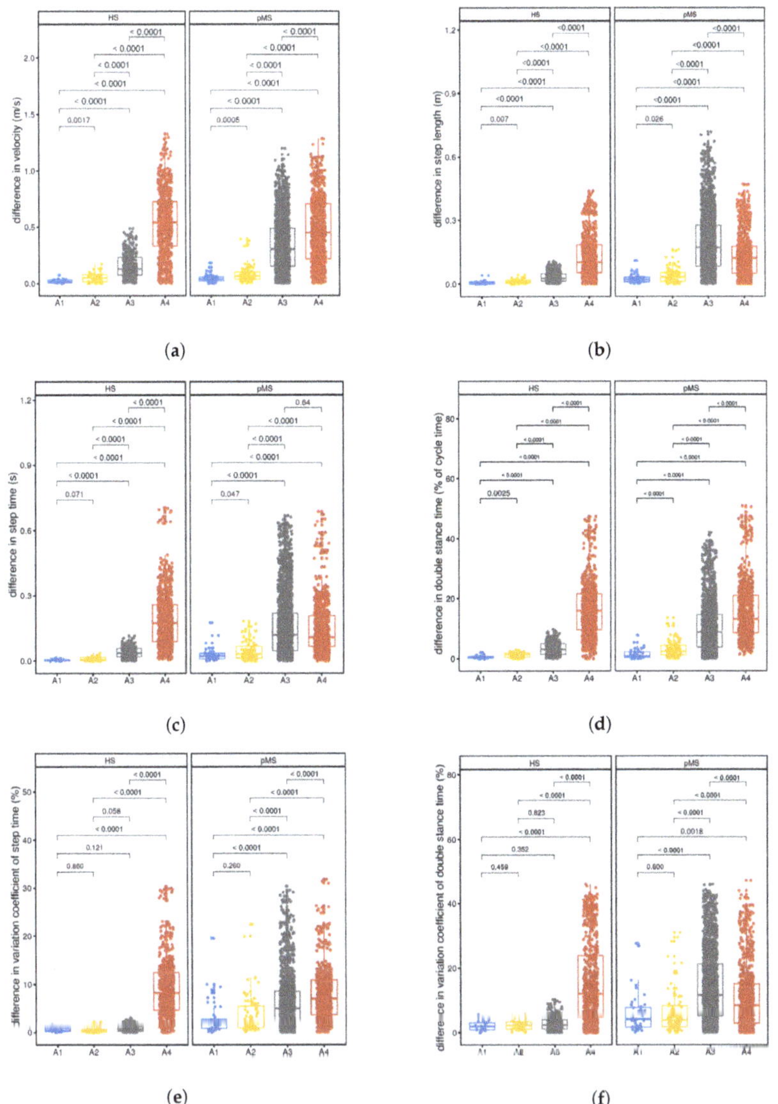

Figure 5. Intra-individual intra-session prediction (A1) vs. intra-individual inter-session prediction (A2) vs. intra-group inter-individual prediction (A3) vs. inter-group inter-individual prediction (A4) for both cohorts : (**a**) Average walking velocity; (**b**) step time; (**c**) step length; (**d**) double stance time; (**e**) coefficient of variation of step time; (**f**) coefficient of variation of double stance time.

To investigate how SId would correlate to change in these conventional features, SId, as measured for each intra-group comparison (A1, A2, A3), was correlated to variation between the respective train trial and test trial for each of the following conventional gait features: The average walking velocity (Figure 6a), step time (Figure 6b), step length (Figure 6c), double stance time (Figure 6d), coefficient of variation of step time (Figure 6e) and coefficient of variation of double stance time (Figure 6f).

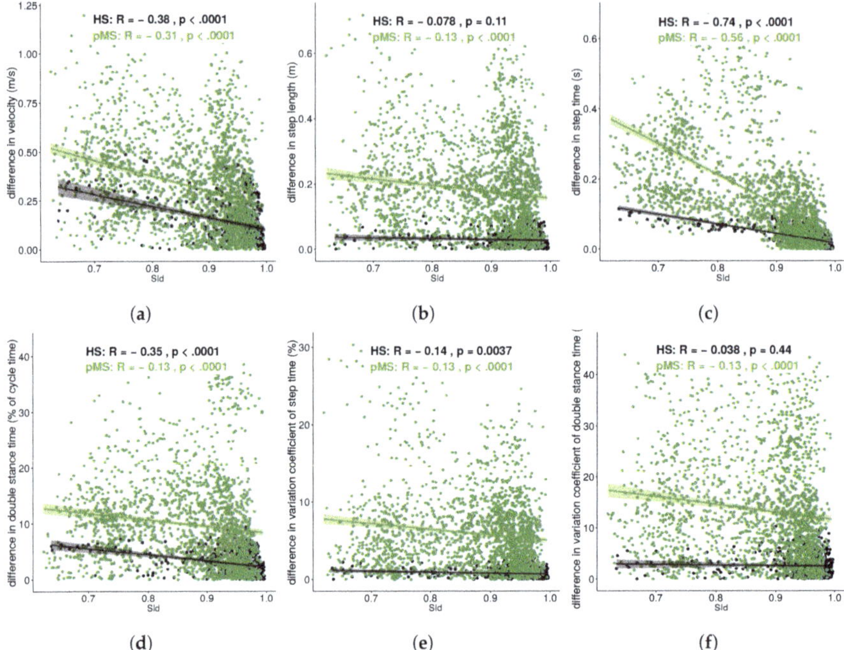

Figure 6. Correlation of SId to conventional features: (**a**) Average walking velocity; (**b**) step length; (**c**) step time; (**d**) double stance; (**e**) coefficient of variation of step time; (**f**) coefficient of variation of double stance time.

For both groups, low correlations were observed for difference in the average walking velocity (Figure 6a) (HS: $r = -0.38$, p-value: < 0.0001; pMS: $r = -0.31$, p-value: < 0.0001), double stance time (Figure 6d) (HS: $r = -0.35$, p-value: < 0.0001; pMS: $r = -0.13$, p-value: < 0.0001) and the variation coefficient of step time (Figure 6e) (HS: $r = -0.14$, p-value: 0.004; pMS: $r = -0.13$, p-value: < 0.0001). Moderate to high correlations were observed for difference in step time (Figure 6c) (HS: $r = -0.74$, p-value: < 0.0001; pMS: $r = -0.56$, p-value: < 0.0001). Additional low correlation was seen for pMS participants for the difference in step length (Figure 6b) ($r = -0.13$, p-value: < 0.0001) and the variation coefficient of double stance time (Figure 6f) ($r = -0.13$, p-value: < 0.0001).

4.3. Correlation to Performance of the Step Detection

Performance and accuracy scores, along with their correlations to SId, are reported in Table 3. In the HS group, SId correlates moderately to the F-measure, $\Delta Start$ and $\Delta Duration$, and weakly to ΔEnd. In the pMS group, SId correlates moderately to the F-measure and strongly to $\Delta Start$ and $\Delta Duration$, while a very low correlation is found with ΔEnd.

Table 3. Correlations between SId and the F-measure and accuracy scores for the step detected. All configurations are pooled together and reported as mean (SD).

	HS ($n = 10$)			pMS ($n = 22$)		
	Value	**Pearson**	***p*-Value**	**Value**	**Pearson**	***p*-Value**
F-measure	0.843 (0.213)	0.560	<0.0001	0.934 (0.130)	0.548	<0.0001
$\Delta Start$	0.18 (0.164)	−0.580	<0.0001	0.154 (0.170)	−0.781	<0.0001
ΔEnd	0.066 (0.087)	−0.306	<0.0001	0.026 (0.035)	−0.084	0.0001
$\Delta Duration$	0.234 (0.209)	−0.548	<0.0001	0.173 (0.179)	−0.771	<0.0001

5. Discussion

This study shows that SId is a valid metric to compare two gait trials both between different subjects or between two visits of a same subject to track changes in gait. In addition, in our small sample of patients, SId seems to give an insight into the performance of the underlying template-based step-detection method.

First, SId showed decreasing values from intra-individual intra-session (A1) to intra-individual inter-session (A2) to intra-group inter-individual (A3) to inter-group inter-individual (A4) trial comparisons for both the HS group and the pMS group. The difference in SId between A1 and A2 was expected for pMS individuals, for which symptoms vary from day to day depending, for instance, on the level of exercise and physical therapy or the weather (Uhtoff effect [47–49]). This higher change in HS participants between trials of different sessions compared to between trials of a same session was also true for conventional features. Average velocity and double stance time, as well as step length in the HS group, both displayed a higher difference when comparing inter-session with intra-session trials. Still, for all features, the difference in A2 remains within the standard error mean for inter-session comparison as found in the literature [50–52]. Furthermore, the hierarchy of variability in gait parameters is also found in the literature in intra-class correlation coefficients for both healthy subjects [19,53] and mixed groups of patients and subjects [10].What is more, SId shows high variability in between-cohort comparisons as compared to intra-cohort comparison for the HS group but not for the pMS group. Two participants from the pMS cohort can then be as distant as one participant from the pMS cohort and one participant from the HS cohort. One explanation is that pMS patients present with a wide range of gait alterations both in terms of the types of symptoms (which can relate to balance deficit, spasticity, decreased muscular strength, etc.) and severity of symptoms. In that regard, it can be observed in Figure 4 and Table 2 that the SId for the detection of steps from HS individuals using steps from pMS individuals seems lower than the detection of steps from pMS individuals using steps from HS individuals, which illustrates the non-symmetrical characteristic of the SId. This difference may be due to the durations of the steps that are different for HS and pMS subjects [23,54,55]. Due the greedy aspect of the matching procedure, it is easier for the algorithm to detect one large step with several small steps than the opposite. Therefore, higher SId values can be achieved by using HS templates to detect pMS steps than the opposite. One other explanation might that the noise level is larger for pMS subjects, thus creating noisy templates that are more difficult to match than HS smoother templates.

Second, as mentioned above, lower SId was associated with increased difference in step time between the train and test trial, a parameter which also showed strong correlation with disease severity as measured using the Expanded Disease Status Scale [16,23,54–56]. The SId has, therefore, potential to give insight into the evolution of the disease, without needing any pre-processing and step detection. However, even though most of them were significant, only low correlations were found for the differences in other conventional features that are usually used to characterize gait. As a matter of fact, very high variability in the difference of conventional features is seen, and one ought to be careful in drawing conclusions before larger and longer studies are carried out.

Third, SId has been shown to provide key information on the underlying step-detection algorithm. One major drawback of automatic step-detection algorithms is that it is tricky to assess their performances in real-life conditions. In particular, when confronted with different types of gait or cohorts, their accuracy may drop, which can have consequences if they are used in a clinical context. As a matter of fact, most of the algorithms designed for a particular type of subject may suffer from degraded performance in other cohorts [57]. Thanks to its construction, a large SId between two trials means that templates used to detect the steps were close to these latter. Very low SId values can therefore be interpreted as a discrepancy between the train and test trial, which is likely to cause a poor step detection. Indeed, we showed in Table 3 a correlation between SId and performance as well as accuracy scores. The SId values are therefore linked to the confidence in the underlying detection algorithm, and could be used to report that the model used in the detection process does not suit the tested data. If several libraries of templates were available (e.g., one for each cohort or one for each gait disability), the SIds could be used to select the most appropriate library and thus improve the step-detection performances. These perspectives shall be investigated in future studies.

Eventually, these results can be applied to a wide range of pMS individuals, with mild as well as severe diseases. Indeed, as patients using walking aids were also included, the conclusions also apply to patients with EDSS 6 and 6.5, which fills a gap in the literature [23]. Comparisons of SId between other populations should be informative to compare distances between gaits of patients disabled by different Neurological illness and participates in the development of a new taxonomy. New matching procedures may also be implemented, for instance, by using Dynamic-Time Warping (DTW), which allows to match time series of different lengths. In particular, the use of this technique dedicated to template matching may be useful in the context of step detection and recognition [58].

Our study has limitations. First, sampling fluctuations may have occurred due to the small sample size, particularly of HS. Recruiting young healthy subjects was difficult due to the necessity of a six month time period between both measurements. In particular, strong variability was found when correlating conventional features with the SId. Even though results were significant, the clouds of points are sparse.

6. Conclusions

In this article, we introduced a novel algorithm for comparing inertial signals of two gait trials. The output parameter, a metric referred to as the similarity index (SId), is comprised between 0 and 1 and reflects how similar two gait trials are. This parameter shows promising results for the longitudinal follow-up of participants, as it is sensitive to changes in gait features. Larger studies are needed to confirm the potential of SId as a predictor of changes and a longer follow-up time could also allow assessment of its prognostic value. Besides, as the SId correlates to the performance and accuracy of the underlying step-detection algorithm, it provides immediate feedback of the detection, which is a key aid for decision making.

Author Contributions: Conceptualization, A.V.-J., L.O., A.M., F.Q., P.-P.V. and D.R.; Data curation, A.M.; Formal analysis, A.V.-J., L.O., A.M. and F.Q.; Methodology, A.V.-J., L.O., A.M., F.Q., P.-P.V. and D.R.; Software, A.V.-J.; Supervision, P.-P.V. and D.R.; Writing—original draft, A.V.-J.; Writing – review and editing, L.O., A.M., F.Q., P.-P.V. and D.R.

Funding: This work was supported by SATT-IDF INNOV.

Acknowledgments: The authors would like to thank Juan Mantilla, Danping Wang, Nicolas Vayatis and Christophe Labourdette for the useful discussions.

Conflicts of Interest: The authors declare no conflict of interest.

References

1. Martinez-Mendez, R.; Sekine, M.; Tamura, T. Detection of anticipatory postural adjustments prior to gait initiation using inertial wearable sensors. *J. Neuroeng. Rehabil.* **2011**, *8*, 17. [CrossRef] [PubMed]
2. Taborri, J.; Palermo, E.; Rossi, S.; Cappa, P. Gait Partitioning Methods: A Systematic Review. *Sensors* **2016**, *16*, 66. [CrossRef] [PubMed]
3. Kotiadis, D.; Hermens, H.; Veltink, P. Inertial Gait Phase Detection for control of a drop foot stimulator. *Med. Eng. Phys.* **2010**, *32*, 287–297. [CrossRef] [PubMed]
4. Vienne, A.; Barrois, R.; Vidal, P.P.; Buffat, S.; Ricard, D. Inertial sensors to assess gait quality in patients with neurological disorders: A systematic review. *Front. Psychol.* **2017**, *8*, 817. [CrossRef] [PubMed]
5. Barrois, R.; Oudre, L.; Moreau, T.; Truong, C.; Vayatis, N.; Buffat, S.; Yelnik, A.; de Waele, C.; Gregory, T.; Laporte, S.; et al. Quantify osteoarthritis gait at the doctor's office: A simple pelvis accelerometer based method independent from footwear and aging. *Comput. Methods Biomech. Biomed. Eng.* **2015**, *18*, 1880–1881. [CrossRef] [PubMed]

6. Barrois, R.; Gregory, T.; Oudre, L.; Moreau, T.; Truong, C.; Aram Pulini, A.; Vienne, A.; Labourdette, C.; Vayatis, N.; Buffat, S.; et al. An Automated Recording Method in Clinical Consultation to Rate the Limp in Lower Limb Osteoarthritis. *PLoS ONE* **2016**, *11*, e0164975. [CrossRef] [PubMed]
7. Barrois, R.; Ricard, D.; Oudre, L.; Tlili, L.; Provost, C.; Vidal, P.P.; Yelnik, A. Observational study of 180° turn using Inertial Measurement Units in post-stroke ambulatory patients. *Ann. Phys. Rehabil. Med.* **2016**, *59S*, e117. [CrossRef]
8. Hsu, W.C.; Sugiarto, T.; Lin, Y.J.; Yang, F.C.; Lin, Z.Y.; Sun, C.T.; Hsu, C.L.; Chou, K.N. Multiple-Wearable-Sensor-Based Gait Classification and Analysis in Patients with Neurological Disorders. *Sensors* **2018**, *18*, 3397. [CrossRef]
9. Tunca, C.; Pehlivan, N.; Ak, N.; Arnrich, B.; Salur, G.; Ersoy, C. Inertial Sensor-Based Robust Gait Analysis in Non-Hospital Settings for Neurological Disorders. *Sensors* **2017**, *17*, 825. [CrossRef]
10. Psarakis, M.; Greene, D.; Cole, M.H.; Lord, S.R.; Hoang, P.; Brodie, M.A.D. Wearable technology reveals gait compensations, unstable walking patterns and fatigue in people with Multiple Sclerosis. *Physiol. Meas.* **2018**, *39*, 075004. [CrossRef]
11. Hilfiker, R.; Vaney, C.; Gattlen, B.; Meichtry, A.; Deriaz, O.; Lugon-Moulin, V.; Anchisi-Bellwald, A.M.; Palaci, C.; Foinant, D.; Terrier, P. Local dynamic stability as a responsive index for the evaluation of rehabilitation effect on fall risk in patients with multiple sclerosis: A longitudinal study. *BMC Res. Notes* **2013**, *6*, 260. [CrossRef] [PubMed]
12. Motl, R.W.; McAuley, E.; Sandroff, B.M. Longitudinal change in physical activity and its correlates in relapsing-remitting multiple sclerosis. *Phys. Ther.* **2013**, *93*, 1037–1048. [CrossRef] [PubMed]
13. Sosnoff, J.J.; Sandroff, B.M.; Motl, R.W. Quantifying gait abnormalities in persons with multiple sclerosis with minimal disability. *Gait Posture* **2012**, *36*, 154–156. [CrossRef] [PubMed]
14. Huisinga, J.M.; Mancini, M.; St. George, R.; Horak, F. Accelerometry reveals differences in gait variability between patients with multiple sclerosis and healthy controls. *Ann. Biomed. Eng.* **2013**, *41*, 1670–1679. [CrossRef] [PubMed]
15. Spain, R.I.; Mancini, M.; Horak, F.B.; Bourdette, D. Body-worn sensors capture variability, but not decline, of gait and balance measures in multiple sclerosis over 18 months. *Gait Posture* **2014**, *39*, 958–964. [CrossRef] [PubMed]
16. Pau, M.; Mandaresu, S.; Pilloni, G.; Porta, M.; Coghe, G.; Marrosu, M.G.; Cocco, E. Smoothness of gait detects early alterations of walking in persons with multiple sclerosis without disability. *Gait Posture* **2017**, *58*, 307–309. [CrossRef]
17. Monticone, M.; Ambrosini, E.; Fiorentini, R.; Rocca, B.; Liquori, V.; Pedrocchi, A.; Ferrante, S. Reliability of spatial–temporal gait parameters during dual-task interference in people with multiple sclerosis. A cross-sectional study. *Gait Posture* **2014**, *40*, 715–718. [CrossRef]
18. Riva, F.; Bisi, M.; Stagni, R. Gait variability and stability measures: Minimum number of strides and within-session reliability. *Comput. Biol. Med.* **2014**, *50*, 9–13. [CrossRef]
19. König, N.; Singh, N.; von Beckerath, J.; Janke, L.; Taylor, W. Is gait variability reliable? An assessment of spatio-temporal parameters of gait variability during continuous overground walking. *Gait Posture* **2014**, *39*, 615–617. [CrossRef]
20. Bautmans, I.; Jansen, B.; Van Keymolen, B.; Mets, T. Reliability and clinical correlates of 3D-accelerometry based gait analysis outcomes according to age and fall-risk. *Gait Posture* **2011**, *33*, 366–372. [CrossRef]
21. Ben Mansour, K.; Gorce, P.; Rezzoug, N. The Multifeature Gait Score: An accurate way to assess gait quality. *PLoS ONE* **2017**, *12*, e0185741. [CrossRef] [PubMed]
22. Givon, U.; Zeilig, G.; Achiron, A. Gait analysis in multiple sclerosis: Characterization of temporal-spatial parameters using GAITRite functional ambulation system. *Gait Posture* **2009**, *29*, 138–142. [CrossRef] [PubMed]
23. Storm, F.A.; Nair, K.P.S.; Clarke, A.J.; Van der Meulen, J.M.; Mazzà, C. Free-living and laboratory gait characteristics in patients with multiple sclerosis. *PLoS ONE* **2018**, *13*, e0196463. [CrossRef] [PubMed]
24. Shanahan, C.J.; Boonstra, F.M.C.; Cofré Lizama, L.E.; Strik, M.; Moffat, B.A.; Khan, F.; Kilpatrick, T.J.; van der Walt, A.; Galea, M.P.; Kolbe, S.C. Technologies for Advanced Gait and Balance Assessments in People with Multiple Sclerosis. *Front. Neurol.* **2018**, *8*, 708. [CrossRef]

25. Caggiari, S.; Mura A.; Leban, B.; Corona, F.; Coghe, G.; Marrosu, M.G.; Cocco, E.; Pau, M. Clinical use of wearable inertial sensors to assess spatialtemporal parameters of gait in people with multiple sclerosis: Correlation with MSWS-12. *Mult. Scler.* **2016**, *22*, 132. [CrossRef]
26. Pearson, O.; Busse, M.; van Deursen, R.; Wiles, C. Quantification of walking mobility in neurological disorders. *QJM* **2004**, *97*, 463–475. [CrossRef] [PubMed]
27. Johansson, J.; Nordström, A.; Nordström, P. Greater Fall Risk in Elderly Women Than in Men Is Associated With Increased Gait Variability During Multitasking. *J. Am. Med. Dir. Assoc.* **2016**, *17*, 535–540. [CrossRef]
28. Spain, R.; George, R.; Salarian, A.; Mancini, M.; Wagner, J.; Horak, F.; Bourdette, D. Body-worn motion sensors detect balance and gait deficits in people with multiple sclerosis who have normal walking speed. *Gait Posture* **2012**, *35*, 573–578. [CrossRef]
29. Motta, C.; Palermo, E.; Studer, V.; Germanotta, M.; Germani, G.; Centonze, D.; Cappa, P.; Rossi, S.; Rossi, S. Disability and Fatigue Can Be Objectively Measured in Multiple Sclerosis. *PLoS ONE* **2016**, *11*, e0148997. [CrossRef]
30. Kempen, J.C.E.; Doorenbosch, C.A.M.; Knol, D.L.; de Groot, V.; Beckerman, H. Newly Identified Gait Patterns in Patients With Multiple Sclerosis May Be Related to Push-off Quality. *Phys. Ther.* **2016**, *96*, 1744–1752. [CrossRef]
31. Filli, L.; Sutter, T.; Easthope, C.S.; Killeen, T.; Meyer, C.; Reuter, K.; Lörincz, L.; Bolliger, M.; Weller, M.; Curt, A.; et al. Profiling walking dysfunction in multiple sclerosis: Characterisation, classification and progression over time. *Sci. Rep.* **2018**, *8*, 4984. [CrossRef] [PubMed]
32. Kelleher, K.J.; Spence, W.; Solomonidis, S.; Apatsidis, D. The characterisation of gait patterns of people with multiple sclerosis. *Disabil. Rehabil.* **2010**, *32*, 1242–1250. [CrossRef] [PubMed]
33. Al-Amri, M.; Nicholas, K.; Button, K.; Sparkes, V.; Sheeran, L.; Davies, J. Inertial Measurement Units for Clinical Movement Analysis: Reliability and Concurrent Validity. *Sensors* **2018**, *18*, 719. [CrossRef] [PubMed]
34. Khurelbaatar, T.; Kim, K.; Lee, S.; Kim, Y.H. Consistent accuracy in whole-body joint kinetics during gait using wearable inertial motion sensors and in-shoe pressure sensors. *Gait Posture* **2015**, *42*, 65–69. [CrossRef] [PubMed]
35. Bilney, B.; Morris, M.; Webster, K. Concurrent related validity of the GAITRite® walkway system for quantification of the spatial and temporal parameters of gait. *Gait Posture* **2003**, *17*, 68–74. [CrossRef]
36. Webster, K.E.; Wittwer, J.E.; Feller, J.A. Validity of the GAITRite walkway system for the measurement of averaged and individual step parameters of gait. *Gait Posture* **2005**, *22*, 317–321. [CrossRef] [PubMed]
37. Menz, H.B.; Latt, M.D.; Tiedemann, A.; Mun San Kwan, M.; Lord, S.R. Reliability of the GAITRite® walkway system for the quantification of temporo-spatial parameters of gait in young and older people. *Gait Posture* **2004**, *20*, 20–25. [CrossRef]
38. van Uden, C.J.; Besser, M.P. Test-retest reliability of temporal and spatial gait characteristics measured with an instrumented walkway system (GAITRite®). *BMC Musculoskelet. Disord.* **2004**, *5*, 13. [CrossRef]
39. McDonough, A.L.; Batavia, M.; Chen, F.C.; Kwon, S.; Ziai, J. The validity and reliability of the GAITRite system's measurements: A preliminary evaluation. *Arch. Phys. Med. Rehabil.* **2001**, *82*, 419–425. [CrossRef]
40. Kuys, S.S.; Brauer, S.G.; Ada, L. Test-retest reliability of the GAITRite system in people with stroke undergoing rehabilitation. *Disabil. Rehabil.* **2011**, *33*, 1848–1853. [CrossRef]
41. Wittwer, J.E.; Webster, K.E.; Andrews, P.T.; Menz, H.B. Test–retest reliability of spatial and temporal gait parameters of people with Alzheimer's disease. *Gait Posture* **2008**, *28*, 392–396. [CrossRef] [PubMed]
42. Polman, C.H.; Reingold, S.C.; Banwell, B.; Clanet, M.; Cohen, J.A.; Filippi, M.; Fujihara, K.; Havrdova, E.; Hutchinson, M.; Kappos, L.; et al. Diagnostic criteria for multiple sclerosis: 2010 revisions to the McDonald criteria. *Ann. Neurol.* **2011**, *69*, 292–302. [CrossRef] [PubMed]
43. Kurtzke, J.F. Rating neurologic impairment in multiple sclerosis: An expanded disability status scale (EDSS). *Neurology* **1983**, *33*, 1444–1452. [CrossRef] [PubMed]
44. Oudre, L.; Barrois-Müller, R.; Moreau, T.; Truong, C.; Vienne-Jumeau, A.; Ricard, D.; Vayatis, N.; Vidal, P.P. Template-Based Step Detection with Inertial Measurement Units. *Sensors* **2018**, *18*, 4033. [CrossRef] [PubMed]
45. Cohen, J. *Statistical Power Analysis for the Behavioral Sciences*, 2nd ed.; Lawrence Erlbaum Associates: New York, NY, USA, 1988.
46. Hinkle, D.E.; Wiersma, W.; Jurs, S.G. *Applied Statistics for the Behavioral Sciences*; Google-Books-ID: 7tntAAAAMAAJ; Houghton Mifflin: Boston, MA, USA, 2003.

47. Frohman, T.C.; Davis, S.L.; Beh, S.; Greenberg, B.M.; Remington, G.; Frohman, E.M. Uhthoff's phenomena in MS—Clinical features and pathophysiology. *Nat. Rev. Neurol.* **2013**, *9*, 535–540. [CrossRef]
48. Davis, F.A.; Michael, J.A.; Tomaszewski, J.S. Fluctuation of motor function in multiple sclerosis related to circadian temperature variations. *Dis. Nervous Syst.* **1973**, *34*, 33–36.
49. Syndulko, K.; Jafari, M.; Woldanski, A.; Baumhefner, R.W.; Tourtellotte, W.W. Effects of Temperature in Multiple Sclerosis: A Review of the Literature. *J. Neurol. Rehabil.* **1996**, *10*, 23–34. [CrossRef]
50. Ade, V.; Schalkwijk, D.; Psarakis, M.; Laporte, M.D.; Faras, T.J.; Sandoval, R.; Najjar, F.; Stubbs, P.W. Between session reliability of heel-to-toe progression measurements in the stance phase of gait. *PLoS ONE* **2018**, *13*, e0200436. [CrossRef]
51. Hsu, C.Y.; Tsai, Y.S.; Yau, C.S.; Shie, H.H.; Wu, C.M. Test-Retest Reliability of an Automated Infrared-Assisted Trunk Accelerometer-Based Gait Analysis System. *Sensors* **2016**, *16*, 1156. [CrossRef]
52. Orlowski, K.; Eckardt, F.; Herold, F.; Aye, N.; Edelmann-Nusser, J.; Witte, K. Examination of the reliability of an inertial sensor-based gait analysis system. *Biomedizinische Technik Biomed. Eng.* **2017**, *62*, 615–622. [CrossRef]
53. Almarwani, M.; Perera, S.; VanSwearingen, J.M.; Sparto, P.J.; Brach, J.S. The test–retest reliability and minimal detectable change of spatial and temporal gait variability during usual over-ground walking for younger and older adults. *Gait Posture* **2016**, *44*, 94–99. [CrossRef] [PubMed]
54. Sandroff, B.M.; Riskin, B.J.; Agiovlasitis, S.; Motl, R.W. Accelerometer cut-points derived during over-ground walking in persons with mild, moderate, and severe multiple sclerosis. *J. Neurol. Sci.* **2014**, *340*, 50–57. [CrossRef] [PubMed]
55. Coulter, E.H.; Miller, L.; McCorkell, S.; McGuire, C.; Algie, K.; Freeman, J.; Weller, B.; Mattison, P.G.; McConnachie, A.; Wu, O.; et al. Validity of the activPAL activity monitor in people moderately affected by Multiple Sclerosis. *Med. Eng. Phys.* **2017**, *45*, 78–82. [CrossRef] [PubMed]
56. Pau, M.; Caggiari, S.; Mura, A.; Corona, F.; Leban, B.; Coghe, G.; Lorefice, L.; Marrosu, M.G.; Cocco, E. Clinical assessment of gait in individuals with multiple sclerosis using wearable inertial sensors: Comparison with patient-based measure. *Mult. Scler. Relat. Disord.* **2016**, *10*, 187–191. [CrossRef] [PubMed]
57. Marschollek, M.; Goevercin, M.; Wolf, K.H.; Song, B.; Gietzelt, M.; Haux, R.; Steinhagen-Thiessen, E. A performance comparison of accelerometry-based step detection algorithms on a large, non-laboratory sample of healthy and mobility-impaired persons. In Proceedings of the Annual International Conference of the IEEE Engineering in Medicine and Biology Society (EMBC), Vancouver, BC, Canada, 20–25 August 2008; pp. 1319–1322.
58. Mantilla, J.; Oudre, L.; Barrois, R.; Vienne, A.; Ricard, D. Template-DTW based on inertial signals: Preliminary results for step characterization. In Proceedings of the Annual International Conference of the IEEE Engineering in Medicine and Biology Society (EMBC), Seogwipo, Korea, 11–15 July 2017.

Article

Comparison of Walking Protocols and Gait Assessment Systems for Machine Learning-Based Classification of Parkinson's Disease

Rana Zia Ur Rehman [1], Silvia Del Din [1], Jian Qing Shi [2], Brook Galna [1,3], Sue Lord [1,4], Alison J. Yarnall [1,5], Yu Guan [6] and Lynn Rochester [1,5,*]

[1] Institute of Neuroscience/Institute for Ageing, Newcastle University, Newcastle Upon Tyne NE4 5PL, UK; rana.zia-ur-rehman@ncl.ac.uk (R.Z.U.R.); silvia.del-din@ncl.ac.uk (S.D.D.); brook.galna@newcastle.ac.uk (B.G.); sue.lord@aut.ac.nz (S.L.); alison.yarnall@ncl.ac.uk (A.J.Y.)

[2] School of Mathematics, Statistics, and Physics, Newcastle University, Newcastle Upon Tyne NE1 7RU, UK; jian.shi@ncl.ac.uk

[3] School of Biomedical, Nutritional and Sport Sciences, Newcastle University, Newcastle Upon Tyne NE1 7RU, UK

[4] Department of Physiotherapy, Auckland University of Technology, Auckland 92006, New Zealand

[5] The Newcastle upon Tyne Hospitals NHS Foundation Trust, Newcastle Upon Tyne NE7 7DN, UK

[6] School of Computing, Newcastle University, Newcastle Upon Tyne NE4 5TG, UK; yu.guan@ncl.ac.uk

* Correspondence: lynn.rochester@ncl.ac.uk; Tel.: +44-(0)-191-208-1291; Fax: +44-(0)-191-208-1251

Received: 1 November 2019; Accepted: 2 December 2019; Published: 5 December 2019

Abstract: Early diagnosis of Parkinson's diseases (PD) is challenging; applying machine learning (ML) models to gait characteristics may support the classification process. Comparing performance of ML models used in various studies can be problematic due to different walking protocols and gait assessment systems. The objective of this study was to compare the impact of walking protocols and gait assessment systems on the performance of a support vector machine (SVM) and random forest (RF) for classification of PD. 93 PD and 103 controls performed two walking protocols at their normal pace: (i) four times along a 10 m walkway (intermittent walk-IW), (ii) walking for 2 minutes on a 25 m oval circuit (continuous walk-CW). 14 gait characteristics were extracted from two different systems (an instrumented walkway—GAITRite; and an accelerometer attached at the lower back—Axivity). SVM and RF were trained on normalized data (accounting for step velocity, gender, age and BMI) and evaluated using 10-fold cross validation with area under the curve (AUC). Overall performance was higher for both systems during CW compared to IW. SVM performed better than RF. With SVM, during CW Axivity significantly outperformed GAITRite (AUC: 87.83 ± 7.81% vs. 80.49 ± 9.85%); during IW systems performed similarly. These findings suggest that choice of testing protocol and sensing system may have a direct impact on ML PD classification results and highlight the need for standardization for wide scale implementation.

Keywords: Parkinson's disease; machine learning; classification; wearables; accelerometer; GAITRite; multi-regression normalization; SVM; random forest classifier

1. Introduction

Parkinson's disease (PD) is a complex neurodegenerative disorder which progresses over time [1] and comprises both motor and non-motor symptoms [2], leading to poor disease management, poorer quality of life [3], and increased health care costs [4]. Early diagnosis of PD is critical for optimal management but remains challenging. Current diagnosis of PD is commonly based on subjective clinical examination (clinical scales) [5] often in conjunction with expensive and time consuming brain imaging techniques. Gait has been shown to act as a marker of global health and has been used to predict morbidity, mortality, falls risk and neurological disorders [6]. Recent work has shown that objective gait quantification of motor impairments can support PD diagnosis, also at an early stage [7,8].

Gait can be objectively quantified via a number of spatial-temporal characteristics and features [9–12]. In order to maximize information for disease classification, analysis of multiple characteristics can be enhanced using machine learning (ML) [6]. The most widely used ML models for PD classification are the support vector machine (SVM) and random forest (RF) [13–19]. Classification accuracy however is inconsistent across studies which may be largely due to methodological differences (e.g., testing protocols, gait assessment systems and normalization of participants' data) [13,14,17]. This leads to difficulty comparing across studies and in turn to select the optimal gait protocol and outcomes for classification purposes. For example, protocols used to measure gait have different durations, distance and speed [10,20,21]. Moreover, gait assessment systems range from gold standards in the field of gait analysis using camera based motion capture and instrumented walkways [11,12] to wearable devices [22]. Practically, wearable sensors such as accelerometers, gyroscopes and magnetometers [23] have advantages as they are not context specific and can be used in the clinic and home [10,24]. This is relevant if gait proves useful for disease classification and clinical use.

For accurate disease classification the differences between participants should also be as low as possible and this requires normalization of the selected features as an important and critical step [16]. Between-participants gait differences are related to demographic characteristics such as the age [25], gender, and BMI [26,27]. Gait characteristics are also speed dependent [28–30] and normalization of gait features with respect to speed is usually performed [16]. Robust normalization processes thus optimize ML models and classification of PD [31].

The effects of walking protocol and gait assessment systems on the performance of ML models and the impacts on disease classification remain unanswered questions. The objective of this study was therefore to investigate the impact of different walking protocols and gait assessment systems on the performance of SVM and RF models for PD classification. We also highlight the strengths and limitations of protocols and devices to guide decision making in future studies. We compared the effect from two different walking protocols at normal pace (four times along a 10 m walkway (intermittent walk-IW); walking for 2 min on a 25 m oval circuit (continuous walk-CW)); and two different gait assessment systems: GAITRite vs. Axivity.

2. Methods

2.1. Participants

Data from the "Incidence of Cognitive Impairment in Cohorts with Longitudinal Evaluation - GAIT" (ICICLE-GAIT) study [11,32] encompassing 93 people with early PD and without dementia at study entry, and 103 healthy controls (HC) were included in this cross-sectional analysis. PD was diagnosed according to the UK Parkinson's Disease Brain Bank criteria [21] by a movement disorder specialist [32]. The study was approved by the "Newcastle and North Tyneside Research Ethics Committee" (REC No. 09/H0906/82). All the participants gave their written informed consent before participating in the study. Experiments were conducted according to the declaration of Helsinki.

2.2. Demographic and Clinical Measures

Demographic characteristics such as age, height, weight, and BMI were recorded for all the participants. Cognition was assessed with the Mini-Mental State Examination (MMSE) [33] and balance confidence was evaluated with the balance self-confidence scale (Activities specific balance confidence scale; ABC) [34]. To assess PD motor severity, Hoehn & Yahr scale score [5] and the modified version of the Movement Disorder Society Unified Parkinson's Disease Rating Scale (MDS-UPDRS)—section III [35] were used. Phenotypes in PD, namely postural instability and gait difficulty (PIGD), indeterminate (ID) and tremor dominant (TD) subtypes, were also calculated from MDS-UPDRS [36]. Levodopa equivalent daily dose (LEDD mg/day) was calculated according to defined criteria [37,38].

2.3. Walking Protocols and Data Collection

Two different protocols were used to assess gait. All PD participants were assessed one hour after medication intake:

(1) Ten meter (m) intermittent walking test (IW). Participants were instructed to walk in a straight line over a 10 m walkway (Figure 1a). They repeated this four times at their preferred speed. The GAITRite mat was placed in the center of the walkway [21].

(2) Two minute continuous walking test (CW). Participants were asked to walk continuously around at 25 m oval circuit at their preferred speed (Figure 1b).

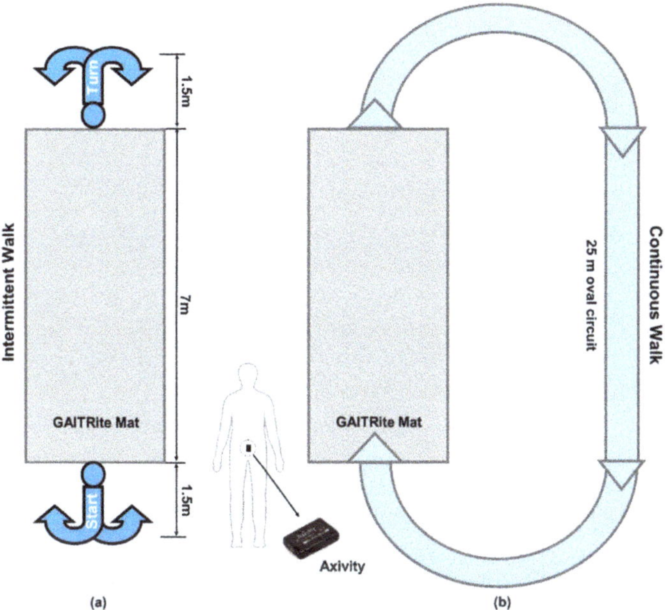

Figure 1. Layout of experimental setup and testing protocols, (**a**) 10 m intermittent walking test (IW); (**b**) 2 min continuous walking test (CW).

2.4. Gait Assessment Systems

Each participant was asked to wear a tri-axial accelerometer (Axivity AX3, dimensions: 23.0 × 32.5 × 7.6 mm) on the lower back (L5), held in place with double sided tape (BSN Medical Limited, Hull, UK) [10]. The monitor measures the vertical, mediolateral and anteroposterior accelerations during walking at 100 Hz sampling frequency (±8 g range, resolution up to 13-bit). Data collected using Axivity were synchronized with real-time clock, and start and stop times of the trials were noted by the experimenter to automatically segment and analyze the accelerometer data via MATLAB®. Gait assessment was also conducted using an instrumented mat (Platinum model GAITRite: 7.0 × 0.6 m) [12]. GAITRite has a spatial accuracy of 1.27 cm and temporal accuracy of one sample (240 Hz, ~4.17 ms).

2.5. Data Processing and Gait Characteristics Extraction

From each testing protocol and gait assessment system, 14 gait characteristics were extracted [10,11]. Methods described in our previous work were used for extracting gait characteristics from the 10 m test and the 2 min test with GAITRite and Axivity [10]. For easy interpretation, these 14 gait characteristics were grouped into five domains (pace, rhythm, variability, asymmetry, and postural control) as described in our previous work [11].

2.6. Statistical Analysis, Gait Normalization and Classification Modeling

Multivariate analysis of variance (MANOVA) was performed on normalized gait characteristics to examine the main effect and interactions of group (PD vs. HC), walking protocols (IW vs. CW) and gait assessment systems (GAITRite vs. Axivity) on the gait characteristics. Independent t-tests were performed to understand the between-group (PD vs. HC) differences of demographic and gait characteristics to include as input to the ML model. Receiver operating characteristics (ROC) analysis was used to measure the discriminative power of each gait characteristic. Pearson's correlation coefficients (r) were used to check the dependency among gait characteristics within each group. Distribution of each gait characteristic was plotted using rain cloud plots [39] for each group, walking protocol and gait assessment system. Gait characteristics were normalized for ML using multiple regression normalization [16] performed with respect to preferred gait speed (step velocity in each walking protocol from each gait assessment system), age, BMI and gender. This gave the ratio of the original and predicted gait characteristics based on the following equations:

$$y_i = \beta_0 + \beta_1 * Gender_i + \beta_2 * Age_i + \beta_3 * BMI_i + \beta_4 * StepVelocity(speed)_i + \epsilon_i \quad (1)$$

where y_i is the gait characteristics from the 5 domains of the conceptual gait model, ith participant. β_0 is the intercept and β is the coefficient of the linear regression. $\epsilon_i \sim N\left(0, \sigma^2\right)$ is the residual for each participant i. For each testing task and sensing system, the model coefficients were estimated using the healthy control participants' data based on Equation (1):

$$y_i = \hat{y}_i + \hat{\epsilon}_i \quad (2)$$

where $\hat{y}_i$ and $\hat{\epsilon}_i$ are the predicted value and residual error for the ith participant. Finally the normalized gait features were obtained by dividing the original independent gait feature with the predicted dependent gait feature by using following equation:

$$y_i^n = \frac{y_i}{\hat{y}_i} \quad (3)$$

where y_i^n is the final normalized gait characteristic for the ith participant and n is normalized. Based on the Taylor's series the expected value of the control group normalized gait characteristics should be 1. The resulting gait characteristics will be unit less due to the division of y_i and $\hat{y}_i$ as these have the same measuring units.

The support vector machine with radial basis function (SVM-RBF) and random forest were used because these are the most widely used ML models for PD classification [13–19,40]. The models were trained on the same conceptual features from both sensing systems to compare the impact of walking protocols and gait assessment systems. 10-fold cross validation repeated 100 times was used for the evaluation of the models. Single measure, area under the curve was used for the model evaluation [41]. Importance of the gait characteristics was identified by extracting the square of the weight of the gait characteristics in the SVM-linear classifier [42,43]. Gait characteristic importance is a unitless number which was used to rank the variables based on their contribution in the classification by SVM model.

This was calculated as the square of the weight calculated in the SVM model for each variable with the following Equation (4):

$$Imporatnce = w^2 = \left(\sum_{k=1}^{k=N} \alpha_k x_k l_k\right)^2 \quad (4)$$

where w^2 gives the importance score and it is the entry wise square of the weight for each gait variable in the model. α_k represents the model parameter trained on data $\{x_k, l_k\}$, where k is 1 to N. N represents the sample size, x_k is each subject data with corresponding label l_k. For ML, standard commands for SVM with different kernels (RBF and linear) and default parameters (slack variable-C:1) were used from SciKit-learn library in Python [44] for comparison among walking protocols and gait assessment systems. Similarly in RF, 100 trees were used for final performance estimation.

3. Results

Demographic characteristics are shown in Table 1. Compared to HCs, PDs had comparable height, weight, and BMI, included proportionally more males; were significantly younger; presented with significantly lower balance confidence (ABC) and poorer cognition (MMSE). Mostly PDs were at mild to moderate stage of the diseases based on the Hoehn & Yahr scale. PD gait was assessed within 23.8 months of clinical diagnosis while taking average 398 mg/day LEDD.

Table 1. Demographic and clinical characteristics of the participants.

Demographics	HC (n = 103) Mean ± SD	PD (n = 93) Mean ± SD	*p*
M/F	49/54	59/34	**0.026**
Age (years)	72.3 ± 6.7	69.2 ± 10.1	**0.012**
Height (m)	1.7 ± 0.09	1.7 ± 0.09	0.623
Mass (kg)	78.6 ± 14.3	78.6 ± 15.9	0.999
BMI (kg/m^2) [1]	27.2 ± 5.6	27.5 ± 4.7	0.750
MMSE (0–30) [2]	28.9 ± 1.9	28.4 ± 1.6	0.102
ABCs (0–100)% [3]	91.2 ± 13.8	80.6 ± 20.7	**<0.001**
LEDD, mg/day [4]		397.7 ± 217.2	
Disease Duration (months)		23.8 ± 4.2	
Hoehn & Yahr (n)		HY I: 8	
		HY II: 71	
		HY III: 14	
MDS-UPDRS III [5]		32.4 ± 10.3	
		(HY I: 17.4 ± 4.5)	
		(HY II: 32.9 ± 9.7)	
		(HY III: 38.1 ± 7.5)	
Motor Phenotype (n)		[6] PIGD: 34	
		[7] ID: 16	
		[8] TD: 43	

[1] BMI: Body Mass Index; [2] MMSE: Mini–Mental State Examination; [3] ABC: Activities specific balance confidence scale; [4] LEDD: Levodopa equivalent daily dose; [5] MDS-UPDRS III: Movement Disorders Unified Parkinson's Disease Rating Scale part III; [6] PIGD: Postural instability and gait disorder phenotype; [7] ID: Indeterminate phenotype; [8] TD: Tremor dominant phenotype. In bold significant p-values ($p < 0.05$).

Table 2 shows the main effects and interaction effect for the group (PD vs. HC), walking protocols and gait assessment systems on gait characteristics. Table 3 shows the mean and standard deviation of raw gait characteristics and the statistical difference for each normalized gait characteristic between PD and HC for the two walking tasks (IW and CW) and two gait assessment systems (GAITRite vs. Axivity). The results for the multi regression normalization are given in the supplementary Tables S1 and S2. Plots of the whole data set to check the distribution, outliers, confidence intervals, and AUC are shown in Figure S1 in the Supplementary Material. Correlations among the gait characteristics are given in Supplementary Figure S2. Gait characteristics were categorized into five domains (pace, rhythm, variability, asymmetry, and postural control) [11] based on a model of gait to help summarize findings.

Table 2. MANOVA to check the effect of walking protocols and gait assessment systems on gait (* indicates interaction).

Effect Assessment on Gait	MANOVA		
	Wilk's Lambda	F	*p*-Value
Group (HC & PD)	0.803	14.198	<0.001
Walking Protocols	0.463	67.337	<0.001
Gait Assessment Systems	0.067	805.792	<0.001
Group * Protocol	0.949	3.092	<0.001
Group * Systems	0.853	9.991	<0.001
Protocols * Systems	0.513	55.168	<0.001

Table 3. Mean comparison among PD and HC for gait characteristics obtained from walking protocols and assessment systems (significant normalized gait characteristics are highlighted in grey color, in bold significant *p*-values ($p < 0.05$) for normalized gait characteristics except step velocity).

Gait Domains	Gait Characteristics	Intermittent Walk (IW)			Continuous Walk (CW)		
		HC (n = 103) Mean ± SD	PD (n = 93) Mean ± SD	*p* Value	HC (n = 103) Mean ± SD	PD (n = 93) Mean ± SD	*p* Value
Gait Characteristics from Axivity							
Pace	Step Velocity (m/s)	1.324 ± 0.153	1.252 ± 0.226	**0.002**	1.283 ± 0.155	1.186 ± 0.262	**0.009**
	Step Length (m)	0.718 ± 0.094	0.717 ± 0.074	**0.010**	0.694 ± 0.121	0.690 ± 0.077	**0.022**
	Swing Time Variability (s)	0.064 ± 0.084	0.123 ± 0.144	**<0.001**	0.037 ± 0.031	0.108 ± 0.082	0.058
Rhythm	Step Time (s)	0.554 ± 0.052	0.614 ± 0.129	**0.001**	0.538 ± 0.046	0.609 ± 0.133	**<0.001**
	Swing Time (s)	0.394 ± 0.047	0.448 ± 0.116	**0.001**	0.386 ± 0.044	0.454 ± 0.125	**<0.001**
	Stance Time (s)	0.705 ± 0.059	0.767 ± 0.141	**0.003**	0.689 ± 0.054	0.763 ± 0.144	**<0.001**
Variability	Step Velocity Variability (m/s)	0.174 ± 0.097	0.196 ± 0.078	0.273	0.137 ± 0.060	0.190 ± 0.076	**<0.001**
	Step Length Variability (m)	0.101 ± 0.060	0.126 ± 0.059	**0.022**	0.072 ± 0.034	0.109 ± 0.044	**<0.001**
	Step Time Variability (s)	0.093 ± 0.103	0.162 ± 0.157	**<0.001**	0.037 ± 0.033	0.114 ± 0.087	**<0.001**
	Stance Time Variability (s)	0.094 ± 0.103	0.166 ± 0.158	**0.001**	0.039 ± 0.033	0.116 ± 0.088	**<0.001**
Asymmetry	Step Time Asymmetry (s)	0.031 ± 0.018	0.051 ± 0.034	0.610	0.021 ± 0.016	0.026 ± 0.025	0.268
	Swing Time Asymmetry (s)	0.023 ± 0.017	0.039 ± 0.028	0.437	0.020 ± 0.018	0.023 ± 0.024	0.592
	Stance Time Asymmetry (s)	0.030 ± 0.019	0.044 ± 0.027	0.771	0.020 ± 0.018	0.024 ± 0.02	0.419
Postural Control	Step length Asymmetry (m)	0.078 ± 0.053	0.119 ± 0.112	0.606	0.066 ± 0.052	0.126 ± 0.128	0.060
Gait Characteristics from GAITRite							
Pace	Step Velocity (m/s)	1.338 ± 0.198	1.194 ± 0.223	**<0.001**	1.301 ± 0.192	1.135 ± 0.218	**<0.001**
	Step Length (m)	0.697 ± 0.084	0.636 ± 0.098	**<0.001**	0.683 ± 0.083	0.616 ± 0.097	**<0.001**
	Swing Time Variability (s)	0.013 ± 0.003	0.016 ± 0.008	**0.327**	0.013 ± 0.004	0.017 ± 0.009	**0.010**
Rhythm	Step Time (s)	0.525 ± 0.045	0.538 ± 0.047	**<0.001**	0.528 ± 0.044	0.548 ± 0.047	**<0.001**
	Swing Time (s)	0.385 ± 0.030	0.382 ± 0.033	**<0.001**	0.385 ± 0.029	0.384 ± 0.031	**0.001**
	Stance Time (s)	0.665 ± 0.068	0.695 ± 0.072	**<0.001**	0.674 ± 0.066	0.714 ± 0.074	**<0.001**
Variability	Step Velocity Variability (m/s)	0.051 ± 0.015	0.047 ± 0.014	0.946	0.050 ± 0.012	0.054 ± 0.014	**0.005**
	Step Length Variability (m)	0.019 ± 0.006	0.020 ± 0.007	**0.008**	0.020 ± 0.006	0.023 ± 0.007	0.338
	Step Time Variability (s)	0.014 ± 0.004	0.016 ± 0.007	0.173	0.014 ± 0.004	0.018 ± 0.006	**0.018**
	Stance Time Variability (s)	0.016 ± 0.005	0.019 ± 0.011	0.260	0.017 ± 0.006	0.023 ± 0.012	**0.011**
Asymmetry	Step Time Asymmetry (s)	0.011 ± 0.008	0.018 ± 0.018	**0.003**	0.012 ± 0.009	0.019 ± 0.022	**0.007**
	Swing Time Asymmetry (s)	0.007 ± 0.006	0.014 ± 0.014	**<0.001**	0.007 ± 0.006	0.014 ± 0.014	**0.003**
	Stance Time Asymmetry (s)	0.007 ± 0.006	0.014 ± 0.014	0.476	0.007 ± 0.006	0.015 ± 0.015	**<0.001**
Postural Control	Step length Asymmetry (m)	0.020 ± 0.016	0.022 ± 0.018	**0.048**	0.019 ± 0.015	0.022 ± 0.020	**0.036**

Firstly, we established the effect of protocol and sensor system on gait characteristics as a first step to evaluate ML performance. There were significant main and interaction effects of pathology, walking protocols, and gait assessment systems on gait as shown in Table 2 and the individual gait characteristics are displayed in Table 3.

For group (PD vs. HC) people with PD had worse gait performance compared to controls irrespective of the protocol or gait assessment system. Grouping variables by domain [11], in general PD pace and rhythm were significantly slower while variability and asymmetry were higher, in both IW and CW protocols for both gait assessment systems.

There was a main effect of walking protocol (IW & CW) on gait characteristics. Performance was typically greater (higher pace, rhythm, variability, and asymmetry) in the IW protocol compared to the CW protocol for both PD and HC. Similarly, there were significant main and interaction effects of assessment systems on gait characteristics. In general, the values from Axivity tended to be higher compared to GAITRite although only asymmetry was significantly different between the systems.

Table 4 shows the contribution of gait characteristics in the classification modelling. A higher importance score indicates a greater contribution of each gait characteristic in the overall classification model. The top 5 Axivity characteristics were from variability, rhythm, and pace domains for both CW and IW. For GAITRite, CW contained gait characteristics from pace, rhythm and asymmetry domains, while for IW pace, rhythm and variability were important. Results without gait normalization are presented in the supplementary Table S3 and Figure S3.

Table 4. Importance of normalized gait characteristics in the classification of PD.

Sensing System	Intermittent Walk		Continuous Walk	
	Characteristic	Importance	Characteristic	Importance
Axivity	Mean Step Length	0.22	Step Velocity Variability	1.10
	Mean Stance Time	0.20	Mean Swing Time	0.72
	Mean Swing Time	0.15	Mean Step Length	0.49
	Swing Time Variability	0.14	Stance Time Variability	0.20
	Mean Step Time	0.07	Step Length Variability	0.12
GAITRite	Mean Step Time	0.23	Mean Step Length	3.80
	Step Velocity Variability	0.22	Mean Step Time	2.72
	Step Length Variability	0.15	Stance Time Asymmetry	1.21
	Swing Time Variability	0.14	Mean Stance Time	1.10
	Mean Step Length	0.09	Swing Time Asymmetry	0.72

Both models (SVM-RBF & RF) behaved in the similar manner for both walking protocols and gait assessment systems, with better performance of Axivity compared to GAITRite in both walking tasks with RF. Overall, SVM-RBF performed better than RF. Therefore for comparison of walking protocols and gait assessment, we only reported the results from SVM-RBF. The results of RF are given in the supplementary Table S4.

Overall, the classification of PD was significantly more accurate with Axivity (<0.001) during the CW test (AUC 87.83 ± 7.81% for Axivity and 80.49 ± 9.85% for GAITRite), while there was no difference ($p = 0.073$) between the systems during the IW test (AUC resulted being 79.09 ± 10.11% for Axivity and 79.90 ± 10.06% for GAITRite) (Figure 2 shows the distribution of the model classification performance, where the x-axis represents the classification performance (AUC), the top x-axis represents walking protocols, and the y-axis represents the gait assessment system). For reference, SVM performance results without gait normalization are presented in the Supplementary Figure S4.

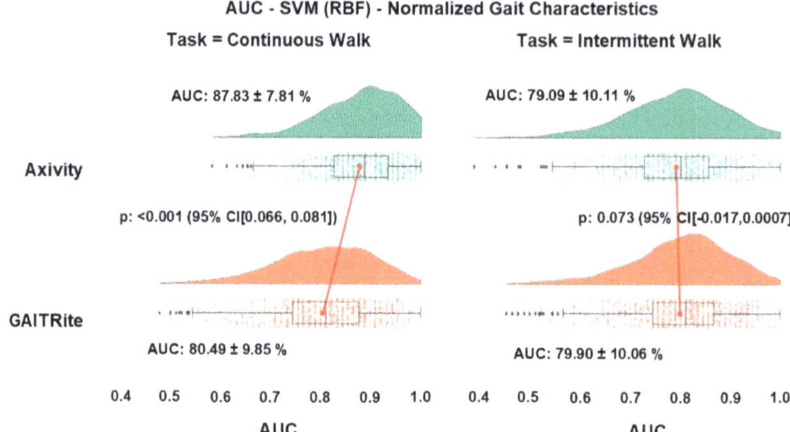

Figure 2. Distribution of SVM classification performance after normalization of gait characteristics.

4. Discussion

To the best of our knowledge this is the first study to investigate the impact of different walking protocols and gait assessment systems on the performance of ML models for classification of PD. Robust normalization techniques were carried out to reduce the effect of demographics and speed on between participant differences within each group (PD and HC). A comprehensive group of 14 gait characteristics were selected based on a validated gait model. Finally, widely used SVM-RBF and RF models were trained for classification of PD and HC. The results show that different walking protocols and gait assessment systems significantly affect gait characteristics and in turn the performance of ML models. Harmonizing methods across multiple levels for comparative purposes is strongly advised to optimize and implement ML in disease classification.

4.1. ML Performance: An Overview

In this study, we found that the combination of CW protocol and Axivity gave the highest PD classification performance. In terms of protocols, ML performance was higher during CW with respect to IW. In terms of systems, Axivity showed a significantly higher AUC (87.83 ± 7.81%) compared to GAITRite (80.49 ± 9.85%) during CW. Similar pattern in results was achieved with RF, where Axivity showed better results compared to GAITRite during both CW & IW. Therefore, walking protocols and gait assessment systems materially impact on ML performance, which makes the comparison of previous ML studies inconclusive. In fact, previous literature has shown that, when using wearables to quantify gait, studies using 2 min CW protocols [18,19] achieved better results compared to those using 10m IW protocols [13,45]. In addition, studies showed that ML models derived from wearable inertial and force feet sensors [14,19,45,46] performed relatively better when compared to studies based on GAITRite data [17].

It's important to underline that many factors can influence ML results: not only walking protocols and gait assessment systems, but also cohort size, disease severity stage of PD, and validation method. However, in the context of this study, we showed that walking protocol and gait assessment have a significant impact on ML performance.

4.2. Effect of Walking Protocols on ML Model and Performance

In general, ML performance was higher during CW with respect to IW. There are a number of possible explanations for this. The gait characteristics included in each ML model were different for CW and IW, which may explain the differences in classification performances (i.e., CW higher AUC than IW). Indeed performance of ML models are influenced by the characteristics included in the model and those characteristics are in turn influenced by the protocol used to assess gait. During IW we observed higher gait performance (e.g., higher pace, etc.) for all characteristics and for both groups (HC and PD). Acceleration and deceleration phases at the beginning and end of each IW increase the dispersion of gait characteristics, especially variability and asymmetry compared to CW where gait was sampled under more steady state conditions (lower variability and asymmetry values for both walking systems).

Another aspect is that even though participants were instructed to walk at their normal preferred pace for both protocols, it is clear that gait performance was faster during the IW compared to CW, and this has been reported previously [21]. The reasons for this are most likely because attention to performance is higher during short intermittent walking tests than a continuous steady state—where walking is performed with less attention and conscious effort. However this will also influence the dispersion of gait characteristics for PD and HC as seen by the standard deviations from Table 3. For accurate classification between groups, this dispersion within each group for each characteristic should be minimized to increase the distance between groups. This explains the need to overcome between subject variability within each group to enhance ML performance.

Collectively, this suggests that the walking protocol should be selected carefully and protocols that capture more steady state gait (in our case CW) may be optimal for classification and therefore early identification of PD.

4.3. Effect of Gait Assessment Systems on ML Model and Performance

In general, Axivity showed significantly higher classification performance compared to GAITRite during CW and comparable performance during IW. Gait characteristics quantified by the two systems showed significant differences. Even if these two systems (GAITRite and Axivity) measured the same spatial-temporal characteristics from the same walking tasks, the mechanism by which gait characteristics are derived is different. GAITRite determines footfalls based on pressure sensors that identify each step from which additional gait characteristics are derived [47]. Axivity, instead, uses accelerometers which detect movement continuously: individual characteristics are then derived from the raw signal. In previous work comparing gait characteristics from Axivity and GAITRite, mean spatiotemporal gait characteristics (such as walking speed, step length and step time) showed high agreement, while variability and asymmetry showed low agreement between the systems [10]. Gait characteristics extracted from GAITRite are more variable (wider dispersion) at slow speed [48]. Conversely, an accelerometer positioned at lower back may mis-detect gait events like initial and final contacts which may impact on gait characteristic quantification [49]. An accelerometer close to the center of mass of the body can capture small variations in body movement (variability) during walking [50] with higher sensitivity compared to GAITRite. Analysis of the current study indicated that Axivity is more sensitive to detect variability and asymmetry, particularly in PD. Collectively all these factors most likely influence: (i) the observed differences in the gait characteristics quantified by each system and (ii) ultimately the performance of the ML models.

The highest classification performance obtained with Axivity during the CW could also simply be due to the higher amount of data available for Axivity vs. GAITRite during the walking protocol: Axivity continuously sampled the entire 2 min while GAITRite sampled only each pass over the mat (e.g., four passes) during the same time frame. This is in part corroborated by the fact that we found comparable performances during the IW, when the amount of data was similar for both the systems. A further explanation for the difference in performance could be related to the inclusion of gradual turns with Axivity during the CW protocol which could have influenced the ML model. Axivity is not able to quantify turning due to the lack of a gyroscope; to try and address this, we measured turning from a sensor with an embedded gyroscope (Opal Mobility lab system APDM Inc., Portland, OR, USA) collected concurrently in a subsample of the same cohort (PD: 31, HC: 49) during CW and IW (Figure 3 shows the probability distribution of turning gait characteristics, where the x-axis represents the corresponding units and the y-axis represents the walking protocols). Turning time and angular velocity were significantly different during IW and so the turning segment of the signal was removed from Axivity analysis to delineate the gait characteristics. There were no significant between-group differences in turning characteristics during CW, and so step data from the turning component was retained for the analysis. However, we can't rule out the possibility that steps from the turning component during CW may have contributed to better classification between groups when using Axivity.

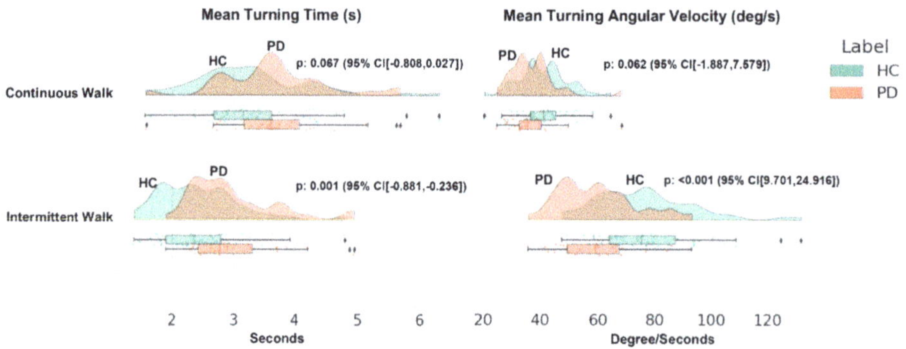

Figure 3. Distribution of turning characteristics.

4.4. Effect of Gait Normalization on ML Performance

From our results, it is clear that participants walked faster during the IW as compared to CW with both gait assessment systems (Table 3). This higher step velocity acts as a function of other gait characteristics [28–30], which can be influenced by its high variability. To find the appropriate walking protocol, multiple regression (MR) normalization using demographics and gait speed was important to reveal important influencing variables and overcome between participant differences among groups. Our findings support this approach, in fact we found that by controlling the effect of speed and demographics, SVM was able to differentiate between PD and HC more accurately. SVM performance increased by 5–7% in CW and 5–9% in the IW in both gait assessment systems (Supplementary Figure S4). The results are in line with previous work where similar gait normalization approaches have been used for better classification [16,31]. Thus, in short walks (IW), normalization may act as a standardized technique to overcome the effect of gait assessment systems. During CW, the effect of gait assessment systems was still significant. Normalization was also important to improve performance of ML—irrespective of protocol or system. This means that, normalization may be important to ensure standardization of walking protocol and gait assessment systems for optimal ML performance.

4.5. Limitations

This study had some limitations. Only two widely used ML models (SVM & RF) [13–19] were used in this study to compare the effect of walking protocols and gait assessment systems. However, future work should explore other classification models such as logistic regression and neural networks. Turning features were not included in this work due to the use of an accelerometer. In order to harmonize gait characteristics, step width and step width variability and a range of time series and frequency based characteristics were not included in the analysis because they could not be calculated from both systems. The inclusion of these additional variables may improve classification for respective systems and should be explored in future studies to investigate their impact on ML models. PD were assessed within 23.8 ± 4.2 months from clinical diagnosis, which is considered relatively early disease. The cohort assessed in this study was relatively young and results may not be applicable or generalizable to older, frailer people with PD with multi-morbidity.

5. Clinical Implications

Based on this study, walking protocols (IW & CW) and gait assessment systems had significant impact on the ML model performance. The extracted characteristics in CW with Axivity gave the highest performance in the classification ML model. Our work emphasizes the importance of the use of standardized walking protocols and wearable devices for ML PD classification purposes, to support clinical decision making. With the recent advancements in this field, this study will help clinicians to understand and select the appropriate walking protocols and gait assessment systems for optimal PD diagnosis. In future studies, such as those looking at prodromal disease, CW assessed with Axivity may give a more accurate reflection of gait changes. For better results, it is recommended to control for demographics and walking speed for gait characteristics normalization in the PD ML classification modeling. Intervention studies seeking to determine changes in particular gait characteristic(s) may be advised to use this methodology.

6. Conclusions

In this study, the impact of different walking protocols (CW & IW) and gait assessment systems (GAITRite & Axivity) on the performance of widely used ML models SVM and RF was investigated. Gait characteristics were normalized with respect to demographic properties and walking speed to overcome the between participants' differences within each group (HC and PD) for each walking protocol (CW vs. IW). Both ML models behaved in similar fashion for both walking protocols and gait assessment systems. Higher performances were achieved with CW compared to IW. Axivity gave higher classification performance compared to GAITRite. The highest PD classification performance was obtained during CW with Axivity (87.83 ± 7.81%). This work supports the idea that direct comparison of various ML studies using different walking protocols and gait assessment systems may not be appropriate. The findings from this study suggest that the choice of the testing protocol and gait assessment systems is important to achieve best classification results, which may have a direct impact on future end points in intervention studies. In conclusion, there is a need for standardization of walking protocols and gait assessment systems for wide scale implementation in clinical gait assessment.

Supplementary Materials: The following are available online at http://www.mdpi.com/1424-8220/19/24/5363/s1, Table S1: Correlation of gender, age, BMI, and step velocity (speed) with gait characteristics before and after normalization, Table S2: Coefficients from the regression model by using the healthy control participants, Table S3: Importance of gait characteristics in the classification of PD before and after gait characteristics normalization, Table S4: Random forest (RF) classification performance after gait normalization, Figure S1: Distribution of the gait characteristics from 5 domains of conceptual gait model with statistical analysis, Figure S2: Correlations among gait characteristics before and after normalization, Figure S3: Contribution of the gait characteristics in the classification modelling in Support Vector Machine, Figure S4: SVM performance before gait characteristics normalization.

Author Contributions: R.Z.U.R. performed data analysis, statistical analysis, drafting and critical revision of the manuscript. S.D.D. helped in data analysis, interpretation of data and critical revision of manuscript for important intellectual content. Y.G. and J.Q.S. provided support for statistical analysis, interpretation, and critical revision of manuscript for important intellectual content. A.J.Y., B.G., and S.L. were involved in interpretation of data and critical revision of the manuscript. L.R. conceptualized and designed the study, helped in interpretation of data, and critically revised the manuscript for important intellectual content.

Funding: This work was supported by "Keep Control" project, which is European Union's Horizon 2020 research and innovation ITN program under the Marie Sklodowska-Curie grant agreement No. 721577. ICICLE-Gait study was supported by Parkinson's UK (J-0802, G-1301) and by the National Institute for Health Research (NIHR) Newcastle Biomedical Research Center (BRC) based at Newcastle Upon Tyne Hospital NHS Foundation Trust and Newcastle University (REC number: 09/H0906/82). The work was also supported by the NIHR/Wellcome Trust Clinical Research Facility (CRF) infrastructure at Newcastle upon Tyne Hospitals NHS Foundation Trust. All opinions are those of the authors and not the funders.

Acknowledgments: The authors would like to thank all the participants and assessors of the ICICLE study, Lisa Alcock and Rachael Lawson for their support.

Conflicts of Interest: The authors declare no conflict of interest. The sponsors had no role in the design, execution, interpretation, or writing of the study.

References

1. Dorsey, E.R.; George, B.P.; Leff, B.; Willis, A.W. The coming crisis: Obtaining care for the growing burden of neurodegenerative conditions. *Neurology* **2013**, *80*, 1989–1996. [CrossRef] [PubMed]
2. Jankovic, J. Parkinson's disease: Clinical features and diagnosis. *J. Neurol. Neurosurg. Psychiatry* **2008**, *79*, 368–376. [CrossRef] [PubMed]
3. Perez-Lloret, S.; Negre-Pages, L.; Damier, P.; Delval, A.; Derkinderen, P.; Destee, A.; Meissner, W.G.; Schelosky, L.; Tison, F.; Rascol, O. Prevalence, determinants, and effect on quality of life of freezing of gait in parkinson disease. *JAMA Neurol.* **2014**, *71*, 884–890. [CrossRef] [PubMed]
4. von Campenhausen, S.; Winter, Y.; e Silva, A.R.; Sampaio, C.; Ruzicka, E.; Barone, P.; Poewe, W.; Guekht, A.; Mateus, C.; Pfeiffer, K.-P. Costs of illness and care in parkinson's disease: An evaluation in six countries. *Eur. Neuropsychopharmacol.* **2011**, *21*, 180–191. [CrossRef]
5. Hoehn, M.M.; Yahr, M.D. Parkinsonism: Onset, progression, and mortality. *Neurology* **1998**, *50*, 318. [CrossRef]
6. Buckley, C.; Alcock, L.; McArdle, R.; Rehman, R.Z.U.; Del Din, S.; Mazzà, C.; Yarnall, A.J.; Rochester, L. The role of movement analysis in diagnosing and monitoring neurodegenerative conditions: Insights from gait and postural control. *Brain Sci.* **2019**, *9*, 34. [CrossRef]
7. Mirelman, A.; Bonato, P.; Camicioli, R.; Ellis, T.D.; Giladi, N.; Hamilton, J.L.; Hass, C.J.; Hausdorff, J.M.; Pelosin, E.; Almeida, Q.J. Gait impairments in parkinson's disease. *Lancet Neurol.* **2019**, *18*, 697–708. [CrossRef]
8. Del Din, S.; Elshehabi, M.; Galna, B.; Hobert, M.A.; Warmerdam, E.; Suenkel, U.; Brockmann, K.; Metzger, F.; Hansen, C.; Berg, D. Gait analysis with wearables predicts conversion to parkinson disease. *Ann. Neurol.* **2019**, *86*, 357–367. [CrossRef]
9. Del Din, S.; Hickey, A.; Hurwitz, N.; Mathers, J.C.; Rochester, L.; Godfrey, A. Measuring gait with an accelerometer-based wearable: Influence of device location, testing protocol and age. *Physiol. Meas.* **2016**, *37*, 1785. [CrossRef]
10. Del Din, S.; Godfrey, A.; Rochester, L. Validation of an accelerometer to quantify a comprehensive battery of gait characteristics in healthy older adults and parkinson's disease: Toward clinical and at home use. *IEEE J. Biomed. Health Inf.* **2016**, *20*, 838–847. [CrossRef]
11. Lord, S.; Galna, B.; Verghese, J.; Coleman, S.; Burn, D.; Rochester, L. Independent domains of gait in older adults and associated motor and nonmotor attributes: Validation of a factor analysis approach. *J. Gerontol. Ser. A* **2013**, *68*, 820–827. [CrossRef] [PubMed]
12. Lord, S.; Galna, B.; Rochester, L. Moving forward on gait measurement: Toward a more refined approach. *Mov. Disord.* **2013**, *28*, 1534–1543. [CrossRef] [PubMed]
13. Klucken, J.; Barth, J.; Kugler, P.; Schlachetzki, J.; Henze, T.; Marxreiter, F.; Kohl, Z.; Steidl, R.; Hornegger, J.; Eskofier, B. Unbiased and mobile gait analysis detects motor impairment in parkinson's disease. *PLoS ONE* **2013**, *8*, e56956. [CrossRef] [PubMed]
14. Caramia, C.; Torricelli, D.; Schmid, M.; Munoz-Gonzalez, A.; Gonzalez-Vargas, J.; Grandas, F.; Pons, J.L. Imu-based classification of parkinson's disease from gait: A sensitivity analysis on sensor location and feature selection. *IEEE J. Biomed. Health Inf.* **2018**, *22*, 1765–1774. [CrossRef] [PubMed]
15. Tahir, N.M.; Manap, H.H. Parkinson disease gait classification based on machine learning approach. *J. Appl. Sci.* **2012**, *12*, 180–185. [CrossRef]
16. Wahid, F.; Begg, R.K.; Hass, C.J.; Halgamuge, S.; Ackland, D.C. Classification of parkinson's disease gait using spatial-temporal gait features. *IEEE J. Biomed. Health Inf.* **2015**, *19*, 1794–1802. [CrossRef]
17. Djurić-Jovičić, M.; Belić, M.; Stanković, I.; Radovanović, S.; Kostić, V.S. Selection of gait parameters for differential diagnostics of patients with de novo parkinson's disease. *Neurol. Res.* **2017**, *39*, 853–861. [CrossRef]
18. Pham, T.D.; Yan, H. Tensor decomposition of gait dynamics in parkinson's disease. *IEEE Trans. Biomed. Eng.* **2018**, *65*, 1820–1827.
19. Alam, M.N.; Garg, A.; Munia, T.T.K.; Fazel-Rezai, R.; Tavakolian, K. Vertical ground reaction force marker for parkinson's disease. *PLoS ONE* **2017**, *12*, e0175951. [CrossRef]
20. Galna, B.; Lord, S.; Burn, D.J.; Rochester, L. Progression of gait dysfunction in incident parkinson's disease: Impact of medication and phenotype. *Mov. Disord.* **2015**, *30*, 359–367. [CrossRef]

21. Galna, B.; Lord, S.; Rochester, L. Is gait variability reliable in older adults and parkinson's disease? Towards an optimal testing protocol. *Gait Posture* **2013**, *37*, 580–585. [CrossRef] [PubMed]
22. Godinho, C.; Domingos, J.; Cunha, G.; Santos, A.T.; Fernandes, R.M.; Abreu, D.; Gonçalves, N.; Matthews, H.; Isaacs, T.; Duffen, J.; et al. A systematic review of the characteristics and validity of monitoring technologies to assess parkinson's disease. *J. Neuroeng. Rehabil.* **2016**, *13*.
23. Qiu, H.; Rehman, R.Z.U.; Yu, X.; Xiong, S. Application of wearable inertial sensors and a new test battery for distinguishing retrospective fallers from non-fallers among community-dwelling older people. *Sci. Rep.* **2018**, *8*, 16349. [CrossRef] [PubMed]
24. Godfrey, A.; Del Din, S.; Barry, G.; Mathers, J.C.; Rochester, L. Instrumenting gait with an accelerometer: A system and algorithm examination. *Med Eng. Phys.* **2015**, *37*, 400–407. [CrossRef] [PubMed]
25. Morris, M.E. Locomotor training in people with parkinson disease. *Phys. Ther.* **2006**, *86*, 1426–1435. [CrossRef]
26. Herman, T.; Giladi, N.; Gurevich, T.; Hausdorff, J. Gait instability and fractal dynamics of older adults with a "cautious" gait: Why do certain older adults walk fearfully? *Gait Posture* **2005**, *21*, 178–185. [CrossRef]
27. Kang, H.G.; Dingwell, J.B. Separating the effects of age and walking speed on gait variability. *Gait Posture* **2008**, *27*, 572–577. [CrossRef]
28. Andriacchi, T.; Ogle, J.; Galante, J. Walking speed as a basis for normal and abnormal gait measurements. *J. Biomech.* **1977**, *10*, 261–268. [CrossRef]
29. Kirtley, C.; Whittle, M.W.; Jefferson, R. Influence of walking speed on gait parameters. *J. Biomed. Eng.* **1985**, *7*, 282–288. [CrossRef]
30. Schwartz, M.H.; Rozumalski, A.; Trost, J.P. The effect of walking speed on the gait of typically developing children. *J. Biomech.* **2008**, *41*, 1639–1650. [CrossRef]
31. Kamruzzaman, J.; Begg, R.K. Support vector machines and other pattern recognition approaches to the diagnosis of cerebral palsy gait. *IEEE Trans. Biomed. Eng.* **2006**, *53*, 2479–2490. [CrossRef] [PubMed]
32. Khoo, T.K.; Yarnall, A.J.; Duncan, G.W.; Coleman, S.; O'Brien, J.T.; Brooks, D.J.; Barker, R.A.; Burn, D.J. The spectrum of nonmotor symptoms in early parkinson disease. *Neurology* **2013**, *80*, 276–281. [CrossRef] [PubMed]
33. Tombaugh, T.N.; McIntyre, N.J. The mini-mental state examination: A comprehensive review. *J. Am. Geriatr. Soc.* **1992**, *40*, 922–935. [CrossRef] [PubMed]
34. Powell, L.E.; Myers, A.M. The activities-specific balance confidence (abc) scale. *J. Gerontol. Ser. A Biol. Sci. Med Sci.* **1995**, *50*, M28–M34. [CrossRef] [PubMed]
35. Goetz, C.G.; Tilley, B.C.; Shaftman, S.R.; Stebbins, G.T.; Fahn, S.; Martinez-Martin, P.; Poewe, W.; Sampaio, C.; Stern, M.B.; Dodel, R. Movement disorder society-sponsored revision of the unified parkinson's disease rating scale (mds-updrs): Scale presentation and clinimetric testing results. *Mov. Disord. Off. J. Mov. Disord. Soc.* **2008**, *23*, 2129–2170. [CrossRef] [PubMed]
36. Stebbins, G.T.; Goetz, C.G.; Burn, D.J.; Jankovic, J.; Khoo, T.K.; Tilley, B.C. How to identify tremor dominant and postural instability/gait difficulty groups with the movement disorder society unified parkinson's disease rating scale: Comparison with the unified parkinson's disease rating scale. *Mov. Disord.* **2013**, *28*, 668–670. [CrossRef]
37. Lawson, R.A.; Yarnall, A.J.; Duncan, G.W.; Breen, D.P.; Khoo, T.K.; Williams-Gray, C.H.; Barker, R.A.; Collerton, D.; Taylor, J.-P.; Burn, D.J. Cognitive decline and quality of life in incident parkinson's disease: The role of attention. *Parkinsonism Relat. Disord.* **2016**, *27*, 47–53. [CrossRef]
38. Tomlinson, C.L.; Stowe, R.; Patel, S.; Rick, C.; Gray, R.; Clarke, C.E. Systematic review of levodopa dose equivalency reporting in parkinson's disease. *Mov. Disord.* **2010**, *25*, 2649–2653. [CrossRef]
39. Allen, M.; Poggiali, D.; Whitaker, K.; Marshall, T.R.; Kievit, R. Raincloud plots: A multi-platform tool for robust data visualization. *PeerJ Prepr.* **2018**, *6*, e27137v1. [CrossRef]
40. Rehman, R.Z.U.; Del Din, S.; Guan, Y.; Yarnall, A.J.; Shi, J.Q.; Rochester, L. Selecting clinically relevant gait characteristics for classification of early parkinson's disease: A comprehensive machine learning approach. *Sci. Rep.* **2019**, *9*, 17269. [CrossRef]
41. Bradley, A.P. The use of the area under the roc curve in the evaluation of machine learning algorithms. *Pattern Recognit.* **1997**, *30*, 1145–1159. [CrossRef]
42. Guyon, I.; Elisseeff, A. An introduction to variable and feature selection. *J. Mach. Learn. Res.* **2003**, *3*, 1157–1182.

43. Guyon, I.; Weston, J.; Barnhill, S.; Vapnik, V. Gene selection for cancer classification using support vector machines. *Mach. Learn.* **2002**, *46*, 389–422. [CrossRef]
44. Pedregosa, F.; Varoquaux, G.; Gramfort, A.; Michel, V.; Thirion, B.; Grisel, O.; Blondel, M.; Prettenhofer, P.; Weiss, R.; Dubourg, V. Scikit-learn: Machine learning in python. *J. Mach. Learn. Res.* **2011**, *12*, 2825–2830.
45. Cuzzolin, F.; Sapienza, M.; Esser, P.; Saha, S.; Franssen, M.M.; Collett, J.; Dawes, H. Metric learning for parkinsonian identification from imu gait measurements. *Gait Posture* **2017**, *54*, 127–132. [CrossRef]
46. Arora, S.; Venkataraman, V.; Donohue, S.; Biglan, K.M.; Dorsey, E.R.; Little, M.A. High accuracy discrimination of parkinson's disease participants from healthy controls using smartphones. In Proceedings of the 2014 IEEE International Conference on Acoustics, Speech and Signal Processing (ICASSP), Florence, Italy, 4–9 May 2014; pp. 3641–3644.
47. Webster, K.E.; Wittwer, J.E.; Feller, J.A. Validity of the gaitrite®walkway system for the measurement of averaged and individual step parameters of gait. *Gait Posture* **2005**, *22*, 317–321. [CrossRef]
48. Bilney, B.; Morris, M.; Webster, K. Concurrent related validity of the gaitrite®walkway system for quantification of the spatial and temporal parameters of gait. *Gait Posture* **2003**, *17*, 68–74. [CrossRef]
49. Panebianco, G.P.; Bisi, M.C.; Stagni, R.; Fantozzi, S. Analysis of the performance of 17 algorithms from a systematic review: Influence of sensor position, analysed variable and computational approach in gait timing estimation from imu measurements. *Gait Posture* **2018**, *66*, 76–82. [CrossRef]
50. Buganè, F.; Benedetti, M.G.; D'Angeli, V.; Leardini, A. Estimation of pelvis kinematics in level walking based on a single inertial sensor positioned close to the sacrum: Validation on healthy subjects with stereophotogrammetric system. *Biomed. Eng. Online* **2014**, *13*, 146. [CrossRef]

Article

Gait Asymmetry Post-Stroke: Determining Valid and Reliable Methods Using a Single Accelerometer Located on the Trunk

Christopher Buckley [1], M. Encarna Micó-Amigo [1], Michael Dunne-Willows [2], Alan Godfrey [3], Aodhán Hickey [4], Sue Lord [1,5], Lynn Rochester [1,6], Silvia Del Din [1] and Sarah A. Moore [1,7,8,*]

1 Institute of Neuroscience/Institute for Ageing, Newcastle University, Newcastle Upon Tyne NE4 5PL, UK; christopher.buckley2@newcastle.ac.uk (C.B.); maria.mico-amigo@newcastle.ac.uk (M.E.M.-A.); sue.lord@aut.ac.nz (S.L.); lynn.rochester@ncl.ac.uk (L.R.); silvia.del-din@newcastle.ac.uk (S.D.D.)
2 EPSRC Centre for Doctoral Training in Cloud Computing for Big Data, Newcastle University, Newcastle Upon Tyne NE4 5PL, UK; m.dunne-willows@newcastle.ac.uk
3 Department of Computer and Information Science, Northumbria University, Newcastle upon Tyne NE1 8ST, UK; alan.godfrey@northumbria.ac.uk
4 Department of Health Intelligence, HSC Public Health Agency, Belfast BT2 7ES, Northern Ireland; Aodhan.Hickey@hscni.net
5 Auckland University of Technology, 55 Wellesley St E, Auckland 1010, New Zealand
6 The Newcastle upon Tyne Hospitals NHS Foundation Trust, Newcastle Upon Tyne NE7 7DN, UK
7 Institute of Neuroscience (Stroke Research Group), Newcastle University, 3-4 Claremont Terrace, Newcastle upon Tyne NE2 4AE, UK
8 Stroke Northumbria, Northumbria Healthcare NHS Foundation Trust, Rake Lane, North Shields, Tyne and Wear NE29 8NH, UK
* Correspondence: s.a.moore@newcastle.ac.uk; Tel.: +44-191-208-3837

Received: 2 November 2019; Accepted: 17 December 2019; Published: 19 December 2019

Abstract: Asymmetry is a cardinal symptom of gait post-stroke that is targeted during rehabilitation. Technological developments have allowed accelerometers to be a feasible tool to provide digital gait variables. Many acceleration-derived variables are proposed to measure gait asymmetry. Despite a need for accurate calculation, no consensus exists for what is the most valid and reliable variable. Using an instrumented walkway (GaitRite) as the reference standard, this study compared the validity and reliability of multiple acceleration-derived asymmetry variables. Twenty-five post-stroke participants performed repeated walks over GaitRite whilst wearing a tri-axial accelerometer (Axivity AX3) on their lower back, on two occasions, one week apart. Harmonic ratio, autocorrelation, gait symmetry index, phase plots, acceleration, and jerk root mean square were calculated from the acceleration signals. Test–retest reliability was calculated, and concurrent validity was estimated by comparison with GaitRite. The strongest concurrent validity was obtained from step regularity from the vertical signal, which also recorded excellent test–retest reliability (Spearman's rank correlation coefficients (rho) = 0.87 and Intraclass correlation coefficient (ICC_{21}) = 0.98, respectively). Future research should test the responsiveness of this and other step asymmetry variables to quantify change during recovery and the effect of rehabilitative interventions for consideration as digital biomarkers to quantify gait asymmetry.

Keywords: stroke; asymmetry; accelerometer; gait; trunk; reliability; validity

1. Introduction

Hemiparesis after stroke typically results in reduced walking speed, an asymmetrical gait pattern, and a reduced ability to make gait adjustments that consequentially limit community ambulation and physical activity [1–4]. Reduction in both predisposes an already at risk population to further cardiometabolic disease [5,6]. Therefore, the improvement of gait is a worthwhile and common target for interventions after stroke. Gait asymmetry, if not addressed early in the recovery process, can prolong and increase gait impairment due to compensatory mechanisms, leading to an increasingly asymmetric gait pattern [7]. The latter is inefficient and requires increased energy expenditure. Consequently, falls risk increases, further reducing levels of physical activity [8]. In order to quantify asymmetry and its improvement from targeted rehabilitative interventions, it is essential to have both valid and reliable tools that are able to quantify movement quality/compensatory strategies of the whole body during gait.

Tests such as the 10 m walk [9] and scales such as the Dynamic Gait Index [10] are used to measure gait after stroke. Although useful and practical for application to clinical settings, these tests are susceptible to subjectivity and not specifically designed to capture the cardinal symptoms of gait after stroke, such as asymmetry. Instrumented walkways can objectively measure asymmetry and have shown excellent intra and inter-rater reliability in subacute stroke [11]. Practically, they are costly and need a controlled dedicated environment with a trained specialist to operate; therefore, they are mainly limited to research settings [12]. From a biomechanical perspective, they limit the number of steps collected per trial and solely obtain information of the participant's footfall. They are not designed to measure the movement of the whole body, where synergistic compensatory movement strategy information may be quantified such as compensatory movements of the pelvis [8,13]. Traditionally, gaining this information would rely on three-dimensional motion analysis systems. However, due to the even higher cost, required experience, and time to use relative to instrumented mats, their application is also limited to research settings [12]. Therefore, a need exists for a valid tool that is capable of quantifying whole body asymmetry, while also being feasible for routine clinical adoption.

Wearable accelerometers are a relatively low-cost alternative that are capable of measuring human movement from a variety of contexts while capturing parameters that are difficult to quantify from clinical inspection by the human eye [1,14]. Previous attempts to quantify measures of asymmetry indicative of spatiotemporal information of the feet with accelerometers have shown their feasibility, but also poor concurrent validity with reference standards of Gaitrite [1]. Therefore, the development of algorithms to capture the complex nature of asymmetry post-stroke has been encouraged [1]. Numerous asymmetry variables exist that have been obtained from cyclical acceleration signals during gait such as variables derived from the frequency domain [15,16]. These variables vary according to the complexity of the sensor, the number of sensors used, their location, and the population on which they were tested [17–19]. Relative to the discreet spatiotemporal movement of the feet equivalents, variables quantifying asymmetry from the cyclical signals of the lower back better classified post-stroke gait from controls [16,18,20]. Their advantage stems from considering the acceleration as a complete waveform, not neglecting temporal information outside of the time domain, which may enable a more complete description of the signal and a better characterisation of gait post-stroke [17].

Previously, studies quantifying asymmetry from acceleration signals of the trunk during post-stroke gait typically focus on differences from a control group, adopt a minimal data set of variables, and to our knowledge do not report the concurrent validity or reliability to reference standards. Knowledge of the most robust asymmetry variables that are capable of quantifying similar information to reference standards using clinically feasible tools is important to further the field. This study compares the validity and test–retest reliability of a wide range of novel acceleration-derived variables to quantify asymmetry post-stroke from a single sensor located on the trunk.

2. Materials and Methods

2.1. Study Design and Setting

This cross-sectional study was undertaken in the gait laboratory at the Clinical Ageing Research Unit, Campus for Ageing and Vitality, Newcastle upon Tyne, UK.

2.2. Participants

The study was approved by the Greater Manchester West Research and Ethics Committee (NRES Committee Northwest-Greater Manchester West 15/NW/0731). All subjects gave informed written consent for the study according to the Declaration of Helsinki.

Inclusion criteria: Community-dwelling stroke survivor; at least one month post-stroke onset; mild to moderate gait deficit defined by clinical observation of gait asymmetry including reduced stance time, increased swing time in the affected limb and/or reduced gait speed/balance problems; no changes in gait-related ability over the past month based on self-report and able to walk 10 m with/without a stick.

Exclusion criteria: Medical problems other than stroke impacting on gait e.g., osteoarthritis. Participants were recruited via advertisement or therapist referral. All eligible participants were consecutively invited to participate in the study.

2.3. Demographic and Clinical Measures

The following data were collected at baseline: age, gender, height and weight, date of stroke, stroke type (Oxford Community Stroke Project Classification [21]), stroke impairment (National Institute of Health Stroke Scale [22]), presence of hemiplegia (clinical observation by two independent experienced clinicians), walking stick use, ankle foot orthosis (AFO) use.

2.4. Test Protocol

Participants were asked to walk at their preferred pace in a straight line for 4 × 10 m intermittent trials (see Figure 1). The trials were repeated on two occasions (Time 1 and Time 2) one week apart (±2 days). A GaitRite instrumented walkway was positioned in the walk path (dimensions were 7.0 m × 0.6 m, spatial accuracy of 1.27 cm and temporal accuracy of one sample (240 Hz, ~4.17 ms) (GaitRite: Platinum model GaitRite, software version 4.5, CIR systems, NJ, USA)). The participants wore an AX3 wearable sensor located at their fifth lumbar vertebrae (L5). The AX3 is a single tri-axial accelerometer-based wearable (AX3, Axivity, York, UK https://axivity.com/, cost ≈ £100, dimensions 23.0 mm × 32.5 mm × 7.6 mm). The AX3 weighs 11 g and has a memory of 512 Mb. AX3 data capture occurs with a sampling frequency of 100 Hz (16-bit resolution) at a range of ±8 g. Recorded AX3 accelerations were stored locally on the device's internal memory and downloaded upon the completion of each session.

2.5. Asymmetry Variables

Acceleration-derived asymmetry variables were selected based upon their ability to represent levels of asymmetry from signals measured from a single accelerometer located at the trunk. The variables that were selected as representative of asymmetry were the harmonic ratio [16], autocorrelation [20], gait symmetry index [18], and phase plot analysis [23–25] (described in more detail below). Four spatiotemporal variables extracted from GaitRite were selected as measures of asymmetry as defined by Lord et al. [26]. The spatiotemporal asymmetry variables included step time asymmetry, stance time asymmetry, swing time asymmetry, and step length asymmetry, and these were calculated as the absolute difference between consecutive left and right steps.

2.6. Description of Acceleration-Derived Variables

All data analysis relating to the raw acceleration signals was performed using MATLAB (version 9.4.0, R2018a). For a full description for the algorithm and data segmentation techniques applied to the accelerometer data, please see references [27,28]. In brief, the vertical acceleration underwent continuous wavelet transformation to estimate the initial contact and final contact in the gait cycle [28]. To ensure that the steady-state gait was analyzed, the initial and final three steps were removed from the signal. Prior to the calculation of additional variables, the acceleration signals were realigned to the earth's gravitational constant [29,30] and a low-pass Butterworth filter with a cut-off frequency of 20 Hz. A full description of the following variables and the required algorithms is the supplied by the provided references. Additionally, they have been summarised in Appendix A.

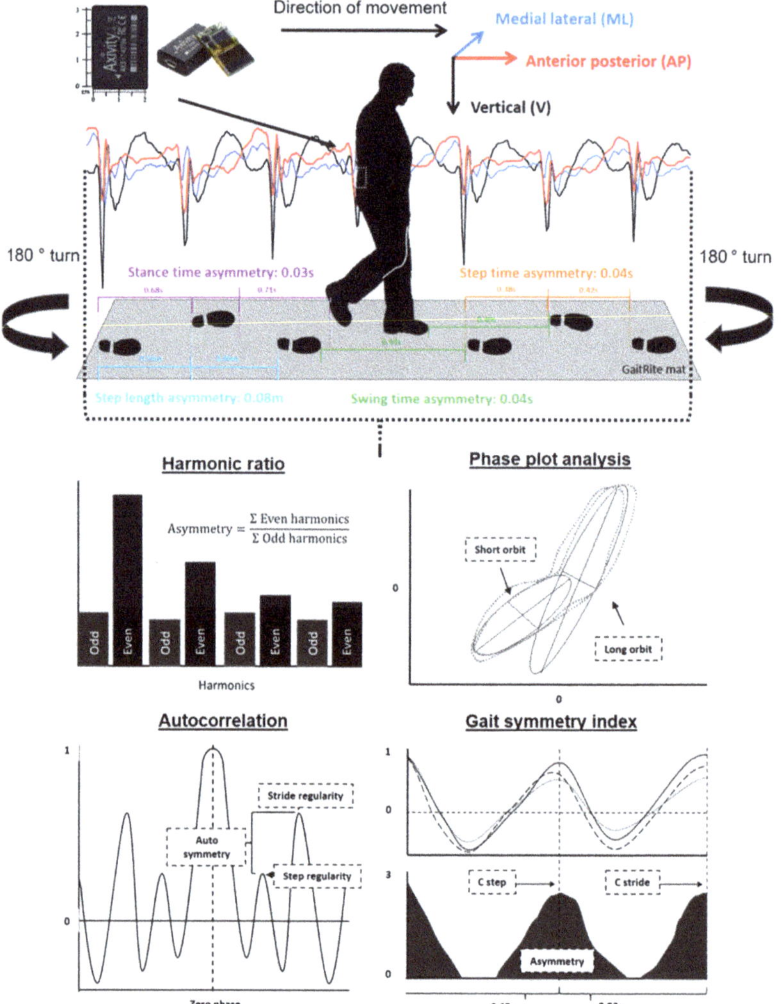

Figure 1. Indication of the instrumentation and the protocol used to collect the acceleration signal and the asymmetry parameters from the GaitRite mat. Also pictured is the acceleration-derived asymmetry variables and the means for the calculation of asymmetry following the processing of the raw acceleration signal.

2.6.1. Harmonic Ratio

The harmonic ratio (HR) describes the step-to-step symmetry within a stride from calculating a ratio of the odd and even harmonics of a signal following fast Fourier transformation [16,31]. This method has been shown previously to reflect increased asymmetry for those post-stroke relative to age and speed-matched controls [16].

2.6.2. Autocorrelation

The unbiased autocorrelation was also calculated due to its ability to reflect the step and stride regularity and the symmetry between the two (autocorrelation symmetry) [20,32,33]. Previously, it has been shown as better capable to characterise hemiplegic gait relative to footfall variables [20,32].

2.6.3. Gait Symmetry Index

The gait symmetry index (GSI) is a more recently proposed variable, which was calculated based upon the concept of the summation of the biased autocorrelation from all three components of movement and a subsequent calculation of step and stride timing asymmetry [18]. It has been shown to be more sensitive than and highly correlated with levels of asymmetry measured with two sensors located at the feet of participants post-stroke [18].

2.6.4. Phase Plot Analysis

Phase plot analysis (aka Poincaré analysis) was performed on vertical components of the acceleration signal [23–25]. This method has had previous applications within electrocardiogram studies. It works by plotting periodic signals as a function of their past values. The resulting ellipses or orbits and the properties thereof can then assess asymmetries in the associated gait. Phase plot analysis also offers the ability to assess intra step correlation i.e., the correlation of signals from immediately successive step cycles, which necessarily corresponds to left-versus-right asymmetry.

2.6.5. Measures Indicative of Stability

Although not indicative of asymmetry, the root mean square of the acceleration signal (Acc RMS) and also its first time derivative (Jerk RMS) were calculated for their potential to highlight synergistic compensatory strategies during gait post-stroke [13,16]. Their test–retest reliability needs to be established in the literature.

2.7. Statistical Analysis

Analysis was completed using SPSS v25 (IBM). The normality of data was tested with a Shapiro–Wilk test. Descriptive statistics (median and interquartile range) were calculated for gait characteristics measured by AX3 and GaitRite. Concurrent validity between the AX3 acceleration-derived variables and those of the GaitRite at Time 1 were tested using Spearman's rank correlation coefficients (RHO). For the AX3 acceleration-derived variables, the test–retest reliability between Time 1 and 2 was established using Spearman's rank correlation coefficients (RHO), intraclass correlation coefficient (ICC_{21}), and limits of agreement (LoA) expressed as a percentage of the mean of the two variables and the 95% LoA. For all analyses, statistical significance was set at $p < 0.05$. Predefined acceptance ratings for ICC_{21} were set at excellent (≥900, 0.0%–4.9%), good (0.750–0.899, 5.0%–9.9%), moderate (0.500–0.749, 10.0%–49.9%), and poor (50.0%) [1,34]. The selection for the most robust variable was based upon the variable with the highest Spearman rank correlation coefficient with the asymmetry variable obtained from the GaitRite while also recording an ICC_{21} greater than 0.8 for test–retest reliability.

3. Results

Twenty-five participants were recruited to the study. Data for two participants who wore a fixed plastic AFO were removed from the analysis, because individual data analysis (including video observations) revealed that the step detection applied were not appropriate for these two participants due to a lack of possible plantar flexion. This was not the case for the remaining participants, as the video analysis confirmed the step detection algorithm was effective to detect both heel strike and toe off [1]. Demographic information for the remaining 23 participants is displayed in Table 1.

Table 1. Participant characteristics.

Demographics (n = 23)	
Gender (male/female)	19/4
Age (years)	63 ± 11
Body mass index	26 ± 4
Stroke characteristics	
Time since stroke (months)	66 ± 48 (range 5–201)
Stroke subtype (OCSP)	
Total anterior circulation	11
Partial anterior circulation	6
Lacunar	3
Posterior circulation	3
Stroke impairment	
NIHSS score (0–40)	4 ± 3 (range 0–11)
NIHSS lower limb score (0–4)	1 ± 0.7 (range 0–3)
Walking speed (m/s)	0. 9 ± 0.4
Marked hemiplegia (Yes/No)	15/8
Walking aid (number (%))	3 (13%)
Push Aequi ankle foot orthosis (number (%))	4 (17%)

Where appropriate mean and standard deviation are displayed, OCSP (Oxford community Stroke Project), NIHSS (National Institute for Health Stroke Scale).

3.1. Concurrent Validity of the Asymmetry Variables

Figure 2 shows the correlation between the asymmetry variables quantified using a GaitRite mat (step time asymmetry, stance time asymmetry, swing time asymmetry, and step length asymmetry) and the acceleration-derived variables proposed to measure asymmetry. Overall, step time asymmetry correlated most with the acceleration-derived variables. Step regularity (vertical acceleration) had the highest concurrent validity with step time asymmetry (−0.87). Six other variables had high levels of agreement (+0.80) (HR V, step regularity (V), step regularity (AP), orbit eccentricity, orbit width deviation, and intra step correlation). Five correlated with step time asymmetry and orbit width deviation correlated with stance time asymmetry. The smallest correlations were achieved by the outputs of the autocorrelation from the medial lateralcomponent of the signal and also a variety of the outputs from the phase plot analysis.

Acceleration-derived variable	GaitRite Asymmetry variables			
	Step time	Stance time	Swing time	Step length
Harmonic ratio (V)	-0.83 **	-0.70 **	-0.73 **	-0.59 **
Harmonic ratio (ML)	-0.47 *	-0.48 *	-0.44 *	-0.26
Harmonic ratio (AP)	-0.76 **	-0.63 **	-0.67 **	-0.44 *
Step regularity (V)	-0.87 **	-0.72 **	-0.72 **	-0.65 **
Step regularity (ML)	-0.23	-0.17	-0.08	-0.29
Step regularity (AP)	-0.83 **	-0.66 **	-0.68 **	-0.52 *
Stride regularity (V)	-0.76 **	-0.65 **	-0.65 **	-0.56 **
Stride regularity (ML)	-0.35	-0.26	-0.17	-0.39
Stride regularity (AP)	-0.45 *	-0.38	-0.37	-0.38
Autocorrelation symmetry (V)	0.54 **	0.38	0.37	0.43 *
Autocorrelation symmetry (ML)	0.23	0.32	0.28	0.17
Autocorrelation symmetry (AP)	-0.57 **	-0.57 **	-0.53 **	-0.49 *
Gait symmetry index	0.62 **	0.48 *	0.50 *	0.36
Orbit eccentricity	-0.81 **	-0.67 **	-0.68 **	-0.38
Relative orbit inclination	0.49 *	0.40	0.35	0.39
Orbit width deviation	0.80 **	0.80 **	0.78 **	0.49 *
Short half orbit eccentricity	0.78 **	0.69 **	0.69 **	0.48 *
Short half orbit segment angle	0.68 **	0.54 **	0.50 *	0.37
Long half orbit eccentricity	0.61 **	0.59 **	0.57 **	0.20
Long half orbit segment angle	0.79 **	0.67 **	0.67 **	0.28
Intra step correlation	-0.83 **	-0.67 **	-0.67 **	-0.45 *

Key: Indication of Spearman's rank correlation coefficient								
-1	-0.75	-0.5	-0.25	0	0.25	0.5	0.75	1

Figure 2. Indication of the correlation between the asymmetry variables quantified using a GaitRite mat and the variables proposed to measure asymmetry from the acceleration signals from the trunk. Black indicates a strong positive or negative correlation. * and ** denotes significance at the 0.05 and 0.01 level, respectively. V = Vertical acceleration, ML = Medial lateral acceleration, and AP = Anterior posterior acceleration.

3.2. *Test–Retest Reliability of the Variables*

Table 2 demonstrates the test–retest reliability between the wearable variables measured one week apart (Time 1 versus Time 2). The most reliable variables were step regularity (V) and HR (V), both recording an ICC_{21} of 0.98. Taken from the ICC_{21} values, excellent reliability was achieved for 12 out of the 27 variables tested. These came from the majority of autocorrelation outputs except for step regularity (ML), stride regularity (AP), and autocorrelation symmetry (vertical acceleration (V) and medial lateral acceleration (ML)) direction, the GSI, the HR in the V and AP direction, Jerk RMS, and the short half-orbit segment angle form the phase plot analysis. Good reliability was achieved for a further five variables (stride regularity (AP), autocorrelation symmetry (V), relative orbit inclination, short half orbit eccentricity, and long half orbit eccentricity).

Table 2. Test–retest reliability (one week apart) for acceleration-derived variables.

Variables	Median (IQR)		Agreement			
	T1	T2	Median Difference (%)	ICC_{21}	LOA % (95% LoA)	Rho
Harmonic ratio (V)	1.71 (1.37)	1.70 (1.23)	−0.01	0.98 **	1.94 (2.52, 1.36)	0.92 **
Harmonic ratio (ML)	1.38 (0.60)	1.57 (0.72)	0.14	0.71 **	1.56 (2.80, 0.31)	0.71 **
Harmonic ratio (AP)	1.26 (0.97)	1.39 (0.92)	0.10	0.92 **	1.54 (2.34, 0.73)	0.91 **
Step regularity (V)	0.53 (0.47)	0.52 (0.54)	−0.02	0.98 **	0.51 (0.67, 0.34)	0.96 **
Step regularity (ML)	0.42 (0.20)	0.44 (0.18)	0.04	0.73 **	0.44 (0.69, 0.19)	0.61 **
Step regularity (AP)	0.51 (0.43)	0.40 (0.49)	−0.20	0.92 **	0.37 (0.68, 0.07)	0.87 **
Stride regularity (V)	0.70 (0.25)	0.68 (0.27)	−0.03	0.94 **	0.66 (0.85, 0.46)	0.88 **
Stride regularity (ML)	0.59 (0.14)	0.66 (0.20)	0.12	0.93 **	0.57 (0.78, 0.37)	0.73 **
Stride regularity (AP)	0.74 (0.18)	0.75 (0.13)	0.01	0.87 **	0.70 (0.92, 0.48)	0.74 **
Autocorrelation symmetry (V)	0.53 (0.26)	0.52 (0.29)	0.56	0.80 **	0.18 (0.40, −0.03)	0.76 **
Autocorrelation symmetry (ML)	0.10 (0.19)	0.16 (0.25)	0.09	0.59 *	0.19 (0.44, −0.05)	0.49 *
Autocorrelation symmetry (AP)	0.18 (0.15)	0.19 (0.14)	0.61	0.93 **	0.36 (0.62, 0.10)	0.79 **
Gait symmetry index	0.21 (0.37)	0.35 (0.43)	−0.02	0.92 **	0.47 (0.70, 0.23)	0.82 **
Orbit eccentricity	7.79 (6.27)	8.32 (15.13)	0.00	0.72 **	0.97 (1.04, 0.91)	0.70 **
Relative orbit inclination	0.01 (0.01)	0.01 (0.01)	0.07	0.76 **	11.02 (28.02, −5.99)	0.60 **
Orbit width deviation	0.01 (0.02)	0.00 (0.02)	−0.07	0.66 **	0.01 (0.05, −0.02)	0.65 **
Short half orbit eccentricity	5.32 (6.35)	4.12 (5.31)	−0.38	0.73 **	0.02 (0.07, −0.03)	0.87 **
Short half orbit segment angle	0.02 (0.05)	0.01 (0.04)	−0.23	0.95 **	7.74 (15.28, 0.20)	0.57 **
Long half orbit eccentricity	5.20 (10.73)	5.61 (6.55)	−0.16	0.79 **	0.04 (0.13, −0.05)	0.59 **
Long half orbit segment angle	0.89 (0.41)	0.88 (0.20)	0.08	0.45	7.77 (26.32, −10.78)	0.57 **
Intra step correlation	1.05 (0.04)	1.05 (0.04)	−0.01	0.58 *	0.78 (1.29, 0.28)	0.68 **
Acceleration RMS (V)	0.18 (0.09)	0.17 (0.06)	0.00	0.03	1.03 (1.24, 0.83)	0.41
Acceleration RMS (ML)	0.25 (0.15)	0.24 (0.15)	−0.06	0.90 **	0.17 (0.24, 0.10)	0.68 **
Acceleration RMS (AP)	8.53 (8.00)	8.57 (7.47)	−0.04	0.20	0.26 (0.62, −0.10)	0.21
Jerk RMS (V)	6.29 (1.18)	6.36 (1.15)	0.01	0.96 **	9.32 (13.49, 5.14)	0.93 **
Jerk RMS (ML)	6.22 (4.89)	6.42 (6.88)	0.01	0.97 **	7.39 (10.67, 4.11)	0.90 **
Jerk RMS (AP)	1.71 (1.37)	1.70 (1.23)	0.03	0.96 **	7.26 (11.23, 3.28)	0.92 **

* and ** denotes significance at the 0.05 and 0.01 level, respectively. V = Vertical acceleration, ML = Medial lateral acceleration, and AP = Anterior posterior acceleration, RMS = root mean square.

3.3. *Selection of the Most Robust Variable*

Table 3 highlights the variables that best correlated with spatiotemporal gait variables calculated from GaitRite while also achieving an ICC_{21} greater than 0.8 for test-retest reliability. For the GaitRite variables of asymmetry, step regularity (V) achieved the highest concurrent validity due to its correlation with step time asymmetry (RHO = 0.87 and ICC_{21} = 0.98 **). The second highest concurrent validity was the HR in the vertical direction, which correlated with swing time asymmetry (RHO = 0.73 and ICC_{21} = 0.98 **).

Table 3. Indication of what wearable sensor variable recorded the highest Spearman's rank correlation coefficient with each variable obtained by the GaitRite mat. The Spearman's rank correlation coefficient between the two devices and the intraclass correlation coefficient is displayed for each variable.

	GaitRite Variable	Acceleration Derived Variable	Spearman's Rank Correlation Coefficient (RHO)	ICC_{21} (Test–Retest)
Asymmetry	Step time (s)	Step regularity (V)	0.87	0.98 **
	Swing time (s)	Harmonic ratio (V)	0.73	0.98 **
	Stance time (s)	Step regularity (V)	0.72	0.98 **
	Step length (m)	Step regularity (V)	0.65	0.98 **

** denotes significance at the 0.01 level. V = Vertical acceleration.

4. Discussion

This study examined the concurrent validity and reliability of a comprehensive range of asymmetry variables derived from a single accelerometer located on the trunk and identified step regularity as the most robust outcome. Step regularity showed strong concurrent validity and excellent test–retest reliability when compared with GaitRite outcomes reflecting asymmetry. This contrasts with previous work based on the AX3 sensor, which achieved poor to moderate criterion validity (Spearman's rank correlation coefficient of RHO = 0.01 to 0.601) for variables engineered to replicate spatiotemporal asymmetry variables calculated from GaitRite [1]. Although clinically more challenging to interpret than traditional spatiotemporal variables, our results support the adoption of novel variables to quantify asymmetry as robust digital variables for measuring asymmetrical gait post stroke.

With one exception (HR correlation with swing time asymmetry), variables calculated from performing an autocorrelation procedure on the original acceleration signal were more strongly correlated with GaitRite asymmetry. Hodt–Billington and colleagues [20] found that autocorrelation variables taken from the trunk were better at discriminating gait post-stroke from controls relative to GaitRite variables of asymmetry. The strength of the autocorrelation procedure may stem from analysing continuous successive steps. Complex measures such as gait asymmetry are not simply portrayed within a single discreet gait cycle; this concept has been highlighted before, whereby continuous measures have been described to highlight different asymmetry causes, symptoms, and gait strategies such as particular compensatory techniques [17]. Data from our study indicate that participants with high asymmetry produced poor forward propulsion from the affected limb, instead of relying on the more dominant limb to achieve progression at the end of each stride. This can be observed by the lack of step regularity and its diminution relative to stride regularity in the AP, ML, and V directions, replicating the gait strategy described by Balasubramanian et al. [35]. The autocorrelation method is well designed to reflect this synergistic gait strategy, which might explain the high correlation found from this sample of participants. However, this strategy will likely vary among a broader range of participants and throughout recovery. Other methods may better reflect true levels of asymmetry at different stages of recovery from acute, early subacute, late subacute, and chronic stroke, meaning that they should still be considered as potential variables [17,20].

Previously, Iosa et al. [16] assessed symmetry together with upright gait stability post-stroke and showed that relative to speed-matched controls, higher instabilities (Acceleration RMS) and reduced symmetry of trunk movements (as measured using the HR) were recorded. In this study, HR in the vertical direction was the only HR variable that performed favourably to autocorrelation variables due to its correlation with swing time asymmetry (RHO = −0.73) while also recording excellent reliability (ICC_{21} = 0.98). Since we did not assess control subjects, we could not determine the best measure to characterise gait post-stroke and highlight the compensatory mechanisms adopted relative to healthy controls. This is a broader aim for ongoing work. However, it has been previously highlighted that compensation strategies may be beneficial to increase gait ability, but this occurs at the compromise of stability. Thus, variables such as Acceleration and Jerk RMS should always be considered in addition to variables directly linked to asymmetry, aiming to provide a more holistic description of gait patterns [13,16]. Future research should explore this relationship so that a holistic, multivariate wearable approach can better assess gait strategies during recovery post-stroke. This potentially would quantify what movements are beneficial to gait, while also highlighting the impact of compensation strategies, consequently quantifying separate movements that can be targeted for rehabilitation.

Although previously suggested as a variable representative of asymmetry in stroke [18], the GSI performed relatively poorer to the previously discussed variables, despite also being based on the autocorrelation (biased) of accelerometry. This was unexpected, as GSI theoretically is designed to detect the asymmetry within temporal footfall parameters. Equally, the autocorrelation symmetry variables did not perform better than step regularity alone, despite being designed to the capture the difference between step and stride regularity and therefore the symmetry between them. Potentially, the GSI and the autocorrelation symmetry did not quantify the synergistic movement strategy that the step regularity variable was suited to highlight and the reason for its favourable concurrent validity. The GSI and the autocorrelation symmetry variables may be better suited to highlight different compensatory synergies at different stages of recovery such as during acute, early subacute, late subacute, and chronic stages, and therefore should not be neglected in future research.

Select phase plot variables achieved RHO values greater than 0.8 when compared to GaitRite asymmetry values and also demonstrated good to excellent reliability, therefore highlighting their ability to quantify symmetry post-stroke. Adaption to the algorithms to the other directional components other than vertical and comparison with controls would better test their application as a biomarker. Similar to the other variables capable of quantifying movements in the AP and ML direction, there is the possibility that they can highlight a new domain of asymmetry separate from the asymmetry footfall asymmetry variables captured by GaitRite. Future research should explore this upper and lower body relationship post-stroke to examine the similarities and differences during gait and determine if added value is obtained [36,37].

All data were collected in a controlled environment; however, wearable technology is not limited by the testing environment and for improved ecological validity; obtaining data from the participant's community is desired [38]. To this goal, future research should utilise the variables tested in the laboratory in the participant's free-living environment. For free-living gait, the majority of walking bouts for people with Parkinson's disease and older adults have been found to be below 10 s, and it has been inferred that these bouts are when the participants are indoors [39]. One limitation with autocorrelation is that it relies on successive steps in a straight line. For free-living data, variables such as the HR may be more useful during these short walking bouts due to their ability to be calculated from a single stride in addition to successive steps [31,40]. Future research should assess the ability of these variables to accurately and reliably quantify asymmetry during short walking bouts or if tested refined spaces, as for this population, the median (and interquartile range) bout length was 16.3 (6.2) seconds for data collected over seven days [1].

4.1. Limitations

The relatively small sample size and limited heterogeneity with respect to time post-stroke did not allow us to determine what variables are the best at quantifying asymmetry for a more general sample or recovery stage-specific populations [41]. Future work is required on a larger sample size that ranges in time since stroke to discover what variables are the most capable to perform as objective biomarkers over all stages of recovery as one variable may not be appropriate for all, and compensatory strategies may change between the different stages of stroke recovery. Equally, future research should confirm that these results are replicable with different accelerometers with differing sampling frequencies, ranges, and resolutions. Further limitations stem from the reliance of the step detection algorithm. Data from two participants was not analysed due to their use of a fixed AFO that impacted on heel strike and the performance of the algorithm, which was based on the detection of initial and final contact within the gait cycle. Future research should integrate/develop step detection algorithms for participants requiring fixed AFOs to broaden application. Alternatively, the variables should be developed so that the cyclical nature of a signal may divide gait cycles (similar to the method used for phase plots) as opposed to methods that rely on detecting the initial and final contact of the foot.

4.2. Applications

These results provide evidence that asymmetry can accurately and reliably be calculated using a single accelerometer. Although much work is needed for accelerometers to be routinely adopted [42,43], these results give evidence that asymmetry can be objectively quantified using a tool applicable for many purposes. Consequently, the variables tested here may then act as a digital biomarker to quantify the impact of targeted interventions proposed to improve gait timing mechanisms and gait asymmetry (e.g., auditory rhythmical cueing) [44]. Accelerometers provide a potentially low burden method for clinicians to collect data from a variety of environments, increasing the ability to objectively quantify asymmetry during stroke rehabilitation. Alongside application within the clinic, accelerometer data can be collected on gait asymmetry in naturalistic environments, thus removing the Hawthorn effect/observer bias associated with clinical testing. With increased development, these variables may provide continuous asymmetry focussed feedback for self-progress specific to each participant during rehabilitation.

5. Conclusions

Gait asymmetry after stroke can be measured robustly using a single wearable sensor on the trunk. Step regularity is the most valid and reliable asymmetry outcome, which is quantified by performing autocorrelation on the vertical component of the signal. The variables tested performed favourably to previous studies that also used GaitRite as the reference. Consequently, their adoption, in addition to other wearable-derived spatiotemporal variables of gait, are encouraged as they provide a more holistic description of gait that appears to indicate compensatory movement post-stroke. Future research is encouraged on larger populations where asymmetry is expected, during recovery/interventions to identify which wearable variables are biomarkers for gait asymmetry and compensatory mechanisms during gait. This will allow for increased accuracy in determining effective interventions.

Author Contributions: Conceptualisation, C.B., S.D.D., L.R. and S.A.M.; methodology, C.B., M.E.M.-A., M.D.-W., A.G., A.H., S.L., L.R., S.D.D., and S.A.M.; software, C.B., S.D.D., M.E.M.-A., M.D.-W., A.G., and A.H.; validation, C.B., A.H., A.G., and S.D.D.; formal analysis, C.B., S.D.D., and S.A.M.; investigation, C.B., A.G., A.H., L.R., S.D.D., and S.A.M.; resources, S.A.M., and L.R.; data curation, C.B., M.E.M.-A., M.D.-W., A.G., A.H., S.D.D.; writing—original draft preparation, C.B.; writing—review and editing, C.B., M.E.M.-A., M.D., A.G., A.H., S.L., L.R., S.D.D., and S.A.M.; visualisation, C.B., S.D.D.; supervision, S.A.M., S.D.D., L.R., and S.L.; project administration, S.A.M.; funding acquisition, S.A.M. and L.R. All authors have read and agreed to the published version of the manuscript.

Funding: A Medical Research Council Centenary award supported the delivery of this research. SM is supported by Health Education England and the National Institute for Health Research (HEE/NIHR ICA Programme Clinical Lectureship, Dr. Sarah Anne Moore, ICA-CL-2015-01- 012). LR and SDD are supported by the NIHR Newcastle Biomedical Research Centre (BRC) based at Newcastle upon Tyne and Newcastle University. The work was also supported by the NIHR/Wellcome Trust Clinical Research Facility (CRF) infrastructure at Newcastle upon Tyne Hospitals NHS Foundation Trust. The views expressed in this publication are those of the author(s) and not necessarily those of the NHS, the NIHR or the Department of Health.

Acknowledgments: We would like to thank the following for their contribution: Patients who took part in the study; Staff from local NHS trusts who assisted with recruitment to the study and lastly, Lisa Alcock for her assistance during in data collection.

Conflicts of Interest: The authors declare no conflict of interest.

Data Sharing: Data cannot be shared publically but can be available upon reasonable request to the corresponding author, as per local data sharing policies.

Appendix A

Appendix A.1. Acceleration-Derived Variable Definitions

Table A1. Indication for the variables used from the signal-derived variables and their respective definitions.

Variable	Definition
Harmonic ratio (V, ML, AP)	The step-to-step symmetry within a stride from calculating a ratio of the odd and even harmonics of a signal following fast Fourier transformation.
Step regularity (V, ML, AP)	Estimated as the normalized unbiased autocovariance for a lag of one step time. Thus, this feature reflects the similarity between subsequent steps of the acceleration pattern over a step. Values of this feature close to 1.0 (maximum possible value) reflect repeatable patterns between subsequent steps.
Stride regularity (V, ML, AP)	Estimated as the normalized unbiased autocovariance for a lag of one stride time. Thus, this feature reflects the similarity between subsequent strides of the acceleration pattern over a stride cycle.
Autocorrelation symmetry (V, ML, AP)	Difference between step and stride regularity designed to quantify the level of symmetry between them and indicative of symmetry during a straight walk.
Gait symmetry index	Calculated based upon the concept of the summation of the biased autocorrelation from all three components of movement and a subsequent calculation of step and stride timing asymmetry.
Orbit eccentricity (V)	Average eccentricity of all fully fitted ellipses.
Relative orbit inclination (V)	Average angle subtended by alternating fitted ellipses within a bout of gait.
Orbit width deviation (V)	Standard deviation of minor axes lengths of all fully fitted ellipses. Analogous to Principle Component Analysis (second component).
Short half orbit eccentricity (V)	Difference in eccentricity of two ellipses fitted to each half-cycle of a full orbit in the phase plot. Averaged over all orbits in a bout's phase plot.
Short half orbit segment angle (V)	Difference in inclination of two ellipses fitted to each half-cycle of a full orbit in the phase plot. Averaged over all orbits in a bout's phase plot.
Long half orbit eccentricity (V)	Difference in eccentricity of two ellipses fitted to each half-cycle of a full orbit in the phase plot. Averaged over all orbits in a bout's phase plot.
Long half orbit segment angle (V)	Difference in inclination of two ellipses fitted to each half cycle of a full orbit in the phase plot. Averaged over all orbits in a bout's phase plot.
Intra step correlation (V)	Average correlation of acceleration signal corresponding to step i with that of step i-1. I.e., a lag-1 autocorrelation where a single lag is one step cycle's duration.
Acceleration RMS (V, ML, AP)	The calculation of the root mean square of the acceleration signal.
Jerk RMS (V, ML, AP)	The calculation of the root mean square of the first time derivative of the acceleration signal (jerk).

Appendix A.2. Explanation and Equation for Each Acceleration Derived Variable for Asymmetry

Appendix A.2.1. Harmonic Ratio

The harmonic ratio is a measure based upon the premise that a stride contains two steps and therefore, during continuous walking, accelerations should repeat in multiples of two. The variable quantifies how well these accelerations are repeated in each stride compared to when accelerations do not repeat and are therefore out of phase. Therefore, the ratio of in and out-of-phase accelerations is a measure of how symmetric the participant is walking. To calculate the harmonic ratio, it is required to evaluate the harmonic content of the acceleration signal using the stride frequency from the analysis of frequency components. Following a fast Fourier transform (using the FFT function in MATLAB), a ratio be can created from the first 20 harmonics extracted from the Fourier series. Due to the AP and V components of the signals being biphasic, the ratio for these components is determined by the sum of the even harmonics (in phase movement) divided by the sum of the odd harmonics (out-of-phase movement).

$$HR_{AP,\,V} = \frac{\Sigma \text{ Amplitudes of even harmonics}}{\Sigma \text{ Amplitudes of odd harmonics}}$$

For the ML component of the signal due to only showing only one dominant acceleration peak within a stride cycle (whereby the odd harmonics are in-phase and even harmonic out-of-phase), the opposite is performed.

$$HR_{ML} = \frac{\Sigma \text{ Amplitudes of odd harmonics}}{\Sigma \text{ Amplitudes of even harmonics}}$$

As a gait measure, a higher harmonic ratio indicates a better symmetry between steps within a single stride For the AP and V components.

Appendix A.2.2. Autocorrelation

Autocorrelation is calculated taking the complete signal of the time when the participant was in contact with the GaitRite mat. Plots of an autocorrelation estimate are used to inspect the structure of a cyclic component within a time series. To do this, the generic unbiased autocorrelation function of the sample sequence x(i) was computed using the below equation:

$$Ad(m) = \frac{1}{N - |m|} \sum_{i=1}^{N-|m|} x(i) \cdot x(i+m)$$

where N is the number of samples and m is the time lag expressed as number of samples.

Since phase shifts can be performed with identical results in both positive and negative directions relative to the original time series, an autocorrelation plot is conventionally organized symmetrically with the zeroth shift located centrally. This central value was used to normalize the signal so that its maxima was one. For a time series of trunk accelerations during walking, autocorrelation coefficients can be produced to quantify the peak values at the first and second dominant period, representing phase shifts equal to one step and one stride, respectively (see Figure 1 as an example). A tailored MATLAB code was used to detect these peaks, particularly using the signals power density to determine the windows in which the peaks would occur. For the symmetry between the step and stride regularity, the absolute difference was calculated as a measure of asymmetry instead of the ratio, which is more conventionally used. This was because the between-step and between-stride autocorrelations may approach zero if the regularity between neighboring steps or neighboring strides is low.

Appendix A.2.3. Gait Symmetry Index (GSI)

Differently from the aforementioned autocorrelation measures, the gait symmetry index (GSI) uses a second-order Butterworth low-pass filter with the cut-off frequency of 10 Hz to filter the complete time series and then uses the biased version of the autocorrelation function as displayed below:

$$Ad(m) = \frac{1}{N} \sum\nolimits_{i=1}^{N-|m|} x(i) \cdot x(i+m).$$

The maximum time lag was 4 s (400 samples), which approximates 2.5 times a single stride duration in post hemiplegic stroke patients. This window length was chosen to capture the repetition of stride cycles in very slow walking. A coefficient of stride cycle repetition (Cstride) was the sum of the positive autocorrelation coefficients of the three axes as a function of the equation displayed below:

$$\text{Cstride(t)} = \text{ADv(t)} + \text{ADml(t)} + \text{ADap(t)}; \quad \text{if AD(t)} < 0, \text{AD(t)} = 0.$$

The coefficient of step repetition (Cstep) was the norm of autocorrelation coefficients as a function of the equation displayed below:

$$\text{Cstep(t)} = \sqrt{\text{ADv(t)} + \text{ADml(t)} + \text{ADap(t)}}; \text{ if AD(t)} < 0, \text{ AD(t)} = 0.$$

One stride time (Tstride) equals t when the Cstride had the maximum value. The hypothesis was that in a perfect symmetric gait pattern, two consecutive steps have the same step duration of 0.5 × Tstride. Thus, the maximum value of Cstep was set at $\sqrt{3}$ when the autocorrelation coefficient of each acceleration axis was 1 at zero-lag (t = 0). The gait symmetry index (GSI) was Cstep (0.5 × Tstride) normalized to its value at zero-lag, as indicated in the below equation:

$$\text{Cstep(t)} = \text{Cstep}(0.5 * \text{Tstride}) / \sqrt{3}.$$

Appendix A.2.4. Phase Plot Analysis

To create an ellipse to apply the following models, the vertical acceleration signal was first transformed to a horizontal–vertical coordinate system and filtered with a low-pass fourth order Butterworth filter at 20 Hz. Following piecewise integration, the full vertical excursion signal must be restored via concatenation of the resultant integrals. Here, the phase shift is introduced. We restore two such vertical excursion signals, one of which is exactly one step cycle lagged behind the other i.e.,:

$$PP1(tt) = PP0(tt - nn)$$

where n is the number of data points comprising a step interval in the vertical excursion signal and $PP1$ and $PP0$ are the lagged and original vertical excursion signal, respectively.

The following conic model is fitted to the two-dimensional phase plot data. This fitting is performed on each orbit in turn.

$$ax^2 + by^2 + cxy + dx + ey + f = 0$$

In the case of ellipse fitting to phase plot data, x and y are taken to be $PP1$ and $PP0$

The above model defines an ellipse subject to the following constraint.

$$c^2 - 4ab < 0$$

This constraint is used to ensure that an elliptical conic is fitted to the data as opposed to a hyperbola or parabola. The model defined by the conic equation can be fitted using ordinary least squares to find an estimate of $\hat{A} = \left(\hat{a}, \hat{b}, \hat{c}, \hat{d}, \hat{e}\right)$. f is set equal to 1 to avoid a trivial solution.

The above form of an ellipse does not lend itself well to geometric interpretation, so the following parameterisation is implemented:

$$\frac{(x-g)^2}{r_1^2} + \frac{(y-k)^2}{r_2^2} = 1.$$

However, this form does not account for inclined ellipses. To account for the significant inclination of ellipses, the following rotated coordinate system is introduced:

$$x' = (x - g)\cos(\theta) + (y - k)\sin(\theta)$$
$$y' = (y - k)y\cos(\theta) + (x - g)\sin(\theta).$$

This form of ellipse and rotated coordinate system ensure more straightforward interpretation of the ellipses and more intuitive feature extraction.

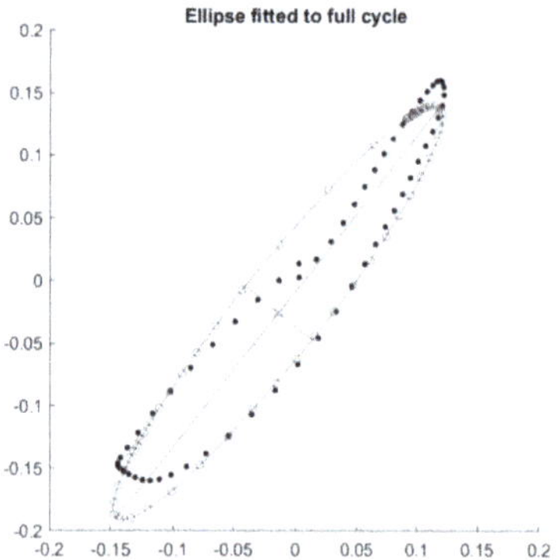

Figure A1. A single orbit with a fitted conic (ellipse).

This Figure A1 shows one such ellipse fitted to a single orbit of a phase plot. From this ellipse, we can extract features relating to the eccentricity and inclination. In general, phase plots consist of many orbits and their respective fitted ellipses (Figure A2). Further features can be extracted by assessing the relative inclination of ellipses from alternating orbits. In general, these inclinations oscillate about the value $\theta = \frac{\pi}{4}$.

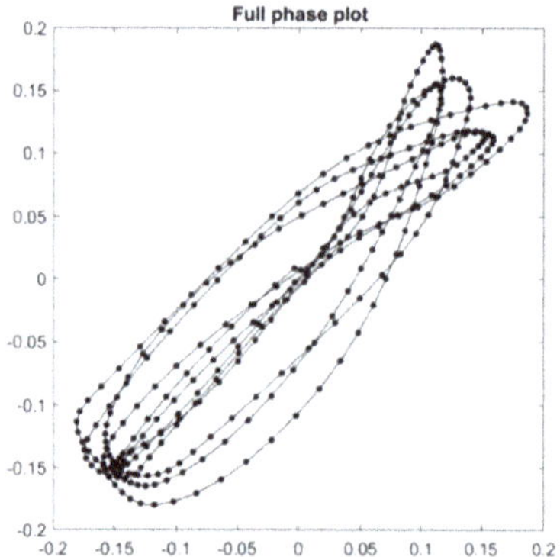

Figure A2. Complete phase plot comprising 7 continuous gait cycles.

Features extracted from ellipses fitted to entire orbits are considered primary features. Ellipses can be fitted to partial orbits; for example, two separate ellipses can be fitted to both halves of an orbit where the orbit in question is halved according to its major/minor axes. This leads to four additional ellipses fitted to each orbit of a phase plot (Figure A3). As an example, take the two ellipses fitted to either half of the shown orbit following halving via the minor axis (Figure A3, lower two figures). Features are extracted from these ellipses by extracting their relative characteristics e.g., their inclination relative to the other, the ratio of their areas, etc. Features extracted from ellipses fitted to partial orbits in this way are considered secondary phase plot features.

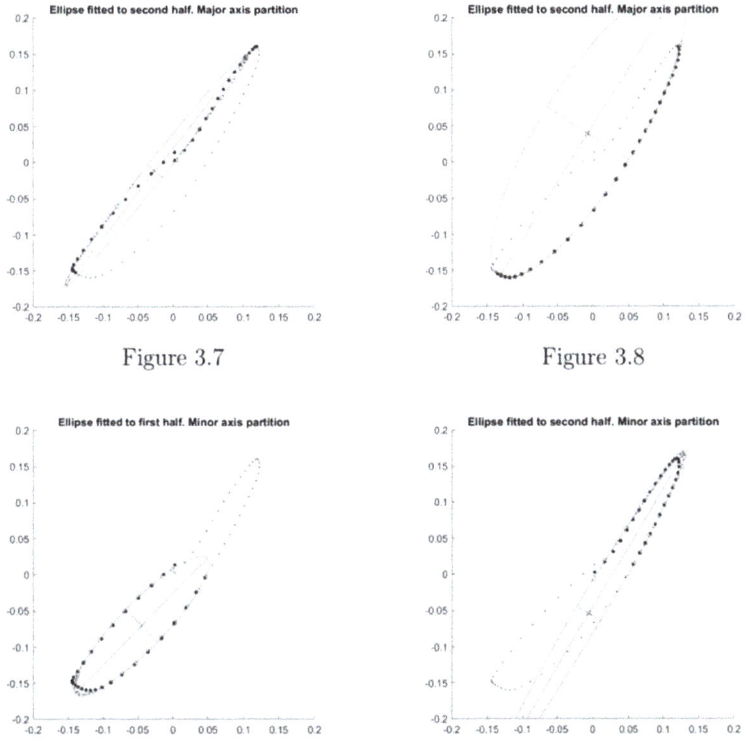

Figure A3. Indication of the different conic (ellipses) fitted to the major/minor axis and the first/second halves.

References

1. Moore, S.A.; Hickey, A.; Lord, S.; Del Din, S.; Godfrey, A.; Rochester, L. Comprehensive measurement of stroke gait characteristics with a single accelerometer in the laboratory and community: A feasibility, validity and reliability study. *J. Neuroeng. Rehabil.* **2017**, *14*, 130. [CrossRef] [PubMed]
2. Gallanagh, S.; Quinn, T.J.; Alexander, J.; Walters, M.R. Physical Activity in the Prevention and Treatment of Stroke. *ISRN Neurol.* **2011**, *2011*, 1–10. [CrossRef] [PubMed]
3. Patterson, K.K.; Gage, W.H.; Brooks, D.; Black, S.E.; McIlroy, W.E. Evaluation of gait symmetry after stroke: A comparison of current methods and recommendations for standardization. *Gait Posture* **2010**, *31*, 241–246. [CrossRef] [PubMed]
4. Fini, N.A.; Holland, A.E.; Keating, J.; Simek, J.; Bernhardt, J. How physically active are people following stroke? Systematic review and quantitative synthesis. *Phys. Ther.* **2017**, *97*, 707–717. [CrossRef]
5. Bull, F.; Goenka, S.; Lambert, V.; Pratt, M. *Physical Activity for the Prevention of Cardiometabolic Disease*; The International Bank for Reconstruction and Development/The World Bank: Washington, DC, USA, 2017; ISBN 9781464805189.
6. Lee, I.M.; Shiroma, E.J.; Lobelo, F.; Puska, P.; Blair, S.N.; Katzmarzyk, P.T.; Alkandari, J.R.; Andersen, L.B.; Bauman, A.E.; Brownson, R.C.; et al. Effect of physical inactivity on major non-communicable diseases worldwide: An analysis of burden of disease and life expectancy. *Lancet* **2012**, *380*, 219–229. [CrossRef]
7. Patterson, K.K.; Parafianowicz, I.; Danells, C.J.; Closson, V.; Verrier, M.C.; Staines, W.R.; Black, S.E.; McIlroy, W.E. Gait Asymmetry in Community-Ambulating Stroke Survivors. *Arch. Phys. Med. Rehabil.* **2008**, *89*, 304–310. [CrossRef]
8. Balaban, B.; Tok, F. Gait Disturbances in Patients with Stroke. *PM R* **2014**, *6*, 635–642. [CrossRef]
9. Wade, D.T. Measurement in neurological rehabilitation. *Curr. Opin. Neurol. Neurosurg.* **1992**, *5*, 682–686.

10. Lin, J.-H.; Hsu, M.-J.; Hsu, H.-W.; Wu, H.-C.; Hsieh, C.-L. Psychometric Comparisons of 3 Functional Ambulation Measures for Patients with Stroke. *Stroke* **2010**, *41*, 2021–2025. [CrossRef]
11. Wong, J.S.; Jasani, H.; Poon, V.; Inness, E.L.; McIlroy, W.E.; Mansfield, A. Inter- and intra-rater reliability of the GAITRite system among individuals with sub-acute stroke. *Gait Posture* **2014**, *40*, 259–261. [CrossRef]
12. Buckley, C.; Alcock, L.; McArdle, R.; Ur Rehman, R.Z.; Del Din, S.; Mazzà, C.; Yarnall, A.J.; Rochester, L. The role of movement analysis in diagnosing and monitoring neurodegenerative conditions: Insights from gait and postural control. *Brain Sci.* **2019**, *9*, 34. [CrossRef] [PubMed]
13. Iosa, M.; Fusco, A.; Giovanni, M.; Paolicci, S. Development and decline of upright gait stability. *Front. Aging Neurosci.* **2014**, *6*, 14. [CrossRef] [PubMed]
14. Wright, R.L.; Brownless, S.B.; Pratt, D.; Sackley, C.M.; Wing, A.M. stepping to the Beat: Feasibility and Potential efficacy of a home-Based auditory-cued step Training Program in chronic stroke. *Front. Neurol.* **2017**, *8*, 412. [CrossRef] [PubMed]
15. Buckley, C.; Galna, B.; Rochester, L.; Mazzà, C. Upper body accelerations as a biomarker of gait impairment in the early stages of Parkinson's disease. *Gait Posture* **2018**, *71*, 289–295. [CrossRef] [PubMed]
16. Iosa, M.; Bini, F.; Marinozzi, F.; Fusco, A.; Morone, G.; Koch, G.; Martino Cinnera, A.; Bonnì, S.; Paolucci, S. Stability and Harmony of Gait in Patients with Subacute Stroke. *J. Med. Biol. Eng.* **2016**, *36*, 635–643. [CrossRef] [PubMed]
17. Viteckova, S.; Kutilek, P.; Svoboda, Z.; Krupicka, R.; Kauler, J.; Szabo, Z. Gait symmetry measures: A review of current and prospective methods. Biomed. Signal Process. *Control* **2018**, *42*, 89–100.
18. Zhang, W.; Smuck, M.; Legault, C.; Ith, M.A.; Muaremi, A.; Aminian, K. Gait Symmetry Assessment with a Low Back 3D Accelerometer in Post-Stroke Patients. *Sensors* **2018**, *18*, 3322. [CrossRef]
19. Huang, X.; Mahoney, J.M.; Lewis, M.M.; Guangwei, D.; Piazza, S.J.; Cusumano, J.P. Both coordination and symmetry of arm swing are reduced in Parkinson's disease. *Gait Posture* **2012**, *35*, 373–377. [CrossRef]
20. Hodt-Billington, C.; Helbostad, J.L.; Moe-Nilssen, R. Should trunk movement or footfall parameters quantify gait asymmetry in chronic stroke patients? *Gait Posture* **2008**, *27*, 552–558. [CrossRef]
21. Bamford, J.; Sandercock, P.; Dennis, M.; Burn, J.; Warlow, C. Classification and natural history of clinically identifiable subtypes of cerebral infarction. *Lancet* **1991**, *337*, 1521–1526. [CrossRef]
22. Brott, T.; Adams, H.P.; Olinger, C.P.; Marler, J.R.; Barsan, W.G.; Biller, J.; Spilker, J.; Holleran, R.; Eberle, R.; Hertzberg, V.; et al. Measurements of acute cerebral infarction: A clinical examination scale. *Stroke* **1989**, *20*, 864–870. [CrossRef] [PubMed]
23. Esser, P.; Dawes, H.; Collett, J.; Howells, K. Insights into gait disorders: Walking variability using phase plot analysis, Parkinson's disease. *Gait Posture* **2013**, *38*, 648–652. [CrossRef] [PubMed]
24. Dunne-Willows, M.; Watson, P.; Shi, J.; Rochester, L.; Del Din, S. A Novel Parameterisation of Phase Plots for Monitoring of Parkinson's Disease. In Proceedings of the 2019 41st Annual International Conference of the IEEE Engineering in Medicine and Biology Society, Berlin, Germany, 23–27 July 2019; IEEE Explore: Berlin, Germany, 2019.
25. Brennan, M.; Palaniswami, M.; Kamen, P. Do existing measures of Poincare plot geometry reflect nonlinear features of heart rate variability? *IEEE Trans. Biomed. Eng.* **2001**, *48*, 1342–1347. [CrossRef] [PubMed]
26. Lord, S.; Galna, B.; Rochester, L. Moving forward on gait measurement: Toward a more refined approach. *Mov. Disord.* **2013**, *28*, 1534–1543. [CrossRef] [PubMed]
27. Del Din, S.; Godfrey, A.; Rochester, L. Validation of an Accelerometer to Quantify a Comprehensive Battery of Gait Characteristics in Healthy Older Adults and Parkinson's Disease: Toward Clinical and at Home Use. *IEEE J. Biomed. Health Inform.* **2016**, *20*, 838–847. [CrossRef] [PubMed]
28. McCamley, J.; Donati, M.; Grimpampi, E.; Mazzà, C. An enhanced estimate of initial contact and final contact instants of time using lower trunk inertial sensor data. *Gait Posture* **2012**, *36*, 316–318. [CrossRef]
29. Buckley, C.; Galna, B.; Rochester, L.; Mazzà, C. Quantification of upper body movements during gait in older adults and in those with Parkinson's disease: Impact of acceleration realignment methodologies. *Gait Posture* **2017**, *52*, 265–271. [CrossRef]
30. Moe-Nilssen, R. A new method for evaluating motor control in gait under real-life environmental conditions. Part 1: The instrument. *Clin. Biomech.* **1998**, *13*, 328–335. [CrossRef]
31. Bellanca, J.L.; Lowry, K.A.; Vanswearingen, J.M.; Brach, J.S.; Redfern, M.S. Harmonic ratios: A quantification of step to step symmetry. *J. Biomech.* **2013**, *46*, 828–831. [CrossRef]

32. Moe-Nilssen, R.; Helbostad, J. Interstride trunk acceleration variability but not step width variability can differentiate between fit and frail older adults. *Gait Posture* **2005**, *21*, 164–170. [CrossRef]
33. Tura, A.; Raggi, M.; Rocchi, L.; Cutti, A.G.; Chiari, L. Gait symmetry and regularity in transfemoral amputees assessed by trunk accelerations. *J. Neuroeng. Rehabil.* **2010**, *7*, 4. [CrossRef] [PubMed]
34. Fleiss, J.L. *The Design and Analysis of Clinical Experiments*; John Wiley & Sons, Inc.: Hoboken, NJ, USA, 1999; ISBN 0471820474.
35. Balasubramanian, C.K.; Bowden, M.G.; Neptune, R.R.; Kautz, S.A. Relationship between step length asymmetry and walking performance in subjects with chronic hemiparesis. *Arch. Phys. Med. Rehabil.* **2007**, *88*, 43–49. [CrossRef] [PubMed]
36. Boström, K.J.; Dirksen, T.; Zentgraf, K.; Wagner, H. The Contribution of Upper Body Movements to Dynamic Balance Regulation during Challenged Locomotion. *Front. Hum. Neurosci.* **2018**, *12*, 8. [CrossRef] [PubMed]
37. Mahaki, M.; Bruijn, S.M.; van Dieën, J.H. The effect of external lateral stabilization on the use of foot placement to control mediolateral stability in walking and running. *PeerJ* **2019**, *7*, e7939. [CrossRef] [PubMed]
38. Van de Port, I.; Punt, M.; Meijer, J.W. Walking activity and its determinants in free-living ambulatory people in a chronic phase after stroke: A cross-sectional study. *Disabil. Rehabil.* **2018**, *16*, 1–6. [CrossRef] [PubMed]
39. Del Din, S.; Godfrey, A.; Galna, B.; Lord, S.; Rochester, L.; Del-Din, S.; Godfrey, A.; Galna, B.; Lord, S.; Rochester, L. Free-living gait characteristics in ageing and Parkinson's disease: Impact of environment and ambulatory bout length. *J. Neuroeng. Rehabil.* **2016**, *13*, 1–12. [CrossRef]
40. Tamburini, A.P.; Storm, F.; Buckley, C.; Bisi, C.; Stagni, R.; Mazzà, C.; Tamburini, P.; Storm, F.; Buckley, C.; Bisi, M.C.; et al. Moving from laboratory to real life conditions: Influence on the assessment of variability and stability of gait. *Gait Posture* **2018**, *59*, 248–252. [CrossRef]
41. Bernhardt, J.; Hayward, K.S.; Kwakkel, G.; Ward, N.S.; Wolf, S.L.; Borschmann, K.; Krakauer, J.W.; Boyd, L.A.; Carmichael, S.T.; Corbett, D.; et al. Agreed Definitions and a Shared Vision for New Standards in Stroke Recovery Research: The Stroke Recovery and Rehabilitation Roundtable Taskforce. *Neurorehabil. Neural Repair* **2017**, *31*, 793–799. [CrossRef]
42. Espay, A.J.; Bonato, P.; Nahab, F.B.; Maetzler, W.; Dean, J.M.; Klucken, J.; Eskofier, B.M.; Merola, A.; Horak, F.; Lang, A.E.; et al. Technology in Parkinson's disease: Challenges and opportunities. *Mov. Disord.* **2016**, *31*, 1272–1282. [CrossRef]
43. Patterson, M.R.; Whelan, D.; Reginatto, B.; Caprani, N.; Walsh, L.; Smeaton, A.F.; Inomata, A.; Caulfield, B. Does external walking environment affect gait patterns? In Proceedings of the 2014 36th Annual International Conference of the IEEE Engineering in Medicine and Biology Society, Chicago, IL, USA, 26–30 August 2014; Volume 2014, pp. 2981–2984.
44. Yoo, G.E.; Kim, S.J. Rhythmic Auditory Cueing in Motor Rehabilitation for Stroke Patients: Systematic Review and Meta-Analysis. *J. Music Ther.* **2016**, *53*, 149–177. [CrossRef]

Article

Is a Wearable Sensor-Based Characterisation of Gait Robust Enough to Overcome Differences Between Measurement Protocols? A Multi-Centric Pragmatic Study in Patients with Multiple Sclerosis

Lorenza Angelini [1,2,*], Ilaria Carpinella [3], Davide Cattaneo [3], Maurizio Ferrarin [3], Elisa Gervasoni [3], Basil Sharrack [4], David Paling [5], Krishnan Padmakumari Sivaraman Nair [2,4] and Claudia Mazzà [1,2]

1 Department of Mechanical Engineering, University of Sheffield, Sheffield S1 3JD, UK; c.mazza@sheffield.ac.uk

2 Insigneo Institute for in silico Medicine, University of Sheffield, Sheffield S1 3JD, UK; siva.nair@nhs.net

3 IRCCS Fondazione Don Carlo Gnocchi, 20121 Milan, Italy; icarpinella@dongnocchi.it (I.C.); dcattaneo@dongnocchi.it (D.C.); mferrarin@dongnocchi.it (M.F.); egervasoni@dongnocchi.it (E.G.)

4 Academic Department of Neuroscience, Sheffield NIHR Neuroscience BRC, Sheffield Teaching Hospital NHS Foundation Trust, Sheffield S10 2JF, UK; basil.sharrack@nhs.net

5 Sheffield Institute of Translational Neuroscience, Sheffield Teaching Hospitals NHS Foundation Trust, Sheffield S10 2JF, UK; david.paling@nhs.net

* Correspondence: L.Angelini@sheffield.ac.uk; Tel.: +44-0774-640-8284

Received: 6 November 2019; Accepted: 18 December 2019; Published: 21 December 2019

Abstract: Inertial measurement units (IMUs) allow accurate quantification of gait impairment of people with multiple sclerosis (pwMS). Nonetheless, it is not clear how IMU-based metrics might be influenced by pragmatic aspects associated with clinical translation of this approach, such as data collection settings and gait protocols. In this study, we hypothesised that these aspects do not significantly alter those characteristics of gait that are more related to quality and energetic efficiency and are quantifiable via acceleration related metrics, such as intensity, smoothness, stability, symmetry, and regularity. To test this hypothesis, we compared 33 IMU-based metrics extracted from data, retrospectively collected by two independent centres on two matched cohorts of pwMS. As a worst-case scenario, a walking test was performed in the two centres at a different speed along corridors of different lengths, using different IMU systems, which were also positioned differently. The results showed that the majority of the temporal metrics (9 out of 12) exhibited significant between-centre differences. Conversely, the between centre differences in the gait quality metrics were small and comparable to those associated with a test-retest analysis under equivalent conditions. Therefore, the gait quality metrics are promising candidates for reliable multi-centric studies aiming at assessing rehabilitation interventions within a routine clinical context.

Keywords: multiple sclerosis; gait metrics; wearable sensors; test-retest reliability; sampling frequency; accelerometry; autocorrelation; harmonic ratio; six-minute walk

1. Introduction

Multiple sclerosis (MS) is a chronic demyelinating disease of the central nervous system affecting 2.3 million people worldwide [1]. MS is the major non-traumatic cause of disability in young and middle-aged adults [2], with a significant negative impact on independence and social participation [3]. Walking impairment is one of the most common functional deficits due to MS, even in the early stages of the disease [4]. Importantly, nearly 70% of people with MS (pwMS) reported that walking difficulty is the most challenging aspect of their condition [5].

Given the high impact of gait impairment on pwMS, different rehabilitation interventions focused on improving locomotion are currently applied to improve the quality of life in this population [6]. The effects of these interventions, together with the progression of the disease, are usually assessed in clinical practice using clinical scales, such as the expanded disability status scale (EDSS) [7] or timed tests, such as the timed up and go test (TUG) [8], the timed 25-foot walk test (T25FW) [9], and the 6-minute walk test (6MWT) [10]. Although widely used, these tests suffer from some limitations. Firstly, they assess only the time taken to execute the test (e.g., TUG and T25FW) or the distance travelled in a given time (6 min for the 6MWT), without providing objective measures of the different components and characteristics of the task that could be useful to describe *how* the performance is possibly impaired [11]. Secondly, these clinical tests have a relatively limited sensitivity to change [9,12,13] and a flooring effect [9,14] that makes it difficult to detect possible alterations in minimally impaired pwMS [15–17].

Instrumental methods may partly overcome these limitations by providing additional quantitative information for a more complete characterisation of walking, which can be useful to tailor the rehabilitative intervention and objectively assess its effects [11,18]. In particular, wearable inertial measurement units (IMUs), including accelerometers, gyroscopes, and magnetometers, represent cost-effective tools to perform objective assessments of walking in pwMS outside movement analysis labs [19,20], and even during free-living and community contexts [21,22]. IMUs have been widely used to analyse different locomotor tasks in pwMS, such as straight-line over ground [17,23–27] and treadmill walking [28], standing up, walking, turning, and sitting down (e.g., the TUG) [15,29], walking with head turns and over/around obstacles [30,31], walking while texting [32], and stairway walking [33]. During these tests, several parameters have been extracted from IMUs, including spatio-temporal parameters [15,24,27,28,31,32,34], indexes of gait variability and stability [17,23,24,26,31,33], trunk sway metrics [15,23,30,34], and angular variables [15,25,27,34]. Nonetheless, what does not yet clearly emerge from current literature on pwMS is which of these could be more reliably adopted within the clinical context.

Besides the issue of identifying among the above metrics those that are more capable of characterising the disease progression, hence providing similar results for patients with similar clinical conditions, and that have the sensitivity to detect changes associated with clinical interventions, the clinical adoption of specific gait metrics also requires accounting for a number of pragmatic limitations associated with testing conditions. These include an understanding of which output is more robust to testing site characteristics (e.g., corridor lengths, lightening, noise, etc.), adopted measuring instruments and their configuration (e.g., brand, location on the body, sampling frequency) [35–37], type of gait test (e.g., a single pass, a 1-minute or a 6MWT), or instructions given to patients (e.g., self-selected or fast walking speed, use or not use of an assistive device) [28,38–45]. All these aspects are particularly difficult to standardise in a busy clinical environment and most likely occur in combination with each other.

The aim of this study was to identify those gait metrics that provide equivalent assessment of pwMS with similar characteristics in terms of age, gender, and gait disability, despite these being tested in different centres and in non-standardised conditions. Our hypothesis was that while pwMS might be able to adjust their gait in terms of spatio-temporal parameters in response to different testing conditions (e.g., if asked to increase their speed), they would not be able to control those aspects of gait more related to its overall quality and energetic efficiency [46,47]. As a result, metrics extracted directly from the acceleration signals and representative of intensity, smoothness, stability, symmetry, and regularity were expected to be more robust to differences in the test settings. To verify this hypothesis, we compared retrospective data from two matched cohorts of pwMS, which were collected by two independent hospitals using protocols that differed for: (i) brand, size, and sampling frequency of the IMUs; (ii) IMU positioning; (iii) subject instructing; (iv) length of the path. As a term of reference, we also compared differences in IMU-based metrics between the two centres (between-centre differences) to those observable between two sessions performed by the same centre (between-day test-retest reliability).

2. Materials and Methods

2.1. Participants

Two research centres, one located in Italy (centre A) and one in the United Kingdom (centre B), provided retrospective IMU data collected while pwMS walked back and forth for 6 min along a hospital corridor. The patients' level of disability was assessed with the EDSS scale, scored by an experienced neurologist. Patients were excluded if not free from any orthopaedic and/or musculoskeletal and neurological disorders other than MS that may have affected their gait and balance. Since there were no restrictions for MS subtypes, both patients with relapsing remitting MS who were relapse-free for 30 days prior to assessment (centre A) and patients with secondary progressive MS (centre B) were included in the study. Thirteen pwMS were selected from each data set to form two cohorts, with individual patients matched if having the same age, gender, EDSS score, and type of assistive device (Table 1). As a result of this matching, the sample size, percentage of females, EDSS score distribution, number of pwMS who required an assistive device, and type of assistive device used during the walking test were the same in the two centres. The average walking speed was calculated as the total distance walked during the test divided by the duration of the walking trial.

Table 1. Clinical characteristics of people with multiple sclerosis for centre A and centre B. Abbreviations: expanded disability status scale (EDSS); people with multiple sclerosis (pwMS); Mann-Whitney U (MWU) statistic; *p*-value (*p*); chi-square (X^2).

	Centre A (n = 13)	Centre B (n = 13)	Statistics
Age [years]	51 (35–63)	57 (34–64)	U = 58, $p = 0.18$
Gender [men/women]	3/10	3/10	$X^2(1) = 0.00$, $p = 1.00$
EDSS score (0–10)	4.5 (2.0–6.5)	4.5 (2.5–6.5)	U = 83, $p = 0.93$
Mild (2.0–2.5)	1	1	
Moderate (3.0–4.5)	6	6	
Severe (5.0–6.5)	6	6	
Assistive devices			
Walker	1 pwMS	1 pwMS	–
Cane	2 pwMS	2 pwMS	–
Walking speed [m/s]	1.1 (0.5–1.4)	0.7 (0.4–1.0)	U = 31, $p < 0.01$ *

Values are median (range) or numbers. * $p < 0.05$.

pwMS from centre B repeated the instrumented walking test on a second visit, which was held 7–14 days after the first test at the same time of the day. The testing procedures were also kept constant between the two sessions. These data were used to assess between-day test-retest reliability.

Institutional review boards or ethics committees at the institutions in each country approved the separate protocols (NRES Committee Yorkshire & The Humber-Bradford Leeds (reference 15/YH/0300) and Ethical Committee of Don Carlo Gnocchi Foundation, Milan, Italy, references 29-03-2017 and 13-02-2019). Written informed consent was provided by all subjects. Data were collected in accordance with the International Declaration of Helsinki.

2.2. Experimental Protocol

Acceleration and angular velocity data from three IMUs, located at the fifth lumbar vertebra and around the right and left ankles, were recorded in both centres while pwMS walked back and forth for 6 min along a straight corridor free of obstacles and other people. If needed, they could use an assistive device and take short resting breaks while standing. Each IMU was manually aligned along the anatomical antero-posterior (AP), medio-lateral (ML), and vertical (V) axes.

The differences between the experimental protocols followed by centre A and centre B were: (i) device manufacturers and sampling frequency used to record acceleration and angular velocity signals; (ii) ankle IMU position; (iii) length of the walkway; (iv) instructions given to participants (Figure 1). Specifically, Xsens IMUs (unit weight 16 g, unit size 47 mm × 30 mm × 13 mm; MTw, Xsens, NL) with a sampling frequency of 75 Hz were used in centre A and OPAL IMUs (unit weight 22 g, unit size 48.5 mm × 36.5 mm × 13.5 mm; OPAL, APDM Inc., Portland, OR, USA) with a sampling frequency of 128 Hz were used in centre B. The IMUs around both ankles were placed laterally in centre A and frontally in centre B. PwMS were requested to walk at their maximum speed along a 30-meter straight corridor in centre A and at preferred comfortable speed along a 10-meter straight corridor in centre B.

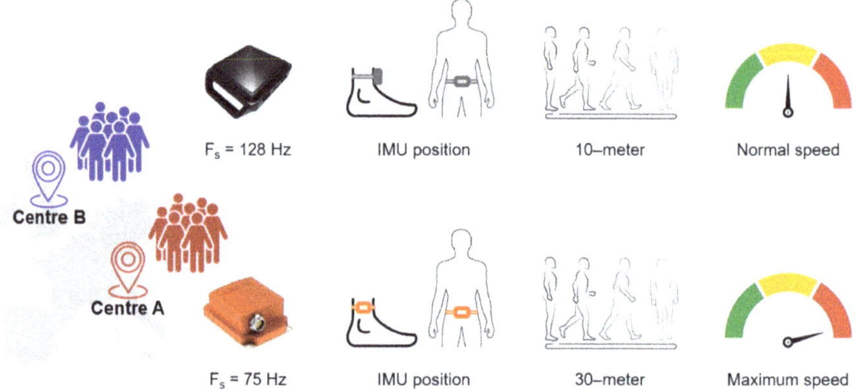

Figure 1. Experimental protocols followed by centre A (red) and centre B (blue).

2.3. Data Processing

Data processing routines were developed in Matlab® (MATLAB R2019b, MathWorks, Inc., Natick, MA, USA). A total of 33 IMU-based metrics were included in this analysis. IMU signals collected in centre B were down sampled from 128 Hz to 75 Hz to match data from centre A, and the influence of down sampling was investigated by comparing the outcome metrics from centre B as obtained before and after the down sampling. Data from the lumbar IMU were reoriented to a horizontal-vertical coordinate system [48] and filtered with a 10 Hz cut-off, zero phase, low-pass Butterworth filter.

The turning motion and resting breaks were detected and removed from IMU signals to isolate steady-state walking bouts, which were used to compute the metrics of interest. The approach proposed by Salarian, et al. [49] was adapted to determine 180° turns, which appear in the V component of the lumbar angular velocity, $\omega_z(t)$, as peaks of a given duration. The turning onset and offset were identified from the trunk rotation angle around the V axis, $\theta_z(t)$, obtained after integrating the $\omega_z(t)$ signal. The turning components were evidenced in $\theta_z(t)$ as steep positive or negative gradients, whereas walking components were evidenced as small oscillations round a flat line. Specifically, $\theta_z(t)$ was first smoothed using a weighted least-squares linear regression. Abrupt change points and their locations were then searched in $\theta_z(t)$ using a predefined Matlab® function based on the minimisation of a linear computational cost function [50]. Resting breaks were automatically detected by checking in 2-s window increments if: (i) the norm of the lumbar IMU angular velocity was less than 0.5 rad/s; (ii) the norm of the lumbar IMU acceleration was within ±10% of 9.81 m/s^2 [51]. A 2-s window was considered motionless if more than 50% of its samples fulfilled both criteria mentioned above.

Twelve gait metrics were extracted from the angular velocities recorded from the ankle IMUs and 21 were extracted from the lumbar IMU accelerations. Following the suggestions of Lord, et al. [52] and Buckley, et al. [53], these metrics were organised in independent gait domains (e.g., rhythm, variability, asymmetry, intensity, stability, smoothness, symmetry, and regularity).

Initial and final foot contact instances, referred to as gait events (GE), were identified for each steady-state walking bout as local minimum values of the ML angular velocity recorded from ankle IMUs of both legs [54]. These minima occur just before and after the instant of maximum ML angular velocity. Once the GE were determined, stride, step, swing and stance durations (representing rhythm domain) were separately estimated for left and right sides. Variability (i.e., within-subject combined standard deviation of left and right; variability domain) and asymmetry (i.e., absolute difference between the mean of left and right time series; asymmetry domain) of these metrics were also computed, applying the established formula in Galna, et al. [55] and Godfrey, et al. [56].

From processing the filtered acceleration signals in time and frequency domain, 21 additional metrics, referred to as gait quality metrics [57], were separately extracted for each acceleration component (AP, ML, and V): (i) intensity as the root mean square (RMS) of each acceleration component around its mean value [44]; (ii) stability as the ratio of the RMS in a given direction to the RMS vector magnitude [58]; (iii) smoothness as the RMS of the jerk [59]; (iv) symmetry represented by the harmonic ratio (HR), defined as the ratio of the sum of the amplitudes of the in-phase harmonics to the sum of the amplitudes of the out-of-phase harmonics [60,61]; (v) regularity as the ensemble of the following three metrics obtained from the unbiased normalised autocorrelation [62]:

$$Step\ regularity = 1st\ peak\ of\ \left(\frac{1}{N-|m|}\sum_{i=1}^{N-|m|} x(i)\cdot x(i+m)\right) \tag{1}$$

$$Stride\ regularity = 2nd\ peak\ of\ \left(\frac{1}{N-|m|}\sum_{i=1}^{N-|m|} x(i)\cdot x(i+m)\right) \tag{2}$$

$$Regularity\ index = \frac{|Stride\ regularity - Step\ regularity|}{mean(Stride\ regularity, Step\ regularity)} \tag{3}$$

All metrics were calculated for the part of signals corresponding to the middle eight steps of each pass along the corridor and then averaged over the whole trial. The choice of eight steps was due to the maximum number of steps which subjects in centre B could walk in completely straight condition. Since centre A adopted a three-times longer path, in order to process the same number of steps, only one walking bout in every three was included for centre B.

2.4. Statistical Analysis

Statistical analyses were performed in R version 3.4.3 [63]. Participant characteristics from centre A and centre B were compared using the independent Mann-Whitney U for age and EDSS scores and Pearson's chi-square for gender. Given the limited sample size and the non-normal distribution of most of the investigated metrics (as a result of the Shapiro-Wilk test), non-parametric tests were performed. The level of significance was taken at 5%. A Wilcoxon signed-rank test was performed to compare the centre B metrics obtained from IMU data sampled at 128 Hz and those down-sampled at 75 Hz.

Between-day test-retest reliability of the metrics was evaluated for centre B through the intra-class correlation coefficients (ICCs) with a 95% confidence interval (CI). ICCs were calculated using a two-way random-effect model and absolute agreement (ICC2,k) [64]. An ICC lower than 0.39 was classified as poor, an ICC between 0.40 and 0.59 as fair, an ICC between 0.60 and 0.74 as moderate, and an ICC greater than 0.75 as excellent [65]. The minimum detectable changes (MDCs), representing the smallest amount of change that can be considered above the bounds of the measurement error and/or within-subject variability, was also computed for each metric at the CI of 95%, according to Equation (4):

$$MDC = 1.96 \cdot \sqrt{2} \cdot SEM = 1.96 \cdot \sqrt{2} \cdot SD \cdot \sqrt{1 - ICC}, \quad (4)$$

where SEM is the standard error of the measurement and SD corresponds to the average of the standard deviations from test and re-test sessions [66].

A Wilcoxon signed-rank test was used to determine if there was a median difference in centre B metrics between the two sessions, whereas an independent Mann-Whitney U test was carried out to compare IMU-based metrics from centre A and centre B.

In all the above tests, if the *p*-value was lower than 0.05, the null hypothesis (e.g., the two population medians were identical) was rejected and the alternative hypothesis accepted. To avoid misinterpretation of the *p*-values and to account for a type II error, the effect size (r) for non-parametric tests was also calculated as follows:

$$r = z / \sqrt{N} \quad (5)$$

where z is the z-score and N is the size of the study (i.e., the number of total observations) on which z is based. Cohen [67] suggested thresholds of 0.1, 0.3, and 0.5 for small, medium, and large effect sizes, respectively.

Median, inter-quartile range, minimum, and maximum values were finally calculated for IMU-based metrics from centre A and centre B (both sessions).

3. Results

3.1. Effect of Sampling Frequency

The results of the comparison between the metrics calculated using the 128 Hz and 75 Hz sampling frequencies are reported in Table 2. The HR, representative of the symmetry domain, was the only metric that significantly differed between the two analyses.

Table 2. Effect of down-sampling of the acceleration and angular velocity signals on the investigated gait metrics. Abbreviations: sampling frequency (F_S), z-score (z), *p*-value (*p*), and effect size (*r*).

Domain	F_s of 128 Hz	F_s of 75 Hz	z	*p*	*r*
Rhythm [s]					
Stride duration	1.20 (1.01–1.74)	1.21 (1.01–1.74)	−0.82	0.41	−0.16
Step duration	0.60 (0.51–0.87)	0.60 (0.50–0.87)	0.00	1.00	0.00
Stance duration	0.75 (0.61–1.18)	0.75 (0.61–1.18)	−1.83	0.07	−0.36
Swing duration	0.44 (0.40–0.58)	0.44 (0.40–0.58)	−1.85	0.06	−0.36
Variability [ms]					
Stride duration	61 (32–100)	63 (32–98)	−1.55	0.12	−0.30
Step duration	46 (20–69)	45 (20–68)	−1.33	0.18	−0.26
Stance duration	65 (34–105)	65 (32–106)	−0.18	0.86	−0.04
Swing duration	29 (23–74)	30 (21–76)	−0.41	0.68	−0.08
Asymmetry [ms]					
Stride duration	2 (0–7)	2 (1–7)	−0.09	0.93	−0.02
Step duration	56 (0–238)	51 (0–242)	−1.49	0.14	−0.29
Stance duration	61 (3–149)	69 (2–130)	−1.58	0.11	−0.31
Swing duration	54 (1–155)	62 (0–138)	−1.33	0.18	−0.26

Table 2. *Cont.*

Domain	F_s of 128 Hz	F_s of 75 Hz	z	p	r
Intensity [m/s²]					
Antero-Posterior	1.10 (0.80–1.96)	1.10 (0.80–1.95)	−0.94	0.34	−0.19
Medio-Lateral	0.93 (0.65–1.41)	0.93 (0.65–1.41)	−1.44	0.15	−0.28
Vertical	1.37 (0.76–3.16)	1.38 (0.75–3.20)	−1.28	0.20	−0.25
Stability [–]					
Antero-Posterior	0.41 (0.37–0.61)	0.41 (0.37–0.61)	−0.30	0.77	−0.06
Medio-Lateral	0.34 (0.25–0.52)	0.34 (0.25–0.52)	−0.89	0.37	−0.18
Vertical	0.58 (0.36–0.62)	0.58 (0.36–0.63)	−0.29	0.77	−0.06
Smoothness [m/s³]					
Antero-Posterior	13.97 (8.86–30.75)	13.95 (8.93–30.82)	0.00	1.00	0.00
Medio-Lateral	13.92 (10.13–28.61)	13.87 (10.19–28.45)	−1.06	0.29	−0.21
Vertical	23.31 (11.06–48.81)	23.21 (10.89–49.38)	−0.75	0.45	−0.15
Symmetry (HR) [–]					
Antero-Posterior	2.94 (1.49–3.73)	2.89 (1.50–3.49)	−2.32	0.02 *	−0.45
Medio-Lateral	0.44 (0.32–0.56)	0.45 (0.32–0.56)	−2.19	0.03 *	−0.43
Vertical	3.01 (1.21–4.84)	2.94 (1.23–4.78)	−3.01	0.00 *	−0.59
Regularity [–]					
Step regularity					
Antero-Posterior	0.60 (0.20–0.85)	0.60 (0.20–0.84)	−1.70	0.09	−0.33
Medio-Lateral	−0.62 (−0.74–−0.37)	−0.60 (−0.73–−0.38)	−1.89	0.06	−0.37
Vertical	0.81 (0.32–0.95)	0.80 (0.32–0.94)	−1.44	0.15	−0.28
Stride regularity					
Anterior-Posterior	0.86 (0.50–0.93)	0.86 (0.50–0.92)	0.00	1.00	0.00
Medio-Lateral	0.77 (0.58–0.85)	0.75 (0.59–0.85)	−1.67	0.09	−0.33
Vertical	0.86 (0.34–0.95)	0.86 (0.34–0.95)	−1.80	0.07	−0.35
Regularity index					
Antero-Posterior	0.37 (0.04–0.82)	0.37 (0.04–0.83)	−1.10	0.27	−0.22
Medio-Lateral	−0.20 (−0.70–−0.08)	−0.20 (−0.66–−0.08)	−0.41	0.68	−0.08
Vertical	0.11 (0.02–0.59)	0.11 (0.02–0.59)	0.00	1.00	0.00

Values are median (range). * $p < 0.05$.

3.2. *Between-Day Test-Retest Reliability*

ICC, SEM, and MDC values for between-day assessment are shown in Table 3 for each metric estimated for pwMS from centre B who completed two testing visits. Overall, 17 out of 33 metrics revealed excellent test-retest reliability (ICC: 0.93–0.98; 95% CI: 0.76–0.93), 11 metrics showed moderate test-retest reliability (ICC: 0.88–0.92; 95% CI: 0.62–0.74), and only 5 metrics exhibited poor to fair test-retest reliability with ICC values between 0.72 and 0.86 and 95% CI between 0.13 and 0.52. The Wilcoxon signed-rank test showed no significant differences in any of the metrics between the two sessions (Figure 2 and Table 4).

Table 3. Intra-class correlation coefficients (ICC) with a 95% confidence interval (CI), standard error of the measurement (SEM), and minimum detectable change (MDC) for the investigated gait metrics.

Domains	ICC	95% CI		SEM	MDC
		Lower	Upper		
Rhythm [s]					
Stride duration	0.97	0.90	0.99	0.04	0.10
Step duration	0.97	0.90	0.99	0.02	0.05
Stance duration	0.96	0.86	0.99	0.03	0.09
Swing duration	0.97	0.91	0.99	0.01	0.03
Variability [ms]					
Stride duration	0.92	0.73	0.97	8	21
Step duration	0.92	0.74	0.98	5	13
Stance duration	0.94	0.80	0.98	6	18
Swing duration	0.95	0.85	0.99	4	11
Asymmetry [ms]					
Stride duration	**0.72**	**0.13**	**0.91**	**1**	**4**
Step duration	0.98	0.93	0.99	10	29
Stance duration	0.90	0.67	0.97	12	33
Swing duration	0.89	0.62	0.97	13	36
Intensity [m/s^2]					
Antero-Posterior	0.97	0.90	0.99	0.06	0.16
Medio-Lateral	0.98	0.93	0.99	0.04	0.11
Vertical	0.97	0.92	0.99	0.10	0.29
Stability [–]					
Antero-Posterior	0.93	0.78	0.98	0.02	0.05
Medio-Lateral	0.93	0.76	0.98	0.03	0.08
Vertical	0.91	0.69	0.97	0.03	0.09
Smoothness [m/s^3]					
Antero-Posterior	0.92	0.73	0.97	2.46	6.83
Medio-Lateral	0.93	0.79	0.98	1.55	4.29
Vertical	0.95	0.82	0.98	2.31	6.41
Symmetry (HR) [–]					
Antero-Posterior	0.95	0.85	0.99	0.14	0.38
Medio-Lateral	**0.75**	**0.15**	**0.92**	**0.04**	**0.10**
Vertical	0.92	0.74	0.98	0.21	0.59
Regularity [–]					
Step regularity					
Antero-Posterior	0.91	0.70	0.97	0.07	0.19
Medio-Lateral	**0.86**	**0.52**	**0.96**	**0.04**	**0.11**
Vertical	0.97	0.92	0.99	0.04	0.10
Stride regularity					
Antero-Posterior	0.88	0.64	0.96	0.05	0.13
Medio-Lateral	**0.85**	**0.50**	**0.96**	**0.04**	**0.10**
Vertical	0.93	0.77	0.98	0.04	0.10
Regularity index					
Antero-Posterior	**0.76**	**0.17**	**0.93**	**0.17**	**0.47**
Medio-Lateral	0.88	0.62	0.96	0.06	0.17
Vertical	0.89	0.63	0.97	0.09	0.24

Inertial measurement unit (IMU)-based gait metrics with poor to fair test-retest reliability are presented in bold.

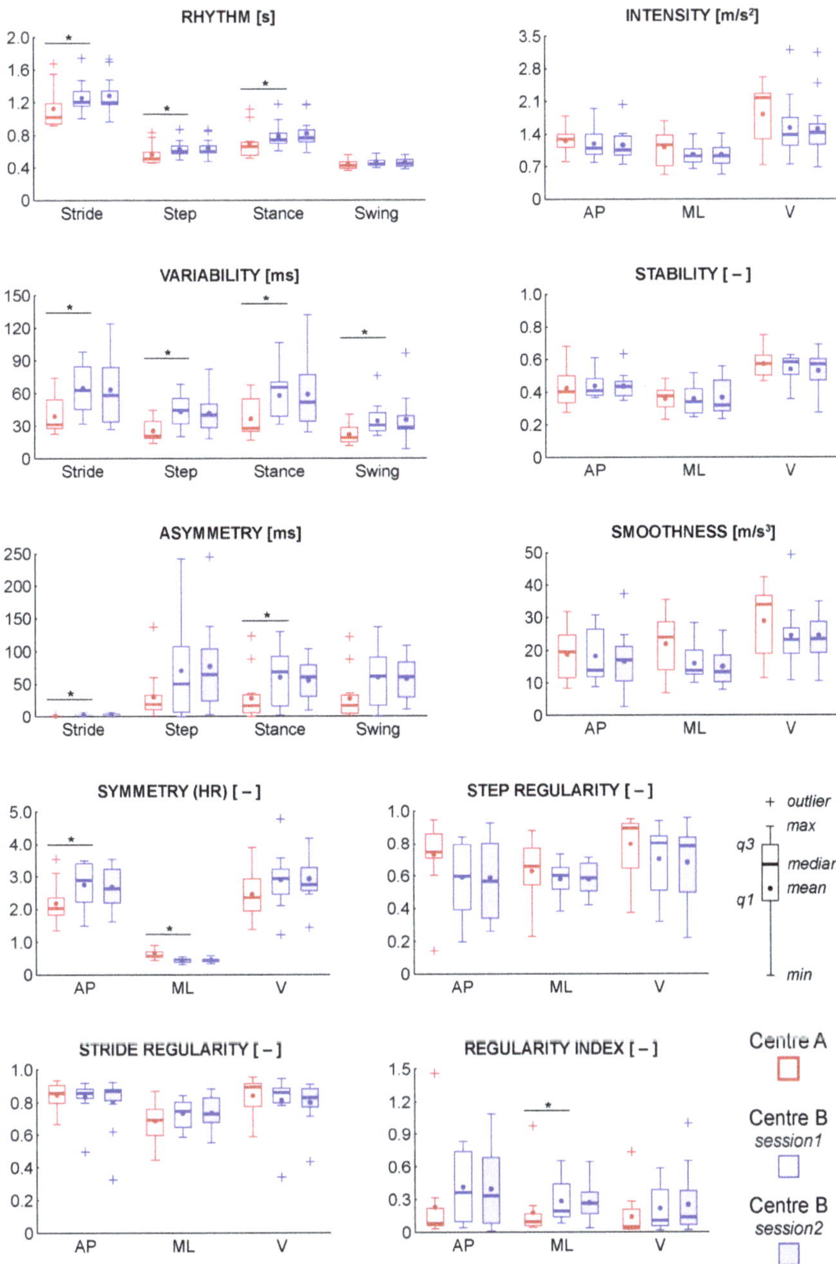

Figure 2. Minimum, first quartile (q1), median, mean, third quartile (q3), and maximum values of each IMU-based metrics relative to centre A (red) and centre B for between-day test-retest assessment (blue empty boxplots and blue filled boxplots). Values larger than q1 + 1.5(q3 + q1) or smaller than q1 − 1.5(q3 − q1) are considered outliers and are represented with crosses (+). * $p < 0.05$. Note that, for graphical convenience, the absolute values have been depicted for the step regularity and regularity index in the ML direction.

Table 4. Descriptive statistics for the investigated gait metrics from centre B (session1 and session2), including the z-score (z), p-value (p), and effect size (r).

Domain	Centre B (session1)	Centre B (session2)	z	p	r
Rhythm [s]					
Stride duration	1.21 (1.01–1.74)	1.20 (0.97–1.74)	−0.70	0.48	−0.14
Step duration	0.60 (0.50–0.87)	0.60 (0.48–0.87)	−0.56	0.58	−0.11
Stance duration	0.75 (0.61–1.18)	0.77 (0.59–1.18)	−1.57	0.12	−0.31
Swing duration	0.44 (0.40–0.58)	0.45 (0.38–0.56)	−1.99	0.05	−0.39
Variability [ms]					
Stride duration	63 (32–98)	58 (27–124)	−0.35	0.72	−0.07
Step duration	45 (20–68)	40 (18–83)	−0.35	0.72	−0.07
Stance duration	65 (32–106)	52 (24–132)	−0.03	0.97	−0.01
Swing duration	30 (21–76)	28 (9–97)	−0.53	0.60	−0.10
Asymmetry [ms]					
Stride duration	2 (1–7)	4 (0–7)	−1.34	0.18	−0.26
Step duration	51 (0–242)	65 (4–245)	−1.30	0.20	−0.25
Stance duration	69 (2–130)	61 (10–104)	−0.52	0.60	−0.10
Swing duration	62 (0–138)	61 (12–109)	−0.38	0.70	−0.08
Intensity [m/s^2]					
Antero-Posterior	1.10 (0.80–1.95)	1.07 (0.76–2.04)	−1.22	0.22	−0.24
Medio-Lateral	0.93 (0.65–1.41)	0.93 (0.53–1.42)	−0.08	0.94	−0.02
Vertical	1.38 (0.75–3.20)	1.43 (0.68–3.14)	−0.38	0.70	−0.08
Stability [–]					
Antero-Posterior	0.41 (0.37–0.61)	0.43 (0.35–0.64)	−0.28	0.78	−0.05
Medio-Lateral	0.34 (0.25–0.52)	0.32 (0.24–0.56)	0.00	1.00	0.00
Vertical	0.58 (0.36–0.63)	0.57 (0.28–0.69)	−0.27	0.79	−0.05
Smoothness [m/s^3]					
Antero-Posterior	13.95 (8.93–30.82)	17.11 (2.76–37.32)	−1.17	0.24	−0.23
Medio-Lateral	13.87 (10.19–28.45)	13.42 (8.04–26.07)	−1.24	0.22	−0.24
Vertical	23.21 (10.89–49.38)	23.43 (10.67–50.74)	−0.41	0.68	−0.08
Symmetry (HR) [–]					
Antero-Posterior	2.89 (1.50–3.49)	2.64 (1.62–3.54)	−0.31	0.75	−0.06
Medio-Lateral	0.45 (0.32–0.56)	0.46 (0.34–0.59)	−0.82	0.41	−0.16
Vertical	2.94 (1.23–4.78)	2.75 (1.45–4.19)	−0.51	0.61	−0.10
Regularity [–]					
Step regularity					
Antero-Posterior	0.60 (0.20–0.84)	0.57 (0.26–0.93)	−0.12	0.91	−0.02
Medio-Lateral	−0.60 (−0.73–−0.38)	−0.58 (−0.71–−0.42)	−0.07	0.94	−0.01
Vertical	0.80 (0.32–0.94)	0.78 (0.22–0.96)	−1.26	0.21	−0.25
Stride regularity					
Anterior-Posterior	0.86 (0.50–0.92)	0.87 (0.33–0.92)	−0.98	0.33	−0.19
Medio-Lateral	0.75 (0.59–0.85)	0.73 (0.55–0.88)	−0.03	0.97	−0.01
Vertical	0.86 (0.34–0.95)	0.83 (0.44–0.91)	−0.52	0.60	−0.10
Regularity index					
Antero-Posterior	0.37 (0.04–0.83)	0.34 (0.01–1.09)	−0.04	0.97	−0.01
Medio-Lateral	−0.20 (−0.66–−0.08)	−0.27 (−0.65–−0.04)	−0.14	0.89	−0.03
Vertical	0.11 (0.02–0.59)	0.14 (0.02–1.00)	−0.43	0.67	−0.08

Values are median (range). * $p < 0.05$.

3.3. Between-Centre Differences

As expected, the comparison between centre A and centre B via the independent Mann-Whitney U test highlighted significant differences for all the temporal metrics (Figure 2 and Table 5; rhythm domain), except for swing duration. Apart from asymmetry of step duration and asymmetry of swing duration, variability and asymmetry of the temporal metrics were significantly lower in centre A compared to centre B (Figure 2 and Table 5; variability and asymmetry domain). However, even though the difference in asymmetry of swing duration between the two centres was non-significant ($U = 48.0$; $p = 0.06$), a fairly moderate effect size was found for this specific metric ($r = 0.37$). Conversely, a consistency between the two centres was found for 18 out of 21 metrics extracted from acceleration signals (Figure 2 and Table 5; intensity, stability, smoothness, symmetry, and regularity domains). Only the differences in the regularity index in the ML direction and in the HR in the AP and ML directions were proved statistically significant between centre A and centre B (Figure 2 and Table 5).

Table 5. Descriptive statistics for the investigated gait metrics from centre A and centre B (session1), including the Mann-Whitney U (MWU) statistic, *p*-value (*p*), and effect size (*r*).

Domain	Centre A	Centre B	U	*p*	*r*
Rhythm [s]					
Stride duration	1.03 (0.92–1.68)	1.21 (1.01–1.74)	43.5	0.04 *	0.41
Step duration	0.51 (0.46–0.84)	0.60 (0.50–0.87)	44.0	0.04 *	0.41
Stance duration	0.66 (0.52–1.12)	0.75 (0.61–1.18)	40.0	0.02 *	0.45
Swing duration	0.43 (0.37–0.56)	0.44 (0.40–0.58)	57.0	0.17	0.28
Variability [ms]					
Stride duration	32 (23–74)	63 (32–98)	26.0	0.00 *	0.59
Step duration	21 (14–45)	45 (20–68)	27.5	0.00 *	0.57
Stance duration	28 (17–68)	65 (32–106)	30.0	0.01 *	0.55
Swing duration	19 (12–41)	30 (21–76)	33.0	0.01 *	0.52
Asymmetry [ms]					
Stride duration	1 (0–4)	2 (1–7)	45.0	0.04 *	0.40
Step duration	19 (1–138)	51 (0–242)	68.0	0.40	0.17
Stance duration	17 (0–123)	69 (2–130)	46.0	0.04 *	0.39
Swing duration	17 (1–122)	62 (0–138)	48.0	0.06	0.37
Intensity [m/s^2]					
Antero-Posterior	1.30 (0.81–1.80)	1.10 (0.80–1.95)	68.0	0.41	0.17
Medio-Lateral	1.17 (0.53–1.69)	0.93 (0.65–1.41)	62.0	0.26	0.23
Vertical	2.17 (0.73–2.62)	1.38 (0.75–3.20)	59.5	0.21	0.25
Stability [–]					
Antero-Posterior	0.40 (0.28–0.68)	0.41 (0.37–0.61)	69.0	0.44	0.16
Medio-Lateral	0.38 (0.23–0.48)	0.34 (0.25–0.52)	83.0	0.96	0.02
Vertical	0.57 (0.47–0.75)	0.58 (0.36–0.63)	72.5	0.55	0.12
Smoothness [m/s^3]					
Antero-Posterior	19.68 (8.56–31.85)	13.95 (8.93–30.82)	82.0	0.92	0.03
Medio-Lateral	24.01 (7.04–35.54)	13.87 (10.19–28.45)	52.5	0.11	0.32
Vertical	33.90 (11.56–42.39)	23.21 (10.89–49.38)	59.0	0.20	0.26
Symmetry (HR) [–]					
Antero-Posterior	2.04 (1.36–3.54)	2.89 (1.50–3.49)	43.5	0.04 *	0.41
Medio-Lateral	0.57 (0.44–0.91)	0.45 (0.32–0.56)	14.0	0.00 *	0.71
Vertical	2.35 (1.39–3.89)	2.94 (1.23–4.78)	52.5	0.11	0.32

Table 5. *Cont.*

Domain	Centre A	Centre B	U	p	r
Regularity [–]					
Step regularity					
Antero-Posterior	0.75 (0.14–0.95)	0.60 (0.20–0.84)	55.5	0.14	0.29
Medio-Lateral	−0.66 (−0.88–−0.23)	−0.60 (−0.73–−0.38)	63.0	0.28	0.22
Vertical	0.89 (0.37–0.95)	0.80 (0.32–0.94)	51.5	0.10	0.33
Stride regularity					
Anterior-Posterior	0.86 (0.67–0.93)	0.86 (0.50–0.92)	81.5	0.90	0.03
Medio-Lateral	0.69 (0.45–0.87)	0.75 (0.59–0.85)	63.0	0.28	0.22
Vertical	0.89 (0.59–0.96)	0.86 (0.34–0.95)	72.5	0.55	0.12
Regularity index					
Antero-Posterior	0.08 (0.03–1.46)	0.37 (0.04–0.83)	50.0	0.08	0.35
Medio-Lateral	−0.10 (−0.98–−0.05)	−0.20 (−0.66–−0.08)	40.5	0.03 *	0.44
Vertical	0.05 (0.02–0.74)	0.11 (0.02–0.59)	51.5	0.09	0.33

Values are median (range). * $p < 0.05$.

4. Discussion

This study aimed to identify comparable gait metrics as quantified from IMU data measured from two different hospital settings on two matched cohorts of pwMS (13 pwMS for each centre, Table 1), under the hypothesis that those metrics associated with the overall balance control and coordination of gait (i.e., gait quality metrics) would be robust, even when obtained from different experimental protocols. Reported results overall corroborated this assumption and showed that between-centre differences for most of these metrics were comparable to those obtained by the same centre in two different sessions.

The small sample size, resulting from the attempt of maximising the cohort match, is certainly a limitation of this study. It is worth noting, in fact, that while some of the investigated gait metrics in centre A (e.g., asymmetry of swing duration from asymmetry domain and regularity index from regularity domain) did not differ significantly from those in centre B, an observed medium effect size suggested the opposite might hold true (Table 5). This is indeed likely to be due to the small sample size and possibly due to the higher inter-subject variability observed in centre B.

Since MS is well known for heterogeneity of symptoms, high day-to-day fluctuations, and a large variability in its course [68], care must be taken before generalising our findings to all pwMS with different levels of gait impairment. Another limitation of this study might lie in the fact that patients recruited by the two centres differed in the subtypes of MS. Nonetheless, Dujmovic, et al. [69] showed that the altered gait pattern in pwMS did not depend on the disease phenotype. Additional studies are of course needed to further investigate this aspect.

The comparison between centre A and centre B implied down-sampling the data from the latter. As expected, this affected only the calculation of HR, which is the only metric based on frequency analysis. In particular, changing sampling frequency from 128 Hz to 75 Hz led to decreased values in the AP and V directions and increased values in the ML direction (Table 2). This is in line with what was previously reported by Riva, et al. [35].

Moderate to excellent between-day test-retest reliability was observed for 28 out of 33 IMU-based metrics with few exceptions, which exhibited poor to fair reliability (Table 3). Additionally, all the investigated metrics were not significantly different between the two sessions (Figure 2 and Table 4), even if some of these results (swing duration in particular) should be interpreted with care, due to the medium effect size. These findings confirmed that sensor-based gait analysis is a reliable tool in pwMS, as also reported in previous test-retest studies on pwMS [34].

Walking speed clearly affected the gait outcomes. In particular, the gait metrics representative of rhythm, variability, and asymmetry domains were evidently lower in centre A compared to centre

B (Figure 2 and Table 5) due to different instructions given to the participants in terms of walking speed (i.e., walk at maximum speed versus walk at self-selected speed). This finding is in agreement with previous studies on pwMS [28] and on people with other neurological conditions, such as Parkinson's disease [70], which observed a reduction of the above metrics with increasing walking speed. The shorter length of the walkway used in centre B could also have contributed to these differences. In fact, Storm, et al. [22] demonstrated that rhythm and variability metrics decreased when walking longer distances (e.g., lower stride duration and lower variability of stride duration). However, the data available for our study did not allow us to separate walking speed and path effects, and further studies should hence be performed to this purpose.

Unlike the temporal metrics, the gait quality metrics appeared to be robust with respect to the notable differences in the experimental gait protocols adopted by the two centres. Among these metrics, in fact, only differences in the regularity index in the ML direction and the HR (representative of symmetry domain) in the AP and ML directions were found to be statistically significant between centre A and centre B (Figure 2 and Table 5). Again, this specific result could be explained both by the different walking speed and by the different lengths of the walkway in the two centres. Indeed, an association between walking speed and HR has been previously showed, both in healthy young [43,44] and older subjects [39]. These authors observed that the HR increased at the self-selected comfortable walking speed and decreased at slower and faster speeds. A similar trend emerged from our analysis, except for the HR in the ML direction, but this specific metric should be handled with care due to its observed low test-retest reliability (Table 3). The low number of steps (i.e., eight steps) used for calculating the HR for each walking bout might also have contributed to reduce robustness and reliability of this metric [57,71]. However, this choice was imposed by the reduced length of the corridor in centre B. Testing the participants along a shorter path also implied a higher number of turns over the 6 min, resulting in a minor validity of the HR as showed in the research by Riva, et al. [35] and by Brach, et al. [40].

While further studies are of course needed to fully validate this hypothesis, our results suggest that, in agreement with what is already reported for other neurological diseases, such as Parkinson's disease [53], the gait quality metrics extracted from the upper body accelerations should not be considered as a simple reflection of gait spatio-temporal features and might bring complementary informative content in quantifying patients' gait ability. Additionally, these metrics have been recently shown to be sensitive to fatigue and pathology progression in pwMS [72] and, as such, they are promising candidates for quantification of disease progression and rehabilitation interventions in these patients.

5. Conclusions

In conclusion, this pragmatic study showed consistency in the gait metrics from two matched groups of pwMS, even when they were assessed in two different hospitals and under notably different gait testing conditions. The identification of such robust gait metrics opens the possibility of comparing retrospective data and paves the way for reliable multi-centre studies to be conducted in routine hospital settings rather than in specialised gait research laboratories. This is essential to allow an increase of sample size and statistical power of clinical trials in which rehabilitation interventions need to be quantitatively assessed.

Author Contributions: Conceptualisation, M.F. and C.M.; methodology, L.A., I.C., D.C., M.F. and C.M.; software, L.A.; formal analysis, L.A.; resources, D.C., E.G., B.S., D.P. and C.M.; writing-original draft preparation, L.A., I.C., M.F. and C.M.; writing-review and editing, L.A., I.C., D.C., M.F., E.G., B.S., D.P., K.P.S.N. and C.M.; visualisation, L.A.; funding acquisition, M.F., B.S., K.P.S.N. and C.M. All authors have read and agreed to the published version of the manuscript.

Funding: This study was co-funded by the NIHR through the Sheffield Biomedical Research Centre (BRC, grant number IS-BRC-1215-20017), by the European Union's Horizon 2020 research and innovation programme and EFPIA via the Innovative Medicine Initiative 2 (Mobilise-D project, grant number IMI22017-13-7-820820), the UK Engineering and Physical Sciences Research Council (Multisim and MultiSim2 project, grant number EP/K03877X/1 and EP/S032940/1), and by the Italian Ministry of Health, Ricerca Corrente (grant number GR-2009-1604984). The views expressed are those of the author(s) and not necessarily those of the NHS, the NIHR, the Department of Health and Social Care, the IMI, the European Union, the EFPIA, or any associated partners.

Acknowledgments: We would like to thank all participants for giving their time to support this research. This study was carried out at the NIHR Sheffield Clinical Research Facility (Sheffield, United Kingdom) and at the IRCCS Don Carlo Gnocchi Foundation (Milan, Italy). The authors would like to acknowledge William Hodgkinson, Craig Smith, and Jessy Moorman Dodd for the support in Sheffield's data collection.

Conflicts of Interest: The authors declare no conflict of interest.

Data Availability: The data used in this paper will be made publicly available (DOI: 10.15131/shef.data.11395641).

References

1. Browne, P.; Chandraratna, D.; Angood, C.; Tremlett, H.; Baker, C.; Taylor, B.V.; Thompson, A.J. Atlas of Multiple Sclerosis 2013: A growing global problem with widespread inequity. *Neurology* **2014**, *83*, 1022–1024. [CrossRef]
2. Pugliatti, M.; Rosati, G.; Carton, H.; Riise, T.; Drulovic, J.; Vécsei, L.; Milanov, I. The epidemiology of multiple sclerosis in Europe. *Eur. J. Neurol.* **2006**, *13*, 700–722. [CrossRef]
3. Cattaneo, D.; Lamers, I.; Bertoni, R.; Feys, P.; Jonsdottir, J. Participation Restriction in People with Multiple Sclerosis: Prevalence and Correlations With Cognitive, Walking, Balance, and Upper Limb Impairments. *Arch. Phys. Med. Rehabil.* **2017**, *98*, 1308–1315. [CrossRef]
4. Martin, C.L.; Phillips, B.A.; Kilpatrick, T.J.; Butzkueven, H.; Tubridy, N.; McDonald, E.; Galea, M.P. Gait and balance impairment in early multiple sclerosis in the absence of clinical disability. *Mult. Scler. J.* **2006**, *12*, 620–628. [CrossRef]
5. LaRocca, N.G. Impact of Walking Impairment in Multiple Sclerosis. *Patient Patient-Cent. Outcomes Res.* **2011**, *4*, 189–201. [CrossRef]
6. Donzé, C. Update on rehabilitation in multiple sclerosis. *La Presse Med.* **2015**, *44*, e169–e176. [CrossRef]
7. Kurtzke, J.F. Rating neurologic impairment in multiple sclerosis. *Neurology* **1983**, *33*, 1444. [CrossRef]
8. Sebastião, E.; Sandroff, B.M.; Learmonth, Y.C.; Motl, R.W. Validity of the Timed Up and Go Test as a Measure of Functional Mobility in Persons with Multiple Sclerosis. *Arch. Phys. Med. Rehabil.* **2016**, *97*, 1072–1077. [CrossRef]
9. Phan-Ba, R.; Pace, A.; Calay, P.; Grodent, P.; Douchamps, F.; Hyde, R.; Hotermans, C.; Delvaux, V.; Hansen, I.; Moonen, G.; et al. Comparison of the timed 25-foot and the 100-meter walk as performance measures in multiple sclerosis. *Neurorehabil. Neural Repair.* **2011**, *25*, 672–679. [CrossRef]
10. Goldman, M.D.; Marrie, R.A.; Cohen, J.A. Evaluation of the six-minute walk in multiple sclerosis subjects and healthy controls. *Mult. Scler. J.* **2008**, *14*, 383–390. [CrossRef]
11. Horak, F.; King, L.; Mancini, M. Role of body-worn movement monitor technology for balance and gait rehabilitation. *Phys. Ther.* **2015**, *95*, 461–470. [CrossRef]
12. Kaufman, M.; Moyer, D.; Norton, J. The significant change for the Timed 25-Foot Walk in the Multiple Sclerosis Functional Composite. *Mult. Scler. J.* **2000**, *6*, 286–290. [CrossRef]
13. Kragt, J.J.; van der Linden, F.A.; Nielsen, J.M.; Uitdehaag, B.M.; Polman, C.H. Clinical impact of 20% worsening on Timed 25-foot Walk and 9-hole Peg Test in multiple sclerosis. *Mult. Scler. J.* **2006**, *12*, 594–598. [CrossRef]
14. Nieuwenhuis, M.M.; Van Tongeren, H.; Sorensen, P.S.; Ravnborg, M. The six spot step test: A new measurement for walking ability in multiple sclerosis. *Mult. Scler. J.* **2006**, *12*, 495–500. [CrossRef]
15. Spain, R.I.; St. George, R.J.; Salarian, A.; Mancini, M.; Wagner, J.M.; Horak, F.B.; Bourdette, D. Body-worn motion sensors detect balance and gait deficits in people with multiple sclerosis who have normal walking speed. *Gait Posture* **2012**, *35*, 573–578. [CrossRef]
16. Liparoti, M.; Della Corte, M.; Rucco, R.; Sorrentino, P.; Sparaco, M.; Capuano, R.; Minino, R.; Lavorgna, L.; Agosti, V.; Sorrentino, G.; et al. Gait abnormalities in minimally disabled people with Multiple Sclerosis: A 3D-motion analysis study. *Mult. Scler. Relat. Disord.* **2019**, *29*, 100–107. [CrossRef]
17. Pau, M.; Mandaresu, S.; Pilloni, G.; Porta, M.; Coghe, G.; Marrosu, M.G.; Cocco, E. Smoothness of gait detects early alterations of walking in persons with multiple sclerosis without disability. *Gait Posture* **2017**, *58*, 307–309. [CrossRef]

18. Muro-de-la-Herran, A.; Garcia-Zapirain, B.; Mendez-Zorrilla, A. Gait analysis methods: an overview of wearable and non-wearable systems, highlighting clinical applications. *Sensors* **2014**, *14*, 3362–3394. [CrossRef]
19. Vienne-Jumeau, A.; Quijoux, F.; Vidal, P.P.; Ricard, D. Wearable inertial sensors provide reliable biomarkers of disease severity in multiple sclerosis: A systematic review and meta-analysis. *Ann. Phys. Rehabil. Med.* **2019**. [CrossRef]
20. Vienne, A.; Barrois, R.P.; Buffat, S.; Ricard, D.; Vidal, P.P. Inertial Sensors to Assess Gait Quality in Patients with Neurological Disorders: A Systematic Review of Technical and Analytical Challenges. *Front. Psychology* **2017**, *8*, 817. [CrossRef]
21. Motl, R.W.; Pilutti, L.; Sandroff, B.M.; Dlugonski, D.; Sosnoff, J.J.; Pula, J.H. Accelerometry as a measure of walking behavior in multiple sclerosis. *Acta Neurol. Scand.* **2013**, *127*, 384–390. [CrossRef]
22. Storm, F.A.; Nair, K.P.S.; Clarke, A.J.; Van der Meulen, J.M.; Mazzà, C. Free-living and laboratory gait characteristics in patients with multiple sclerosis. *PLoS ONE* **2018**, *13*, e0196463. [CrossRef]
23. Huisinga, J.M.; Mancini, M.; St George, R.J.; Horak, F.B. Accelerometry reveals differences in gait variability between patients with multiple sclerosis and healthy controls. *Ann. Biomed. Eng.* **2013**, *41*, 1670–1679. [CrossRef]
24. Moon, Y.; Wajda, D.A.; Motl, R.W.; Sosnoff, J.J. Stride-Time Variability and Fall Risk in Persons with Multiple Sclerosis. *Mult. Scler. Int.* **2015**, *2015*, 7. [CrossRef]
25. Motta, C.; Palermo, E.; Studer, V.; Germanotta, M.; Germani, G.; Centonze, D.; Cappa, P.; Rossi, S.; Rossi, S. Disability and Fatigue Can Be Objectively Measured in Multiple Sclerosis. *PLoS ONE* **2016**, *11*, e0148997. [CrossRef]
26. Engelhard, M.M.; Dandu, S.R.; Patek, S.D.; Lach, J.C.; Goldman, M.D. Quantifying six-minute walk induced gait deterioration with inertial sensors in multiple sclerosis subjects. *Gait Posture* **2016**, *49*, 340–345. [CrossRef]
27. Psarakis, M.; Greene, D.A.; Cole, M.H.; Lord, S.R.; Hoang, P.; Brodie, M. Wearable technology reveals gait compensations, unstable walking patterns and fatigue in people with multiple sclerosis. *Phys. Meas.* **2018**, *39*, 075004. [CrossRef]
28. Moon, Y.; McGinnis, R.S.; Seagers, K.; Motl, R.W.; Sheth, N.; Wright, J.A., Jr.; Ghaffari, R.; Sosnoff, J.J. Monitoring gait in multiple sclerosis with novel wearable motion sensors. *PLoS ONE* **2017**, *12*, e0171346. [CrossRef]
29. Craig, J.J.; Bruetsch, A.P.; Lynch, S.G.; Huisinga, J.M. The relationship between trunk and foot acceleration variability during walking shows minor changes in persons with multiple sclerosis. *Clin. Biomech.* **2017**, *49*, 16–21. [CrossRef]
30. Corporaal, S.H.A.; Gensicke, H.; Kuhle, J.; Kappos, L.; Allum, J.H.J.; Yaldizli, Ö. Balance control in multiple sclerosis: Correlations of trunk sway during stance and gait tests with disease severity. *Gait Posture* **2013**, *37*, 55–60. [CrossRef]
31. Anastasi, D.; Carpinella, I.; Gervasoni, E.; Matsuda, P.N.; Bovi, G.; Ferrarin, M.; Cattaneo, D. Instrumented Version of the Modified Dynamic Gait Index in Patients with Neurologic Disorders. *PM&R* **2019**. [CrossRef]
32. Pau, M.; Corona, F.; Pilloni, G.; Porta, M.; Coghe, G.; Cocco, E. Texting while walking differently alters gait patterns in people with multiple sclerosis and healthy individuals. *Mult. Scler. Relat. Disord.* **2018**, *19*, 129–133. [CrossRef]
33. Carpinella, I.; Gervasoni, E.; Anastasi, D.; Lencioni, T.; Cattaneo, D.; Ferrarin, M. Instrumental Assessment of Stair Ascent in People with Multiple Sclerosis, Stroke, and Parkinson's Disease: A Wearable-Sensor-Based Approach. *IEEE Trans. Neural Syst. Rehabil. Eng.* **2018**, *26*, 2324–2332. [CrossRef]
34. Craig, J.J.; Bruetsch, A.P.; Lynch, S.G.; Horak, F.B.; Huisinga, J.M. Instrumented balance and walking assessments in persons with multiple sclerosis show strong test-retest reliability. *J. Neuroeng. Rehabil.* **2017**, *14*, 43. [CrossRef]
35. Riva, F.; Grimpampi, E.; Mazzà, C.; Stagni, R. Are gait variability and stability measures influenced by directional changes? *Biomed. Eng. Online* **2014**, *13*, 56. [CrossRef]
36. Brønd, J.C.; Arvidsson, D. Sampling frequency affects the processing of Actigraph raw acceleration data to activity counts. *J. Appl. Physiol.* **2015**, *120*, 362–369. [CrossRef]
37. England, S.A.; Granata, K.P. The influence of gait speed on local dynamic stability of walking. *Gait Posture* **2007**, *25*, 172–178. [CrossRef]

38. Brodie, M.A.D.; Menz, H.B.; Lord, S.R. Age-associated changes in head jerk while walking reveal altered dynamic stability in older people. *Exp. Brain Res.* **2014**, *232*, 51–60. [CrossRef]
39. Mazzà, C.; Iosa, M.; Pecoraro, F.; Cappozzo, A. Control of the upper body accelerations in young and elderly women during level walking. *J. Neuroengineering Rehabil.* **2008**, *5*, 30. [CrossRef]
40. Brach, J.S.; McGurl, D.; Wert, D.; Vanswearingen, J.M.; Perera, S.; Cham, R.; Studenski, S. Validation of a measure of smoothness of walking. *J. Gerontology. Ser. A Biol. Sci. Med. Sci.* **2010**, *66*, 136–141. [CrossRef]
41. Helbostad, J.L.; Moe-Nilssen, R. The effect of gait speed on lateral balance control during walking in healthy elderly. *Gait Posture* **2003**, *18*, 27–36. [CrossRef]
42. Rabuffetti, M.; Scalera, M.G.; Ferrarin, M. Effects of Gait Strategy and Speed on Regularity of Locomotion Assessed in Healthy Subjects Using a Multi-Sensor Method. *Sensors* **2019**, *19*, 513. [CrossRef] [PubMed]
43. Latt, M.D.; Menz, H.B.; Fung, V.S.; Lord, S.R. Walking speed, cadence and step length are selected to optimize the stability of head and pelvis accelerations. *Exp. Brain Res.* **2008**, *184*, 201–209. [CrossRef]
44. Menz, H.B.; Lord, S.R.; Fitzpatrick, R.C. Acceleration patterns of the head and pelvis when walking on level and irregular surfaces. *Gait Posture* **2003**, *18*, 35–46. [CrossRef]
45. Lowry, K.A.; Lokenvitz, N.; Smiley-Oyen, A.L. Age- and speed-related differences in harmonic ratios during walking. *Gait Posture* **2012**, *35*, 272–276. [CrossRef]
46. Pecoraro, F.; Mazzà, C.; Cappozzo, A.; Thomas, E.E.; Macaluso, A. Reliability of the intrinsic and extrinsic patterns of level walking in older women. *Gait Posture* **2007**, *26*, 386–392. [CrossRef]
47. Cappozzo, A. Analysis of the linear displacement of the head and trunk during walking at different speeds. *J. Biomech.* **1981**, *14*, 411–425. [CrossRef]
48. Moe-Nilssen, R. A new method for evaluating motor control in gait under real-life environmental conditions. Part 1: The instrument. *Clin. Biomech.* **1998**, *13*, 320–327. [CrossRef]
49. Salarian, A.; Horak, F.B.; Zampieri, C.; Carlson-Kuhta, P.; Nutt, J.G.; Aminian, K. iTUG, a Sensitive and Reliable Measure of Mobility. *IEEE Trans. Neural Syst. Rehabil. Eng.* **2010**, *18*, 303–310. [CrossRef]
50. Killick, R.; Fearnhead, P.; Eckley, I.A. Optimal Detection of Changepoints With a Linear Computational Cost. *J. Am. Stat. Assoc.* **2012**, *107*, 1590–1598. [CrossRef]
51. Palmerini, L.; Rocchi, L.; Mazilu, S.; Gazit, E.; Hausdorff, J.M.; Chiari, L. Identification of Characteristic Motor Patterns Preceding Freezing of Gait in Parkinson's Disease Using Wearable Sensors. *Front. Neurol.* **2017**, *8*, 394. [CrossRef] [PubMed]
52. Lord, S.; Galna, B.; Rochester, L. Moving forward on gait measurement: toward a more refined approach. *Movement Disorders* **2013**, *28*, 1534–1543. [CrossRef] [PubMed]
53. Buckley, C.; Galna, B.; Rochester, L.; Mazzà, C. Upper body accelerations as a biomarker of gait impairment in the early stages of Parkinson's disease. *Gait Posture* **2019**, *71*, 289–295. [CrossRef] [PubMed]
54. Salarian, A.; Russmann, H.; Vingerhoets, F.J.; Dehollain, C.; Blanc, Y.; Burkhard, P.R.; Aminian, K. Gait assessment in Parkinson's disease: toward an ambulatory system for long-term monitoring. *IEEE Trans. Biomed. Eng.* **2004**, *51*, 1434–1443. [CrossRef] [PubMed]
55. Galna, B.; Lord, S.; Rochester, L. Is gait variability reliable in older adults and Parkinson's disease? Towards an optimal testing protocol. *Gait Posture* **2013**, *37*, 580–585. [CrossRef]
56. Godfrey, A.; Del Din, S.; Barry, G.; Mathers, J.C.; Rochester, L. Instrumenting gait with an accelerometer: A system and algorithm examination. *Med. Eng. Phys.* **2015**, *37*, 400–407. [CrossRef]
57. Pasciuto, I.; Bergamini, E.; Iosa, M.; Vannozzi, G.; Cappozzo, A. Overcoming the limitations of the Harmonic Ratio for the reliable assessment of gait symmetry. *J. Biomech.* **2017**, *53*, 84–89. [CrossRef]
58. Sekine, M.; Tamura, T.; Yoshida, M.; Suda, Y.; Kimura, Y.; Miyoshi, H.; Kijima, Y.; Higashi, Y.; Fujimoto, T. A gait abnormality measure based on root mean square of trunk acceleration. *J. Neuroeng. Rehabil.* **2013**, *10*, 118. [CrossRef]
59. Fazio, P.; Granieri, G.; Casetta, I.; Cesnik, E.; Mazzacane, S.; Caliandro, P.; Pedrielli, F.; Granieri, E. Gait measures with a triaxial accelerometer among patients with neurological impairment. *Neurol. Sci.* **2013**, *34*, 435–440. [CrossRef]
60. Gage, S.H. Microscopy in America (1830–1945). *Trans. Am. Microsc. Soc.* **1964**, *83*, 1–125. [CrossRef]
61. Smidt, G.L.; Arora, J.S.; Johnston, R.C. Accelerographic analysis of several types of walking. *Am. J. Phys. Med. Rehabil.* **1971**, *50*, 285–300.
62. Moe-Nilssen, R.; Helbostad, J.L. Estimation of gait cycle characteristics by trunk accelerometry. *J. Biomech.* **2004**, *37*, 121–126. [CrossRef]

63. R Core Team. *R: A Language and Environment for Statistical Computing*; R Foundation for Statistical Computing: Vienna, Austria; Available online: http://cran.fhcrc.org/web/packages/dplR/vignettes/intro-dplR.pdf (accessed on 20 December 2019).
64. Li, L.; Zeng, L.; Lin, Z.-J.; Cazzell, M.; Liu, H. Tutorial on use of intraclass correlation coefficients for assessing intertest reliability and its application in functional near-infrared spectroscopy–based brain imaging. *J. Biomed. Opt.* **2015**, *20*, 050801. [CrossRef] [PubMed]
65. Cicchetti, D.V. Methodological Commentary The Precision of Reliability and Validity Estimates Re-Visited: Distinguishing Between Clinical and Statistical Significance of Sample Size Requirements. *J. Clin. Exp. Neuropsychology* **2001**, *23*, 695–700. [CrossRef]
66. Almarwani, M.; Perera, S.; VanSwearingen, J.M.; Sparto, P.J.; Brach, J.S. The test–retest reliability and minimal detectable change of spatial and temporal gait variability during usual over-ground walking for younger and older adults. *Gait Posture* **2016**, *44*, 94–99. [CrossRef]
67. Cohen, J. CHAPTER 3—The Significance of a Product Moment rs. In *Statistical Power Analysis for the Behavioral Sciences*; Available online: http://www.utstat.toronto.edu/~{}brunner/oldclass/378f16/readings/CohenPower.pdf (accessed on 20 December 2019).
68. Compston, A.; Coles, A. Multiple sclerosis. *Lancet* **2008**, *372*, 1502–1517. [CrossRef]
69. Dujmovic, I.; Radovanovic, S.; Martinovic, V.; Dackovic, J.; Maric, G.; Mesaros, S.; Pekmezovic, T.; Kostic, V.; Drulovic, J. Gait pattern in patients with different multiple sclerosis phenotypes. *Mult. Scler. Relat. Disord.* **2017**, *13*, 13–20. [CrossRef]
70. Cole, M.H.; Sweeney, M.; Conway, Z.J.; Blackmore, T.; Silburn, P.A. Imposed Faster and Slower Walking Speeds Influence Gait Stability Differently in Parkinson Fallers. *Arch. Phys. Med. Rehabil.* **2017**, *98*, 639–648. [CrossRef]
71. Riva, F.; Bisi, M.C.; Stagni, R. Gait variability and stability measures: Minimum number of strides and within-session reliability. *Comput. Biol. Med.* **2014**, *50*, 9–13. [CrossRef]
72. Shema-Shiratzky, S.; Gazit, E.; Sun, R.; Regev, K.; Karni, A.; Sosnoff, J.J.; Herman, T.; Mirelman, A.; Hausdorff, J.M. Deterioration of specific aspects of gait during the instrumented 6-min walk test among people with multiple sclerosis. *J. Neurol.* **2019**, *266*, 3022–3030. [CrossRef]

Brief Report

Functional Evaluation Using Inertial Measurement of Back School Therapy in Lower Back Pain

Claudia Celletti [1], Roberta Mollica [1], Cristina Ferrario [2,3], Manuela Galli [3] and Filippo Camerota [1,*]

1 Physical Medicine and Rehabilitation, Umberto I University Hospital, 00161 Rome, Italy; clacelletti@gmail.com (C.C.); roberta.mollica@uniroma1.it (R.M.)
2 Department of Mechanic, Politecnico di Milano, 20124 Milan, Italy; cristina.ferrario@polimi.it
3 Department of Electronics, Information and Bioengineering (DEIB) Politecnico di Milano, 20133 Milan, Italy; manuela.galli@polimi.it
* Correspondence: filippo.camerota@uniroma1.it

Received: 4 December 2019; Accepted: 16 January 2020; Published: 18 January 2020

Abstract: Lower back pain is an extremely common health problem and globally causes more disability than any other condition. Among other rehabilitation approaches, back schools are interventions comprising both an educational component and exercises. Normally, the main outcome evaluated is pain reduction. The aim of this study was to evaluate not only the efficacy of back school therapy in reducing pain, but also the functional improvement. Patients with lower back pain were clinically and functionally evaluated; in particular, the timed "up and go" test with inertial movement sensor was studied before and after back school therapy. Forty-four patients completed the program, and the results showed not only a reduction of pain, but also an improvement in several parameters of the timed up and go test, especially in temporal parameters (namely duration and velocity). The application of the inertial sensor measurement in evaluating functional aspects seems to be useful and promising in assessing the aspects that are not strictly correlated to the specific pathology, as well as in rehabilitation management.

Keywords: back school; inertial sensor; lower back pain; rehabilitation; stability; timed up and go test

1. Introduction

Lower back Pain (LBP) is a well described and extremely widespread health problem [1]. LBP is a pain that goes from the twelfth rib to the lower gluteal folds; pain can also spread to the lower limbs for one day or more [1]. This condition is the main cause of absence from work and activity limitations in much of the world. The consequence is a heavy economic burden for subjects, families, communities, industry, and governments [2]. Of the 291 conditions studied in the 2010 Global Burden of Disease (GBD) report, LBP had the highest load. LBP is the leading cause of disability globally [3].

The main components to treat this condition are education, reassurance, analgesic drugs, and non-pharmacological therapies. During the treatment, periodic check-ups are recommended based on individual patient needs, such as prognosis, treatment prescribed, and remaining concerns about serious pathological abnormality [4].

Chronic LBP is defined as lower back pain that lasts for over 12 weeks. Generally, one-third of the patients with LBP reported that in the year after an acute episode, lower back pain was of moderate intensity [2]. In patients with chronic back pain, a multidisciplinary approach leads to better results when combined with medical, rehabilitative, and psychological treatments [5].

Among other rehabilitation approaches, back schools (BS) are interventions that comprise an education component and exercises. BS are training programs with lessons given by a therapist to patients or workers, with the aim of treating or preventing lower back pain [6]. Several studies have demonstrated the efficacy of BS in reducing and managing lower back pain [7]. BS, due to the validity of their educational exercises, enhance the quality of life, reduce disability induced by LBP [8,9], and also improve mental well-being.

The aim of this study is to evaluate not only the efficacy of BS therapy in reducing pain but also in functional improvement, an aspect strictly related to pain but normally not evaluated in the studies that focus on assessing pain relief. A new and simple gait evaluation method is used to make the analysis. In particular, stability and ability to perform functional tests, such as the timed "up and go" test, are evaluated in order to verify if a rehabilitation program based on BS therapy is able to improve stability and walking.

2. Materials and Methods

Patients were recruited from the Rehabilitation Ambulatory Service of Umberto I University Hospital. All participants signed informed consent forms after receiving detailed information about the study's aims and procedures for the Declaration of Helsinki.

2.1. Eligibility Criteria

Patients were included in the study if they had lower back pain that had lasted for more than six weeks that was associated with limitations of motion. The presence of vertebral infections; tumoral metastasis; fractures and neoplasm; rheumatological, neurological, or oncological disease; previous back surgery; severe cognitive impairments; or pregnancy was considered an exclusion criterion.

2.2. Intervention

The BS program was supervised by a multidisciplinary professional team. A total of 10 one-hour sessions scheduled 3 times a week were carried out. The adopted rehabilitation program was chosen by considering the effectiveness of the BS on LBP reported in previous studies. The details of the program followed in this study are described below.

The first treatment session was used to provide subjects with basic anatomical knowledge of the spine and its functions; the correct ergonomic positions to be maintained in everyday life were also shown. During the following 9 sessions, the physiotherapists supervised the activities, which consisted of exercises based on diaphragmatic breathing (10 min), self-stretching of the trunk muscles (10 min), strengthening of erector muscles of the spine, abdominal strengthening, and postural exercises. The tasks were divided into 3 sets of 10 repetitions for each one; 3 min of rest was provided between each series. Explanations of the ergonomic position of the spine and how to introduce self-correction in daily life were provided for the whole duration of the treatment.

2.3. Health State: Clinical Evaluations

Patients were evaluated before and after physiotherapy treatment with the following clinical scales:

1. The *numeric rating scale* (NRS) is a rapidly administered 11-point numeric scale used to roughly measure any kind of pain, with a score ranging from 0 (no pain) to 10 (acute pain) [10];
2. The *Oswestry disability index* (ODI), also known as the Oswestry lower back pain disability questionnaire, is considered the "gold standard" of lower back functional outcome tools and consists of 10 sections, with a score varying from 0 to 5 for each one. A low score indicates minimal disability; the disability is more severe for higher scores [11];
3. The *performance-oriented mobility assessment* (POMA) scale was developed by Tinetti in 1986 to assess the mobility and risk of falling of the elderly [12]. This scale was chosen because it is very reliable and widely used. In this study, we used the balance scale of the POMA, which evaluates the positions and changes in position of the subject, assessing stability tasks. Each item is scored on a two- or three-point scale, where the maximum is 18 [13];

4. The *timed up and go test* (TUG) is a clinical test that evaluates the balance and mobility of a subject [14,15]. In the traditional TUG test, a stopwatch is used to measure how long it takes a subject to lift off a chair, walk 3 m, turn 180°, return to the chair, and sit back down.

2.4. Biomechanical Evaluation

Instrumentation

In this study, we evaluated the TUG as both a time test and also using an inertial measurement unit (IMU). The commercial name of the device used is a G-Sensor instrument (BTS SpA, Milan, Italy). The communication with the receiving unit (personal computer) takes place via a Bluetooth connection. The associated software (BTS® G-Studio) is used to acquire, process, and archive data. In the IMU there is a triaxial accelerometer (16 bits/axes, up to 1000 Hz) with different sensitivities (±2, ±4, ±8, ±16 g), a triaxial 16-bit magnetometer (±1200 µT, up to 100 Hz), and a triaxial gyroscope (16 bits/axes, up to 8000 Hz) with multiple sensitivities (±250, ±500, ±1000, ±2000°/s). The G-Sensor is positioned at level L5 using an elastic belt. It is important to keep the power connector facing upwards and the logo outwards to correctly define the reference system (Figure 1a)

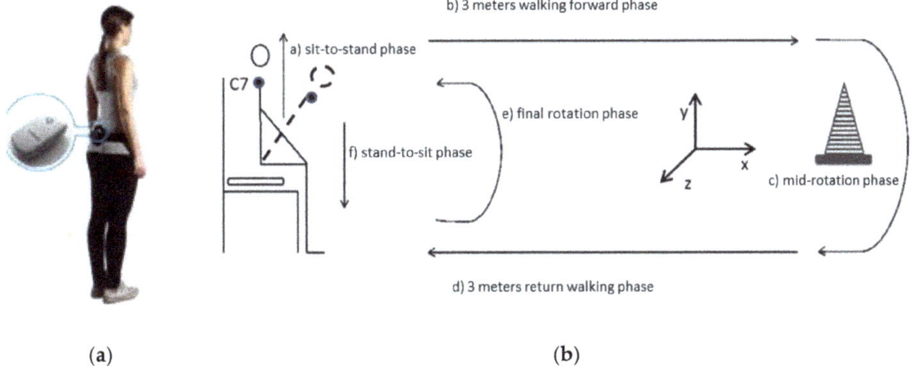

(a) (b)

Figure 1. (**a**) Inertial measurement unit (IMU) position and (**b**) timed up and go test (TUG) phases.

The test begins with patients seated in a standard chair with their arms on either side of their body. After a signal from the clinician, the subject rises from the chair, walks three meters in a straight line at a speed that is normal for them, turns around an obstacle, and finally returns to the chair and sits down. The software used is BTS G-Studio, which has a specific protocol capable of analyzing the TUG test and automatically generates a TUG report with temporal parameters identifying the duration of the different sub-phases [16]. The mathematical method used to identify each sub-phase is the one described in the study by Salarian et al. [17]. Additionally, a detailed description of the practical operation of BTS G-Studio in iTUG analysis, as compared with an optoelectronic system, is provided in the study by Negrini [18].The test can, therefore, be divided into different phases: the first is that of rising from the chair (sit-to-stand sub-phase), walking for 3 m until reaching an obstacle (walking forward sub-phase), turning around the cone (mid-turning sub-phase), walking three m back towards the chair (return walking sub-phase), and then turning and sitting down on the chair (stand-to-sit sub-phase) without using the assistance of their arms, if possible. The test is concluded when the subject is seated again. The final report of the TUG test shows all the spatiotemporal parameters related to the walk for each sub-phase considered: the sit-to-stand, the steady-state gait, the turning, and the turn-to-sit phases [17]. The parameters supplied automatically by the IMU for each trial are: total time duration, sub-phase durations, mean velocity turning (mid-turning and final turning sub-phases), and the maximum trunk flexion angle and its range of motion during sit-to-stand and stand-to-sit sub-phases (Figure 1b).

Furthermore, an instrumental evaluation of stability was carried out using a baropodometric platform (P-Walk BTS Engineering). The stabilometry test measures the oscillations by evaluating the elliptical area containing 95% of sway points, velocities with closed eyes (CE) and opened eyes (OE), and the length of the excursion of the center of pressure. The test we performed had a duration of 30 s, within which the position of the CoP was recorded during quiet standing [19]. Patients were adequately informed about the procedure; the requirements were to maintain a natural standing position with the arms alongside the body, the feet open at an angle of about 30°, and the heels at a distance of about 3 cm. All tests were performed by the same examiner in order to reduce the inter-operator error and to increase the reproducibility of the test; thus, the subjects were given the same information before each test. For each trial condition (EO and EC), three tests were carried out, for which the median scores are reported. Considering the EO condition, subjects were required to stare at a mark fixed at eye level on a wall 1.5 m away.

2.5. Statistical Analysis

The statistical analysis was performed with SPSS software. To verify the normality of the parameters, the Kolmogorov–Smirnov test was used. When the normality assumption was not fulfilled, the median and range (minimum–maximum) were evaluated. The differences between variables were evaluated using the Friedman test for paired samples. The probability level for statistical significance in all tests was set at a $p < 0.05$.

3. Results

Forty-eight patients (mean age 71 ± 13.66) were recruited for this study; 4 patients did not complete the rehabilitation program and were excluded from the study; a total of 44 patients (34 female and 10 male, mean age 70 ± 14.02) were evaluated before and after back school treatment.

We observed a global pain reduction in patients with LBP that attended the back-school program. This reduction was also associated with clinical improvement of stability, as shown by the POMA balance score increase. When the postural analysis data were examined, a variation was not registered when considering the opened eyes test; instead, in the closed eyes test a significant reduction of the length of CoP was registered (Table 1).

Table 1. Clinical scale and instrumental evaluation before and after back school cycle.

		T0 (Median ± s.d.)	T1 (Median ± s.d.)	p	Chi Quadro	df
POMA Balance		12.88 ± 2.00	13.86 ± 1.92	0.000	17.19	1
	NRS	6.11 ± 1.57	4.32 ± 1.99	0.000	33	1
	ODI	30.51 ± 13.29	28.72 ± 14.91	0.60	0.273	1
Stabilometria	Area OE (mm^2)	210.27 ± 1012.07	231.84 ± 1007.86	0.75	0.1	1
	Lenght OE (mm)	115.24 ± 76.58	126.28 ± 99.40	0.15	2.07	1
	Area CE (mm^2)	446.73 ± 2540.10	591.74 ± 3412.65	0.42	0.64	1
	Length OE (mm)	167 ± 308.20	162.33 ± 221.95	0.02	4.9	1
TUG	Total time (s)	13.37 ± 3.86	11.25 ± 2.16	0.00	19.70	1
	Stand up (s)	1.65 ± 0.37	1.47 ± 0.29	0.02	5.15	1
	Sitting (s)	2.20 ± 0.60	1.99 ± 0.42	0.50	0.44	1
	Rotation velocity (°/s)	77.71 ± 19.80	83.23 ± 22.45	0.04	8.52	1

Legend: POMA = performance-oriented mobility assessment; NRS = numeric rating scale; ODI = Oswestry disability index; OE = opened eyes; CE = closed eyes; TUG = timed up and go; s = second.

It is interesting to notice that there was a significant reduction of the total duration of the TUG test, and also of the stand-up and sitting phases (Table 1).

The BS groups showed significant improvement in several instrumental TUG (iTUG) parameters, especially in temporal (duration and velocity) parameters.

The BS treatment significantly reduced the total duration of the task and its sub-phases: the stand-to-sit sub-phase and the sit-to-stand phase, the mean velocity of TUG, and of mid-turning and final turning sub-phases increased at a significant level.

4. Discussion

As far as we know, this is the first paper to evaluate not only the pain aspect of lower back syndrome after treatment, but also the functional aspect that is not strictly related to this pathology (i.e., timed up and go evaluation). The TUG test provided in this study is an instrumented TUG. While the TUG test taken by an expert operator using a stopwatch has excellent reliability, accuracy, and precision, this measure is subjective and operator-dependent (i.e., a less experienced clinician could affect the quality of the measure). The use of the stopwatch in the clinical setting has several limitations: (a) the identification of the start time and the end time are not easily detectable by the operator; (b) the evaluation of the TUG time requires a high level of attention by the operator, which could decrease when many trials are required; (c) the quantification of sub-phases is not possible.

The instrumented TUG analysis is of considerable interest, as it evaluates the various sub-phases of the test (chair transition, straight-ahead gait, and 180° turn); this allows a better understanding of movement strategies. Considering, for example, the 180° turn, there is a variability between subjects with different gaits and with **or** without balance impairment. A further variation is introduced for patients using an assistive device, such as a walker.

Therefore, the IMU technology implementations for the iTUG quantification of pre- and post- specific therapies have several benefits, including additional performance parameters, generation of reports, fast assessment, and that the patient does not need to be undressed. In addition to this, it is important to consider the ability for self-administration at home and in a clinical environment. This could provide more details and insights about patient performance [16]. Although other variables could have been derived using the data provided by the wearable sensor, as the purpose of this work was to analyze the TUG, which is an automatic functional clinical test, the analysis focused mainly on the evaluation of the duration of the task included in the test. It is known that lower back pain is associated with functional impairment. In particular, the opportunity to analyze the different phases of this test using an inertial measurement instrument made it possible to assert that back school therapy may improve back function, increasing the promptness to position changes and speeding up movements. The changes observed with iTUG represent the effect of the reduction of LBP on functional ability. As the patients experience pain during the movement, the biomechanical result is a slow movement and a higher TUG time. After treatment, the patients feel better, experience less pain, and can get out of the chair faster. No changes are evidenced as far as postural acquisition is concerned. In maintaining postural control, pain in the lumbar area has a minor effect in terms of functional limitation, and therefore one can expect to have no obvious variations in postural control.

5. Conclusions

In conclusion, through the quantitative evaluation of the iTUG test, it is proven that the BS could be considered a promising new rehabilitative treatment for LBP in improving motor functional limitations. Moreover, as the IMU sensor can provide data that might provide many more temporal and kinematic measures after successive elaboration, future development of this study should provide additional data for a more detailed analysis, in order to show more important changes in patients' movement patterns after the treatment.

Author Contributions: Conceptualization, C.C. and F.C.; methodology, C.C., F.C., and R.M.; formal analysis, C.C., M.G., and C.F.; investigation, C.C. and F.C.; data curation, C.C. and F.C.; writing—original draft preparation, C.C.; writing—review and editing, F.C., M.G., and C.F.; supervision, R.M. and M.G. All authors have read and agreed to the published version of the manuscript.

Funding: This research received no external funding.

Acknowledgments: We thanks Daniele Scilimati and Laura Venerucci for their contributions in evaluating and treating patients.

Conflicts of Interest: The authors declare no conflict of interest.

References

1. Hoy, D.; March, L.; Brooks, P.; Blyth, F.; Woolf, A.; Bain, C.; Williams, G.; Smith, E.; Vos, T.; Barendregt, J.; et al. The global burden of lower back pain: Estimates from the Global Burden of Disease 2010 study. *Ann. Rheum. Dis.* **2014**, *73*, 968–974. [CrossRef] [PubMed]
2. Rapoport, J.; Jacobs, P.; Bell, N.N.; Klarenbach, S. Refining the measurement of the economic burden of chronic diseases in Canada. *Chronic Dis. Can.* **2014**, *25*, 13–21.
3. Hoy, D.D.; March, L.; Brooks, P.; Woolf, A.; Blyth, F.; Vos, T.; Buchbinder, R. Measuring the global burden of lower back pain. *Best Pract. Res. Clin. Rheumatol.* **2010**, *24*, 155–165. [CrossRef] [PubMed]
4. Maher, C.; Underwood, M.; Buchbinder, R. Non-specific lower back pain. *Lancet* **2017**, *389*, 736–747. [CrossRef]
5. Urits, I.; Burshtein, A.; Sharma, M.; Testa, L.; Gold, P.A.; Orhurhu, V.; Viswanath, O.; Jones, M.R.; Sidransky, M.A.; Spektor, B.; et al. Lower back Pain, a Comprehensive Review: Pathophysiology, Diagnosis, and Treatment. *Curr. Pain Headache Rep.* **2019**, *23*, 23. [CrossRef] [PubMed]
6. Airaksinen, O.; Brox, J.J.; Cedraschi, C.; Hildebrandt, J.; Klaber-Moffett, J.; Kovacs, F.; Mannion, A.F.; Reis, S.; Staal, J.B.; Ursin, H.; et al. Chapter 4. European guidelines for the management of chronic nonspecific lower back pain. *Eur. Spine J.* **2006**, *15* (Suppl. 2), S192–S300. [CrossRef] [PubMed]
7. Heymans, M.M.; van Tulder, M.M.; Esmail, R.; Bombardier, C.; Koes, B.W. Back schools for non-specific low-back pain: A systematic view within the framework of the cochrane collaboration back review group. *Spine* **2005**, *30*, 2153–2163. [CrossRef] [PubMed]
8. Paolucci, T.; Morone, G.; Iosa, M.; Fusco, A.; Alcuri, R.; Matano, A.; Bureca, I.; Saraceni, V.M.; Paolucci, S. Psychological features and outcomes of the back-school treatment in patients with chronic non-specific lower back pain. A randomized controlled study. *Eur. J. Phys. Rehabil. Med.* **2012**, *48*, 245–253. [PubMed]
9. Morone, G.; Paolucci, T.; Alcuri, M.M.; Vulpiani, M.M.; Matano, A.; Bureca, I.; Paolucci, S.; Saraceni, V.M. Quality of life improved by multidisciplinary back school program in patients with chronic non-specific lower back pain: A single blind randomized controlled trial. *Eur. J. Phys. Rehabil. Med.* **2011**, *47*, 533–541. [PubMed]
10. Deschamps, M.; Band, P.P.; Coldman, A.J. Assessment of adult cancer pain: Shortcomings of current methods. *Pain* **1988**, *32*, 133–139. [CrossRef]
11. Fairbank, J.J.; Pynsent, P.B. The Oswestry Disability Index. *Spine* **2000**, *25*, 2940–2953. [CrossRef] [PubMed]
12. Tinetti, M.E. Performance oriented assessment of mobility problems in elderly patients. *J. Am. Geriatr. Soc.* **1986**, *34*, 119–126. [CrossRef] [PubMed]
13. Faber, M.M.; Bosscher, R.R.; van Wieringen, P.C. Clinimetric properties of the performance-oriented mobility assessment. *Phys. Ther.* **2006**, *86*, 944–954. [CrossRef] [PubMed]
14. Lin, M.; Hwang, H.; Hu, M.; Wu, H.; Wang, Y.; Huang, F. Psychometric comparisons of the timed up and go, one-leg stand, functional reach, and tinetti balance measures in community-dwelling older people. *J. Am. Geriatr. Soc.* **2004**, *52*, 1343–1348. [CrossRef] [PubMed]
15. Barry, E.; Galvin, R.; Keogh, C.; Horgan, F.; Fahey, T. Is the timed up and go test a useful predictor of risk of falls in community dwelling older adults: A systematic review and meta-analysis. *BMC Geriatr.* **2014**, *14*, 14. [CrossRef] [PubMed]
16. Kleiner, A.F.R.; Pacifici, I.; Vagnini, A.; Camerota, F.; Celletti, C.; Stocchi, F.; De Pandis, M.F.; Galli, M. Timed Up and Go evaluation with wearable devices: Validation in Parkinson's disease. *J. Body Mov. Ther.* **2018**, *22*, 390–395. [CrossRef] [PubMed]
17. Salarian, A.; Horak, F.F.; Zampieri, C.; Carlson-Kuhta, P.; Nutt, J.J.; Aminian, K. iTUG, a Sensitive and Reliable Measure of Mobility IEEE Trans. *Neural Syst. Rehabil. Eng.* **2010**, *18*, 303–310. [CrossRef] [PubMed]

18. Negrini, S.; Serpelloni, M.; Amici, C.; Gobbo, M.; Silvestro, C.; Buraschi, R.; Borboni, A.; Crovato, D.; Lopomo, N.F. Use of Wearable Inertial Sensor in the Assessment of Timed-Up-and-Go Test: Influence of Device Placement on Temporal Variable Estimation. In Proceedings of the 6th International Conference on Wireless Mobile Communication and Healthcare, MobiHealth 2016, Milan, Italy, 14–16 November 2016; Springer: Berlin/Heidelberg, Germany, 2017; Volume 192, pp. 310–317.
19. Scoppa, F.; Capra, R.; Gallamini, M.; Shiffer, R. Clinical stabilometry standardization: Basic definitions–acquisition interval–sampling frequency. *Gait Posture* **2013**, *37*, 290–292. [CrossRef] [PubMed]

Article

Turning Analysis during Standardized Test Using On-Shoe Wearable Sensors in Parkinson's Disease

Nooshin Haji Ghassemi [1,*], Julius Hannink [1], Nils Roth [1], Heiko Gaßner [2], Franz Marxreiter [2], Jochen Klucken [2] and Björn M. Eskofier [1]

1 Machine Learning and Data Analytics Lab, Department of Computer Science, Friedrich-Alexander-University Erlangen-Nürnberg (FAU), Carl-Thiersch-Strasse 2b, D-91052 Erlangen, Germany
2 Department of Molecular Neurology, University Hospital Erlangen, Schwabachanlage 6, D-91054 Erlangen, Germany
* Correspondence: nooshin.haji@fau.de; Tel.: +49-9131-85-20286

Received: 23 May 2019; Accepted: 9 July 2019; Published: 13 July 2019 

Abstract: Mobile gait analysis systems using wearable sensors have the potential to analyze and monitor pathological gait in a finer scale than ever before. A closer look at gait in Parkinson's disease (PD) reveals that turning has its own characteristics and requires its own analysis. The goal of this paper is to present a system with on-shoe wearable sensors in order to analyze the abnormalities of turning in a standardized gait test for PD. We investigated turning abnormalities in a large cohort of 108 PD patients and 42 age-matched controls. We quantified turning through several spatio-temporal parameters. Analysis of turn-derived parameters revealed differences of turn-related gait impairment in relation to different disease stages and motor impairment. Our findings confirm and extend the results from previous studies and show the applicability of our system in turning analysis. Our system can provide insight into the turning in PD and be used as a complement for physicians' gait assessment and to monitor patients in their daily environment.

Keywords: Parkinson's disease; pathological gait; turning analysis; wearable sensors; mobile gait analysis

1. Introduction

Gait is an important part of mobility that is impaired in neurodegenerative diseases like Parkinson's disease (PD). As the disease progresses, gait fluctuations become more severe. Different locomotor patterns in gait, such as straight walking and turning, require different levels of functioning and coordination. For a person with impaired mobility caused, for example, by PD, turning is challenging and potentially risky, even more than straight walking [1,2]. There have been attempts to identify and characterize turning abnormalities in order to complement the physicians' assessment of pathological gait.

Studies showed that turning deficits are manifested in mild PD even when there are no signs of impairment in straight walking [3]. Difficulty while turning may lead to posture instability and, potentially, even falls [4,5]. Risk of falling is higher during turning compared with straight walking [4,5]. Furthermore, deterioration of motor function during turning can cause progressive episodes of freezing of gait (FoG) [6–8].

Some studies have attempted to utilize the definition of disease stages and motor impairments by UPDRS-III [9] and H&Y [10] clinical scores and objectively assess turning deficits [11–15]. Studies on spatio-temporal parameters quantifying turning have demonstrated decreased speed, longer duration of turning, and a larger number of strides as the disease progresses [3,16–18]. Postural stability also decreases during turning for PD patients in comparison to healthy controls, particularly during fast walking [19].

Outside the clinics and in the majority of standardized clinical tests, a gait sequence includes both straight walking and turning. In order to differentiate between them during the course of a gait, different definitions of turning have been presented in the literature. For example, turning was defined as the movement between two pre-defined points that indicated the initiation and termination of turning [5]. Salarian et al. [17] used mathematical modeling in order to isolate turns from the whole gait sequence. Spatio-temporal parameters extracted from individual strides are different in straight walking compared to turning. Many studies used characteristics and statistics of spatio-temporal gait parameters to define turning [3,6,20]. Without a standard turning definition, studies then presented some clinical validations to support their definitions—for example, they showed that turning parameters were correlated to the established clinical scores [3,6,20].

Gait and turning can be measured by a variety of systems—from accurate but stationary motion capture systems [19] to small wearable sensors [3,17]. The focus of this study is on wearable sensors, since they give the opportunity to perform long-term monitoring of PD patients. Sensor placement plays a crucial factor in designing wearable systems. Many turning studies place the sensors on the upper extremity [3,6,20]. One advantage is that turning is easily detectable in the sensor signals [17]. However, gait disturbances such as FoG cannot be detected clearly from sensors on the upper extremity. Such systems still need additional sensors on the lower extremity in order to quantify turning in terms of spatio-temporal parameters [3,17]. In contrast, sensors on the lower extremity and, in particular, on the shoe provide higher biomechanical resolutions. Panebianco et al. [21] examined different sensor locations and showed that as sensors get closer to the foot, higher accuracy for gait events and parameters can be obtained. Moreover, for long-term monitoring of patients, sensors integrated in the footwear are less obtrusive and stigmatizing.

In order to measure gait, we used wearable sensors mounted on the lateral side of the shoe. In order to isolate turning from gait, we used the statistics of spatio-temporal parameters. The goal of this study is to show the applicability of the system in the objective analysis of turning and to evaluate whether it confirms the findings of other studies. To this end, we first introduce our novel turning isolation algorithm targeting data from a standardized 4×10 m gait test measured with wearable sensors placed on the shoe. Then, we quantify the isolated turnings through several spatio-temporal parameters that proved to be effective in detecting pathological gait [22–24]. Through meticulous statistical analysis, we evaluate the turning abnormalities in a large PD cohort. The value of this objective turning assessment was clinically validated by the correlation of the turn-derived parameters to clinical scores, including motor impairment and disease stages in PD.

2. Methods

2.1. Wearable Measurement System

For our experiments, data was recorded with a Shimmer 2R/3 Inertial measurement unit (IMU) (Shimmer Sensing, Dublin, Ireland), measuring acceleration and angular velocity at 102.4 Hz. Each unit consisted of a tri-axial accelerometer (range Shimmer 2R: ± 6 g, Shimmer 3: ± 8 g) and a tri-axial gyroscope (range Shimmer 2R: $\pm 500°/s$, Shimmer 3: $\pm$ $1000°/s$). The sensor units were mounted laterally on each shoe below the patient's ankle. The measurements from both feet were included in the experiments. Figure 1 shows the sensor placement on the shoe and the axes definition.

2.2. Study Population

We recruited 108 PD patients during their regular visit in the movement disorder outpatient center at the University Hospital Erlangen. Sporadic PD was defined according to the guidelines of the German Association for Neurology (DGN), which are similar to the UK PD Society Brain Bank criteria [25]. Patients had to be able to walk independently (H&Y $<$ 4, UPDRS gait item $<$ 3) [10,26]. All PD patients were clinically (UPDRS-III) and biomechanically (gait analysis) investigated in stable ON medication without the presence of clinically relevant motor fluctuations during the assessments.

We had an exclusion criterion for a severe cognitive impairment. To obtain quantitative gait data from controls, we recruited 42 age-matched controls with no signs of PD and/or other motor impairments. With respect to age, height, and body-mass-index (BMI), PD and control cohorts were matched (see Table 1). Data regarding laterality of the disease can be found in Table 1, where the UPDRS sub-items of rigidity lower and upper extremities were reported. This data shows that patients affected on the right and left sides are almost equally represented in our cohort. Written informed consent was obtained from all participants (IRB-approval-No. 4208, 21.04.2010, IRB, Medical Faculty, Friedrich-Alexander University Erlangen-Nürnberg, Germany).

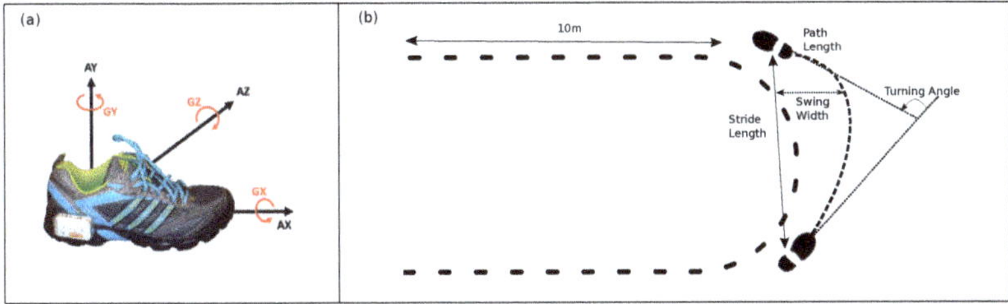

Figure 1. (**a**) Shimmer sensor placement and axes definition. (**b**) Definition of turning angle, stride length, path length, and swing width.

Table 1. Clinical characteristics of patients with Parkinson's disease (PD) and healthy controls.

	PD (N = 108)	Control (N = 42)
Age (years)	57.61 ± 10.42 [36–85]	58.78 ± 11.14 [41–84]
Sex (Male/Female)	74/34	25/17
Height (m)	1.74 ± 0.1	1.73 ± 0.07
BMI	25.81 ± 3.71	26.48 ± 3.76
Hoehn and Yahr stage	2.06 ± 0.84	
I (<1)	28	
II (1-2]	34	
III (2<)	46	
UPDRS-III total	18.24 ± 9.8 [2–50]	
Low [0–12]	36	
[13–22]	38	
High [23<)	34	
Laterality based on Rigidity item (upper and lower extremity)		
No rigidity or both sides	22%	
Right side	42%	
Left side	36%	
Gait item		
0 [0]	34	
1 (0–1]	62	
2 (1–2]	12	
Postural stability item		
0 [0]	46	
1 (0–1]	49	
2 (1–2]	13	

Participants walked freely at a comfortable, self-chosen speed in an obstacle-free and flat environment for 4 × 10 m. After each 10 m of straight walking, participants were instructed to turn 180° at a preferred direction.

2.3. Turning Isolation

The standardized 4 × 10 m walking included four straight gait bouts and three turnings in between each two straight bouts. The goal was to isolate the three turnings from the whole gait sequence. To this end, the gait sequence was segmented to individual strides semi-automatically [27,28]. These strides should then be categorized as straight walking, turning, and transitions between straight walking and turning. In order to differentiate between these categories, we used statistics of spatio-temporal parameters.

The change of azimuth between two successive mid-stances was defined as the turning angle between consecutive strides (see Figure 1). The absolute values of turning angles were considered since the sign of values only showed the direction of the turnings, which is not of importance in our analysis. Similarly to Mariani et al. [20], strides with turning angles larger than 20° were classified as turning.

In order to identify transition strides in a gait sequence [20], again, statistics over turning angles were used, since this parameter is the best indicator of spatial foot movement during turning (see Figure 1). The turning strides with angles larger than 20° were eliminated from the sequence. A gamma distribution was then fitted to the tuning angels from the rest of the strides. We chose gamma distribution due to the fact that the distribution is one-hand tailed, in a way that strides from straight walking mainly centers on the mean. The highest 10% of the distribution was classified as the transition if the strides were adjacent to the turning strides. In fact, the strides in the highest 10% of the distribution were considered as anomalies in the distribution of straight strides. For turning analysis, we only considered turning and transition strides.

2.4. Turning Parameters

After the turning isolation, we had three sets of strides related to three turns in the standardized test. We extracted spatio-temporal parameters from these strides based on the algorithms in previous works [20,22]. The algorithms for obtaining parameters from our wearable sensor-based system were validated previously using a gold standard, such as an optical motion capture system or instrumented walkway. To quantify turning, two sets of parameters were computed for each turning—per-stride parameters and global parameters per-turn.

For the first group, a set of parameters was extracted from each stride: stride time, path length (normalized on patient's height), stride length (normalized on patient's height), stride velocity, and swing width. In turning, it is very likely that a stride has a curved trajectory, rather than a straight line. In such cases, length of movement in the straight line between the beginning and end of a stride is measured as stride length. In addition, path length was introduced to measure curve length between the beginning and end of a stride (see Figure 1). All these parameters were calculated from mid-stance of a stride to the successive mid-stance.

For the global parameters, we calculated the number of strides and total duration per turn. This set of parameters measures characteristics of the whole turn.

2.5. Statistical Analysis

In order to determine whether parameters can distinguish between different groups (controls and three stages of disease (see Table 1)), we applied the one-way analysis of variance (ANOVA). When a significant difference was found, a post hoc analysis was performed using Bonferroni's test to obtain a pairwise comparison between the groups. The significance level was set at $p < 0.05$. For measuring effect sizes, η^2 was defined as the ratio of variability between groups to the total variation in the data that was used. Cutoff values for small, medium, and large effect sizes were set at 0.01, 0.06, and 0.14, respectively, according to Cohen [29]. Statistical analysis and parameter computations were performed using MATLAB R2015a.

3. Results

As the disease progresses, gait impairment associated with deteriorated mobility becomes more prevalent. In this section, we examined whether spatio-temporal parameters that characterize turning were able to reflect gait impairments.

Figures 2 and 3 show spatio-temporal parameters that are characteristic of turning for global and per-stride parameters, respectively. Clinical scores in PD studies determine the severity of gait impairment and disease stages: the H&Y, UPDRS-III score, and the UPDRS-III sub-items for gait and postural instability. Patients with different levels of disease severity (see Table 1) and controls were statistically compared using ANOVA, followed by Bonferroni's post hoc test.

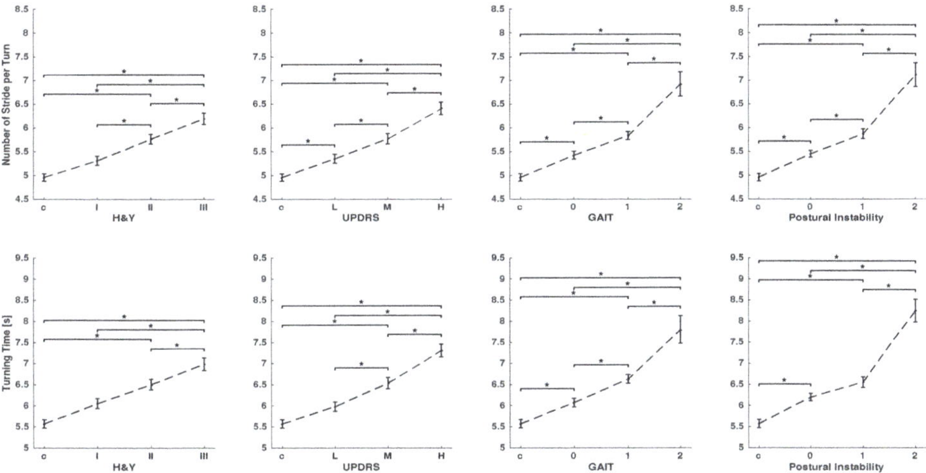

Figure 2. Global parameters characterizing turning: number of strides per-turn and turning time were calculated for controls and PD patients grouped according to H&Y disease stage, UPDRS-III total score, and the single items, gait and postural instability of the UPDRS-III. Group data are displayed as mean ± SEM and were compared using one-way ANOVA followed by Bonferroni's post hoc test, where * indicates $p < 0.05$.

As the disease progresses, stride velocity, path length, stride length, and swing width (per-stride parameters) decreases, and as a result, patients need more strides and time (global parameters) to complete a turn. This can be observed for all clinical scores, although the two sub-items of gait and postural instability are showing larger differences between stages of the disease. Stride time shows no clear change between different groups.

Global parameters showed that PD patients, in contrast to controls, need significantly more time and a larger number of strides to complete a turn (see Figure 2). Number of strides per turn, in particular, shows a significant difference between the control and even early stage of the disease for the UPDRS-III score and its two sub-items. Moreover, there are significant differences between stages of the disease in most comparisons. Per-stride parameters, except stride time, show a significant difference between the controls, mild, and severe stages of the disease for all clinical scores. Stride velocity, stride length, path length, and swing width are able to differentiate disease severity by means of all tested clinical scores (see Figure 3).

To quantify effect sizes, η^2 is reported in Table 2. The effect sizes range from small to large. The largest effect sizes are obtained consistently over all clinical scores with $p < 0.001$ by the global parameters, number of strides per turn, and turning time. Path length showed consistently higher effect sizes than stride length, which suggests that it is a more meaningful parameter for estimation of spatial foot displacement in turning. The effect sizes of per-stride duration are very small.

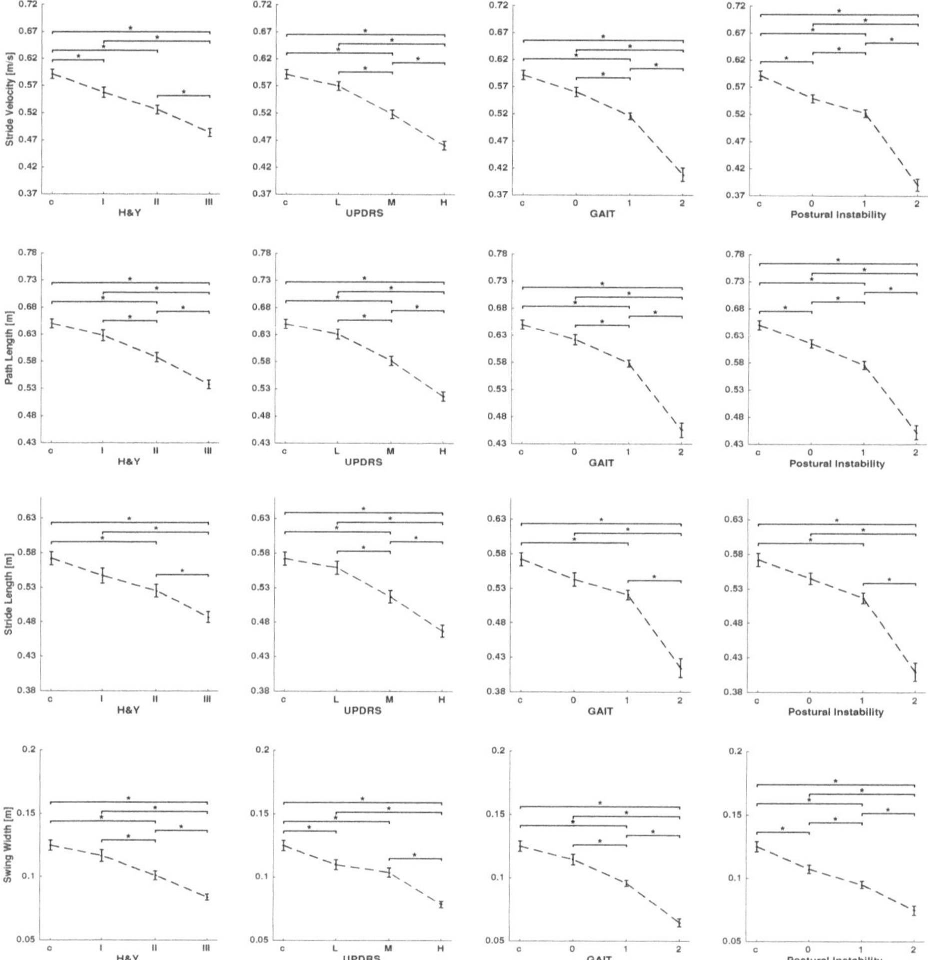

Figure 3. Per-stride parameters characterizing turning: stride velocity, path length, stride length, and swing width were calculated for controls and PD patients who were grouped according to the H&Y disease stage, UPDRS-III score, and the single items, gait and postural instability, of the UPDRS-III. Group data are displayed as mean ± SEM and were compared using one-way ANOVA, followed by Bonferroni's post hoc test, where * indicates $p < 0.05$.

Table 2. ANOVA test: η^2 values for different parameters and clinical scores. Values with * correspond to $p < 0.001$. Bold font indicates values with strong effect sizes.

Parameters	H&Y	UPDRS	Gait	Postural Instability
Number of Strides per-Turn	**0.172 ***	**0.2 ***	**0.202 ***	**0.232 ***
Turning Time	**0.149 ***	**0.199 ***	**0.187 ***	**0.228 ***
Stride Velocity	0.054 *	0.057 *	0.06 *	0.069 *
Path Length	0.054 *	0.054 *	0.06 *	0.063 *
Stride Length	0.03 *	0.03 *	0.034 *	0.038 *
Mid Swing	0.034 *	0.035 *	0.039 *	0.029 *
Stride Time	0.003	0.003	0.002	0.007 *

4. Discussion

The aim of the present study was to investigate whether an on-shoe, sensor-based gait analysis system reflected turning abnormalities and whether it could objectively complement physicians' gait assessments. To this end, we recruited 108 PD patients and 42 age-matched controls, and measured their gait during a 4 $\times$ 10 m walk by using our system. We then isolated the turnings from the whole gait sequence and quantified them using several spatio-temporal parameters. The parameters extracted using an on-shoe wearable system were previously validated against gold-standard systems, such as an optical motion capturing system [30] or instrumented walkway [22], and results indicated their technical validity. The clinical validation that followed turn quantification showed that turning parameters extracted using our measurement system and the turn isolation algorithm can effectively reflect gait abnormalities and be successfully used for the objective assessment of turning.

There have been many studies regarding turning analysis in PD [3,17,20]; yet, there is no unique way to define turning. Turning has been defined using mathematical modeling [17], statistics of spatio-temporal parameters [20], or the path between two pre-defined points [5]. One reason for these diverse turning definitions is that, basically, there is no standard way to determine the start and end of the turning. Common gold standards, such as motion-capture systems or videos, cannot provide a ground truth for turning. Since transitions between straight walking and turning happen gradually, it is inherently difficult to determine a specific start- and end-point for turning. A technical validation seems impractical with the usual gold standards. Nevertheless, a specific definition of turning, supported by some clinical validations that show its usability, can be an asset in objective gait assessment [3,17,20].

Turns can have different lengths, angles, and bases of support. We can expect that different types of turning require different levels of coordination [19]. In this study, we analyzed 180° during the 4 $\times$ 10 m walk test. Turnings with 180° were also analyzed in other standardized tests, like Timed Up and Go (TUG) [16,31,32]. We studied a 4 $\times$ 10 m walk because it includes three turns, which makes it statistically more meaningful to draw any general conclusions from the experiments. Regardless of the type of the turns, the underlying concepts that were used in this study are valid, although the turning isolation algorithm may need some adjustments to distinguish between straight walking and turning in an optimal way.

The findings of this study confirm the results from other studies [11–15], showing that spatio-temporal parameters can manifest gait deficits even in early stages of the disease. Results show that as the total duration of a turn increases, the stride length and velocity decreases and more strides are needed to complete a turn in the PD population. Such changes in parameters were scaled with PD severity. Global parameters of turning, such as the number of strides per turn and the total duration of the turn, can distinguish different groups. This is an important finding for PD studies, because gait problems are difficult to detect by physicians in early stages of the disease, whereas sensor signals can capture subtle differences between a healthy and abnormal gait in the early stages of the disease.

The large effect sizes for global parameters further emphasized the efficiency of these parameters for yielding statistical differences between different groups. Previous studies showed similar results for such global turning parameters [3,16,17]. Per-stride parameters of stride velocity, path length, stride length, and swing width can distinguish the majority of groups, although to a lesser extent in contrast to global parameters. For example, the distinction between controls and early-stage PD patients is more effective in global parameters. Furthermore, the effect sizes for per-stride parameters are in the range of small to medium (see Table 2), which again proves to be less effective than global parameters.

The total duration of turns showed a clear correlation with clinical scores, but such a correlation has not been obtained for per-stride timing. We may be able to explain this by considering two kinds of compensatory actions taken by patients in order to complete the turning. One compensatory action is to take smaller strides, and the other one is having longer pauses in a mid-stance phase in order to secure balance. While the first compensatory action decreases the per-stride time, the latter increases it. These compensatory actions may be different from patient to patient, and a patient may take both of these actions to safely complete a turn. Hence, overall, we cannot see any clear increase or decrease in the per-stride duration; however, the total turn duration did increase, because we may have a decrease of time per stride but patients take more strides that compensate for the decrease in time per stride. Having a long pause at mid-stance phases did not have any effect on stride and path length. These parameters decrease as the disease progresses.

Established clinical scores have no sub-item to assess specific characteristics of turning. Turning is evaluated as part of the gait in general; yet, our findings show that clinical scores reveal turning deficits at different levels. Parameters consistently show a higher correlation with gait and postural instability sub-items than with H&Y and UPDRS-III global scores, both in terms of p-values and effect sizes. Postural instability and gait sub-items are widely used for assessing gait, balance, and risk of falling in PD patients [23]. These two sub-items effectively demonstrate turning abnormalities, even at early stages of the disease (see Figure 2).

Despite the importance, there has not been a study to objectively compare straight walking and turning parameters in order to understand which set of parameters reflects gait abnormalities better. However, parameters quantifying straight walking differentiate between controls and PD patients in more moderate stages of the disease or higher levels of motor impairment [23]. Spatio-temporal parameters characterizing gait abnormalities have been widely used in data-driven applications, from PD diagnosis to disease monitoring [20,33]. However, most of such studies focus only on analyzing straight walking. Our results suggest that turning analysis may improve the performance of data-driven methods in medical applications.

One of the key goals of mobile gait analysis is to monitor patients outside of the clinics. Long-term monitoring of patients during the course of a day can provide better insight into their disease condition, in contrast to time-limited examinations inside the clinics [3]. Moreover, continuous monitoring of patients can be supplemented with preventative strategies for falling and FoG. The fact that turning during standardized tests demonstrates clear signs of deficiency emphasizes that turning analysis needs to be integrated into the long-term gait analysis. Turning isolation during long-term monitoring is even more challenging than in a standardized test, since the strides can be highly variable and different types of turning may happen within the course of a day. Some studies successfully addressed turning analysis in long-term monitoring [3,6], although they did not use on-shoe sensor systems. More research is needed to understand how findings of the current study can be transferred using an on-shoe sensor system to long-term monitoring.

Laterality of PD is another important factor in turning analysis, since turning to the direction of the most affected side is more challenging for patients. However, analyzing the laterality of the disease was beyond the scope of this study—here, the patients were instructed to turn at a convenient speed and preferred direction.

A limitation of our study is that we were, at this stage, not able to analyze the asymmetry between the left and right foot, since the sensors were not synchronized. Even better results may be obtained by an experiment design that takes into account the specific characteristics of PD patients and assessments during OFF medication.

5. Conclusions

Mobile gait analysis using wearable sensors offers elaborate assessments of pathological gait, leading to deeper insight into the motor deficits of PD. A high level of deficiency has been frequently reported for turning in PD. We investigated the feasibility of turning analysis during standardized gait tests using on-shoe wearable sensors. Turning measurements in our experiments clearly demonstrated turning deficits in Parkinson's patients. However, global parameters proved more effective than per-stride parameters. This should be taken into account in designing gait analysis systems, and has an important implication for PD clinical examinations, since physicians can readily assess global parameters. The current result is in alignment with other studies of turning in Parkinson's patients, which proves the feasibility of turning analysis using on-shoe sensor systems. The results of the current study can be applied to studies evaluating turning inside the clinic, and provide useful insight into long-term monitoring outside the clinic.

Author Contributions: Conceptualization, H.G., F.M., N.H.G., J.K. and B.M.E.; Formal analysis, N.H.G.; Funding acquisition, J.K. and B.M.E.; Methodology, H.G., N.H.G.; Resources, H.G. and J.K.; Software, J.H. and N.R., N.H.G.; Supervision, H.G., J.K. and B.M.E.; Writing—original draft, N.H.G.; Writing—review & editing, J.H., N.R., H.G., F.M., J.K. and B.M.E.

Acknowledgments: N. Haji Ghassemi acknowledges financial support from the Bavarian Research Foundation (BFS) and Federal Ministry of Education and Research (BMBF). This work was in part supported by the FAU Emerging Fields Initiative (EFIMoves), Bavarian Ministry for Economy, Regional Development & Energy via the Medical Valley Award 2017 (FallRiskPD Project) and EIT Health innovation project. F. Marxreiter is supported by the interdisciplinary center for clinical research Erlangen (IZKF), clinician scientist program. Björn M. Eskofier gratefully acknowledges the support of the German Research Foundation (DFG) within the framework of the Heisenberg professorship program (grant number ES 434/8-1).

Conflicts of Interest: The authors declare no conflict of interest. The founding sponsors had no role in the design of the study; in the collection, analyses, or interpretation of data; in the writing of the manuscript, and in the decision to publish the results.

References

1. Stack, E.; Ashburn, A. Dysfunctional turning in Parkinson's disease. *Disabil. Rehabil.* **2008**, *30*, 1222–1229. [CrossRef] [PubMed]
2. Crenna, P.; Carpinella, I.; Rabuffetti, M.; Calabrese, E.; Mazzoleni, P.; Nemni, R.; Ferrarin, M. The association between impaired turning and normal straight walking in Parkinson's disease. *Gait Posture* **2007**, *26*, 172–178. [CrossRef] [PubMed]
3. El-Gohary, M.; Pearson, S.; McNames, J.; Mancini, M.; Horak, F.; Mellone, S.; Chiari, L. Continuous monitoring of turning in patients with movement disability. *Sensors* **2014**, *14*, 356–369. [CrossRef] [PubMed]
4. Pickering, R.M.; Grimbergen, Y.A.; Rigney, U.; Ashburn, A.; Mazibrada, G.; Wood, B.; Gray, P.; Kerr, G.; Bloem, B.R. A meta-analysis of six prospective studies of falling in Parkinson's disease. *Mov. Disord.* **2007**, *22*, 1892–1900. [CrossRef] [PubMed]
5. Stack, E.; Jupp, K.; Ashburn, A. Developing methods to evaluate how people with Parkinson's disease turn 180°: An activity frequently associated with falls. *Disabil. Rehabil.* **2004**, *26*, 478–484. [CrossRef] [PubMed]
6. Mancini, M.; Weiss, A.; Herman, T.; Hausdorff, J.M. Turn Around Freezing: Community-Living Turning Behavior in People with Parkinson's Disease. *Front. Neurol.* **2018**, *9*, 18. [CrossRef] [PubMed]
7. Moore, O.; Peretz, C.; Giladi, N. Freezing of gait affects quality of life of peoples with Parkinson's disease beyond its relationships with mobility and gait. *Mov. Disord.* **2007**, *22*, 219–2195. [CrossRef]
8. Bachlin, M.; Plotnik, M.; Roggen, D.; Maidan, I.; Hausdorff, J.M.; Giladi, N.; Troster, G. Wearable assistant for Parkinson's disease patients with the freezing of gait symptom. *IEEE Trans. Inf. Technol. Biomed.* **2010**, *14*, 436–446. [CrossRef]

9. Goetz, C.G.; Tilley, B.C.; Shaftman, S.R.; Stebbins, G.T.; Fahn, S.; Martinez-Martin, P.; Poewe, W.; Sampaio, C.; Stern, M.B.; Dodel, R.; et al. Movement Disorder Society-sponsored revision of the Unified Parkinson's Disease Rating Scale (MDS-UPDRS): Scale presentation and clinimetric testing results. *Mov. Disord.* **2008**, *23*, 2129–2170. [CrossRef]
10. Hoehn, M.M.; Yahr, M.D. Parkinsonism: Onset, progression and mortality. *Neurology* **1967**, *17*, 427–442. [CrossRef]
11. Stack, E.L.; Ashburn, A.A.; Jupp, K.E. Strategies used by people with Parkinson's disease who report difficulty turning. *Park. Rel. Disord.* **2006**, *12*, 87–92. [CrossRef] [PubMed]
12. Mak, M.; Patle, A.; Hui-Chan, C. Sudden turn during walking is impaired in people with Parkinson's disease. *Exp. Brain Res.* **2008**, *190*, 43–51. [CrossRef] [PubMed]
13. Huxham, F.; Baker, R.; Morris, M.E.; Iansek, R. Footstep adjustments used to turn during walking in Parkinson's disease. *Mov. Disord.* **2008**, *23*, 817–823. [CrossRef] [PubMed]
14. Hong, M.; Perlmutter, J.; Earhart, G.A. kinematic and electromyographic analysis of turning in people with Parkinson disease. *Neurorehabil. Neural Repair* **2009**, *23*, 166–176. [CrossRef] [PubMed]
15. Hong, M.; Earhart, G.M. Effects of medication on turning deficits in individuals with Parkinson's disease. *J. Neurol. Phys. Ther.* **2010**, *34*, 11–16. [CrossRef]
16. King, L.; Mancini, M.; Priest, K.; Salarian, A.; Rodrigues-de Paula, F.; Horak, F. Do clinical scales of balance reflect turning abnormalities in people with Parkinson's disease? *J. Neurol. Phys. Ther.* **2012**, *36*, 25–31. [CrossRef] [PubMed]
17. Salarian, A.; Zampieri, C.; Horak, F.; Carlson-Kuhta, P.; Nutt, J.; Aminian, K. Analyzing 180 Degrees Turns Using an Inertial System Reveals Early Signs of Progression of Parkinson's Disease. In Proceedings of the 2009 IEEE Engineering in Medicine and Biology Society, Minneapolis, MN, USA, 3–6 September 2009; pp. 224–227.
18. Huxham, F.; Baker, R.; Morris, M.E.; Iansek, R. Head and trunk rotation during walking turns in Parkinson's disease. *Mov. Disord.* **2008**, *23*, 1391–1397. [CrossRef]
19. Mellone, S.; Mancini, M.; King, L.A.; Horak, F.B.; Chiari, L. The quality of turning in Parkinson's disease: A compensatory strategy to prevent postural instability? *J. Neuroeng. Rehabil.* **2016**, *13*, 39. [CrossRef]
20. Mariani, B.; Jimenez, M.; Vingerhoets, F.; Aminian, K. On-shoe wearable sensors for gait and turning assessment of patients with Parkinson's disease. *IEEE Trans. Biomed. Eng.* **2013**, *60*, 155–158. [CrossRef]
21. Panebianco, G.P.; Bisi, M.C.; Stagni, R.; Fantozzi, S. Analysis of the performance of 17 algorithms from a systematic review: Influence of sensor position, analysed variable and computational approach in gait timing estimation from IMU measurements. *Gait Posture* **2018**, *66*, 76–82. [CrossRef]
22. Rampp, A.; Barth; J., Schülein, S.; Gassmann; K. G.; Klucken, J.; Eskofier, B. M. Inertial sensor-based stride parameter calculation from gait sequences in geriatric patients. *IEEE Trans. Biomed. Eng.* **2014**, *62*, 1089–1097. [CrossRef] [PubMed]
23. Schlachetzki, J.C.; Barth, J.; Marxreiter, F.; Gossler, J.; Kohl, Z.; Reinfelder, S.; Gassner, H.; Aminian, K.; Eskofier, B.M.; Winkler, J.; et al. Wearable sensors objectively measure gait parameters in Parkinson's disease. *PLoS ONE* **2017**, *12*, e0183989. [CrossRef] [PubMed]
24. Gassner, H.; Raccagni, C.; Eskofier, B.M.; Klucken, J.; Wenning, G.K. The diagnostic scope of sensor-based gait analysis in atypical Parkinsonism: Further observations. *Front. Neurol.* **2019**, *10*, 5. [CrossRef] [PubMed]
25. Hughes, A.J.; Daniel, S.E.; Kilford, L.; Lees, A.J. Accuracy of clinical diagnosis of idiopathic parkinson's disease: A clinico-pathological study of 100 cases. *J. Neurol. Neurosurg. Psychiatry* **1992**, *55*, 181–184. [CrossRef] [PubMed]
26. Parkinson, J. *An Essay on the Shaking Palsy*; Neely & Jones: London, UK, 1817.
27. Haji Ghassemi, N.; Hannink, J.; Martindale, C.F.; Gassner, H.; Müller, M.; Klucken, J.; Eskofier, B.M. Segmentation of Gait Sequences in Sensor-Based Movement Analysis: A Comparison of Methods in Parkinson's Disease. *Sensors* **2018**, *18*, 145. [CrossRef]
28. Barth, J.; Oberndorfer, C.; Pasluosta, C.; Schülein, S.; Gassner, H.; Reinfelder, S.; Kugler, P.; Schuldhaus, D.; Winkler, J.; Klucken, J.; et al. Stride segmentation during free walk movements using multi-dimensional subsequence dynamic time warping on inertial sensor data. *Sensors* **2015**, *15*, 6419–6440. [CrossRef]
29. Cohen, J. *Statistical Power Analysis for the Behavioral Sciences*, 2nd ed.; Lawrence Erlbaum Associates: Hillsdale, NJ, USA, 1988.

30. Kanzler, C.M.; Barth, J.; Klucken, J.; Eskofier, B.M. Inertial sensor based gait analysis discriminates subjects with and without visual impairment caused by simulated macular degeneration. *Conf. Proc. IEEE Eng. Med. Biol. Soc.* **2016**, *2016*, 4979–4982.
31. Salarian, A.; Horak, F.B.; Carlson-Kuhta, P.; Nutt, J.; Zampieri, C.; Aminian, K. iTUG, a Sensitive and Reliable Measure of Mobility. *IEEE Trans. Neural Syst. Rehabil. Eng.* **2010**, *18*, 303–310. [CrossRef]
32. Herman, T.; Giladi, N.; Hausdorff, J.M. Properties of the 'timed up and go' test: More than meets the eye. *Gerontology* **2011**, *57*, 203–210.
33. Klucken, J.; Barth, J.; Kugler, P.; Schlachetzki, J.; Henze, T.; Marxreiter, F.; Kohl, Z.; Steidl, R.; Hornegger, J.; Eskofier, B.; et al. Unbiased and Mobile Gait Analysis Detects Motor Impairment in Parkinson's Disease. *PLoS ONE* **2013**, *8*, e56956. [CrossRef]

Article

Gait Quality Assessment in Survivors from Severe Traumatic Brain Injury: An Instrumented Approach Based on Inertial Sensors

Valeria Belluscio [1,2], Elena Bergamini [1], Marco Tramontano [1,2], Amaranta Orejel Bustos [1], Giulia Allevi [2], Rita Formisano [2], Giuseppe Vannozzi [1,*] and Maria Gabriella Buzzi [2]

1 Department of Movement, Human and Health Sciences, University of Rome "Foro Italico", Interuniversity Centre of Bioengineering of the Human Neuromusculoskeletal System, P.zza Lauro de Bosis 15, 00135 Roma, Italy; v.belluscio@studenti.uniroma4.it (V.B.); elena.bergamini@uniroma4.it (E.B.); m.tramontano@hsantalucia.it (M.T.); a.orejelbustos@studenti.uniroma4.it (A.O.B.)

2 IRCSS Fondazione Santa Lucia, Via Ardeatina 306, 00179 Roma, Italy; giulia.allevi@gmail.com (G.A.); r.formisano@hsantalucia.it (R.F.); mg.buzzi@hsantalucia.it (M.G.B.)

* Correspondence: giuseppe.vannozzi@uniroma4.it; Tel.: +39-063673-3522

Received: 6 October 2019; Accepted: 28 November 2019; Published: 3 December 2019

Abstract: Despite existing evidence that gait disorders are a common consequence of severe traumatic brain injury (sTBI), the literature describing gait instability in sTBI survivors is scant. Thus, the present study aims at quantifying gait patterns in sTBI through wearable inertial sensors and investigating the association of sensor-based gait quality indices with the scores of commonly administered clinical scales. Twenty healthy adults (control group, CG) and 20 people who suffered from a sTBI were recruited. The Berg balance scale, community balance and mobility scale, and dynamic gait index (DGI) were administered to sTBI participants, who were further divided into two subgroups, severe and very severe, according to their score in the DGI. Participants performed the 10 m walk, the Figure-of-8 walk, and the Fukuda stepping tests, while wearing five inertial sensors. Significant differences were found among the three groups, discriminating not only between CG and sTBI, but also for walking ability levels. Several indices displayed a significant correlation with clinical scales scores, especially in the 10 m walking and Figure-of-8 walk tests. Results show that the use of wearable sensors allows the obtainment of quantitative information about a patient's gait disorders and discrimination between different levels of walking abilities, supporting the rehabilitative staff in designing tailored therapeutic interventions.

Keywords: wearables; inertial sensors; traumatic brain injury; dynamic balance; gait disorders; gait patterns; head injury; gait symmetry; gait smoothness; acceleration

---

1. Introduction

Head injuries are considered a major health problem as they are associated with high mortality and disability in young adults (<45 years of age) [1,2]. Nearly 70% of all brain-injury cases are males [3], and most events are caused by falls (28%), followed by motor vehicle accidents (20%) and blows (19%) [4]. Traumatic brain injuries (TBI) impress a significant burden on the health care system, due to the need for therapy to address physical, communicative, and psychological problems [5]. Costs are usually more elevated when the traumatic brain injury is considered severe [5]; that is with an initial Glasgow coma scale score (GCS) of 8 or less [6]. Neuropsychological and cognitive impairments, such as anxiety and depression, selective/sustained attention, language, and executive function deficits have been well documented in the literature [7–12]. Less attention has been placed on motor impairments, in striking contrast with available data on other neurological populations, such as stroke and Parkinson's disease ones [13–18]. The available studies mainly focused on impaired balance and altered coordination. Specifically, Rinne and colleagues [19] described that well-recovered men with TBI had impaired balance and agility compared to healthy controls. A recent review performed by Williams and colleagues [20] evidenced that people with TBI walked more slowly than healthy controls, primarily due to reduced step length. A few authors emphasized the impact of post-traumatic parkinsonism or post-traumatic cerebellar syndrome [21,22], two conditions that interfere with walking and balance performances in persons surviving from TBI. Additionally, balance abnormalities have also been reported in terms of increased postural sway during quiet standing or functional tasks, with altered sensory inputs [23–25]. Additionally, gait analysis has been used in few studies: Chou and colleagues [26] showed that people who suffered from TBI usually present a gait pattern with a significantly slower speed and a shorter stride length, confirming previous results [27]. Basford and colleagues [28] reported that gait analysis, balance, and vestibular testing could document subtle biomechanical changes among participants with TBI, suggesting the appropriateness of gait and balance testing in this population, even when motor disorders are not clinically evident.

Taken together, evidence exists for persistent motor deficits after TBI. However, these studies have focused on mild (GCS > 13) and moderate (GCS between 9 and 13) TBI, and to the authors' knowledge, no quantitative information is available about motor ability in people who have incurred a severe TBI (sTBI). An objective characterization of their level of motor impairment could be an important step in the rehabilitation process of this population, helping in obtaining not only physical improvements, but also increasing the independence in daily life and the overall quality of life. This characterization, in order to be helpful and informative, should be ecological and as non-intrusive as possible.

In this framework, attention is growing on miniaturized and wearable instruments that quantify movement patterns in a non invasive way: inertial measurement units (IMUs), embedding accelerometers and gyroscopes, have been widely used in the last two decades since they present many advantages compared to the traditional gait analysis approach based on stereophotogrammetry and force platforms. From the data measured by these units, spatiotemporal gait parameters [29] and stability-related parameters [13,30] can be extracted, allowing fall risk to be assessed [31], and allowing one to differentiate gait patterns between healthy and pathological populations [13,32–34]. However, in the sTBI population, an instrumented approach with IMUs has never been proposed and no information is available about their capability to discriminate among different levels of walking ability, as defined by currently administered clinical scales, such as the dynamic gait index scale [35]. An integrated approach based on the "gold standard" clinical evaluation method which relies on clinical scales and the proposed sensor-based assessment would overcome the limitations of a subjective evaluation, depending on the operator's specific training, helping in revealing changes hardly detectable using clinical scales. In addition, this integration would allow to assess patients in ecological contexts, where they perform tasks more similarly to those of real life, providing objective motor ability characterization.

Given these premises, the aims of the present study were twofold: (i) to quantify gait patterns in sTBI population using a set of wearable inertial sensors; (ii) to investigate the association of the estimated gait quality indices with the level of walking ability and the scores of commonly administered clinical scales. Specifically, spatiotemporal parameters and gait quality indices (dynamic stability, symmetry, and smoothness) were investigated considering clinical performance tests commonly used in the routine assessment [36,37].

The hypothesis is that the instrumental approach could be a valid support to the traditional clinical evaluation in order to obtain quantitative and objective information about sTBI patients' motor impairments, discriminating between different levels of walking abilities, and helping clinicians with defining and evaluating the efficacy of personalized rehabilitation treatments, as previously reported in the literature [38]. Furthermore, the correlation analysis could help with simplifying and facilitating routine evaluation in terms of time-consuming administration of clinical scales, possibly allowing a reduction in the number of scales used, maintaining those necessary to characterize the investigated population/motor task.

2. Materials and Methods

The research was performed at the Santa Lucia Foundation and it was approved by the Local Independent Ethics Committee of Fondazione Santa Lucia IRCCS (Rome, Italy) (protocol number: CE/PROG.700).

2.1. Participants

Twenty healthy subjects (control group, CG) (age: 33.9 ± 9.5 years), 15 males and 5 females, and 20 people who suffered from a sTBI (age: 33.4 ± 10.5 years), 15 males and 5 females, were involved in the study. This sample size complied with the minimum number of participants recommended by a power analysis purposely performed ($\alpha = 0.05$; power ($1-\beta$) = 0.95, effect size d: 0.7) for non parametric comparisons [39]. Exclusion criteria for CG were the presence of any orthopedic, neurological, or other co-morbidities which could have influenced the motor performance. Inclusion criteria for sTBI were: (i) age between 15 and 65 years; (ii) Glasgow coma scale (GCS) score ≤ 8 (used to objectively describe the severity of impaired consciousness at the time of injury) [6]; (iii) level of cognitive functioning (LCF) ≥ 7 [40]; (iv) presence of disturbances in static and dynamic balance; (v) ability to understand verbal commands. Almost all the patients selected suffered from a sTBI as a consequence of a traffic accident (19 out 20 participants), whereas one person suffered from a sTBI due to a fall.

2.2. Procedures

2.2.1. Clinical Assessment

The following clinical scales were administered by an expert physiotherapist to all sTBI participants, to assess static and dynamic balance, ambulation skills, and mobility deficits:

- Dynamic gait index (DGI)—to assess a subject's ability to modify gait in response to changing task demands. It consists of items rated from 0 to 3 (0 = severely impaired; 3 = normal performance), yielding a maximum score of 24 points. A score lower than 19 points has been associated with impairment of gait and fall risk [35,41].
- Berg balance scale (BBS)—to measure 14 different tasks related to balance and postural control. It is scored from 0 to 4, with 0 indicating that the subject is unable to perform the task and 4 that the subject fully meets the most difficult criteria required for the task [42].
- Community balance and mobility scale (CB&M)—to assess specific aspects of balance and mobility which are necessary for independent functioning within the community [43]. This scale includes several challenging tasks and it is based on 19 tests. Higher scores are indicative of better balance and mobility.

To codify for different levels of walking ability, sTBI patients were further divided into two sub-groups, according to their score in the dynamic gait index clinical scale: persons with a score >19 were considered severe (10 people, sTBI-1), while those with a score ≤ 19 were considered very severe (10 people, sTBI-2), according to [35]. The demographic characteristics of each subgroup are reported in Table 1.

Table 1. Demographic and anthropometric characteristics of the control group (CG), severe traumatic brain injury 1 (sTBI-1), and sTBI-2. Mean ± standard deviation values are displayed.

	CG	sTBI-1	sTBI-2
Nr. of Participants	20	10	10
Nr. of Males	15	8	7
Age [Years]	33.9 ± 9.5	33.2 ± 9.6	36.1 ± 13.1
Body Mass [kg]	78.3 ± 14.9	75.9 ± 16.2	71.0 ± 14.7
Body Height [m]	1.78 ± 0.09	1.73 ± 0.11	1.70 ± 0.11
Time Since Trauma [days]	-	308 ± 182	512 ± 476

2.2.2. Motor Assessment

Each participant was asked to perform three different motor tasks in a randomized order: the 10 m walk Test (10mWT), the figure-of-8 walk test (F8WT), and the Fukuda stepping test (FST). All tests were carried out in a fully dedicated quiet area at the Santa Lucia Foundation, where the surface was accurately kept flat, and participants were asked to stay barefoot and to stand upright for at least 5 s at the beginning and at the end of each trial. Tasks were carefully explained and demonstrated by an instructor before the testing. The instructor also gave the patients start and stop commands and stayed close to participants to prevent dizziness and/or falls. A detailed description of the motor tasks is reported below.

10 m Walk Test (10mWT)

The 10mWT is a widely used and recommended test for measuring gait speed in different populations [44]. The experimental protocol of the assessment was selected according to previous studies [13,45]: it consists of walking on a straight 14 m long walkway for three repetitions at the participant's preferred walking pace, with the middle 10 m marked on the floor and considered as steady-state walking for further analysis. The time taken to walk the middle 10 m was measured using a stopwatch and walking speed was calculated by dividing the distance covered (i.e., 10 m) by the time taken.

Figure-of-8 Walk Test (F8WT)

The F8WT requires a person to walk a figure-of-8 shape, as illustrated in Figure 1, marked on the floor with tape, with each circle diameter of 1.66 m (5.44 ft) [46]. Participants were instructed: (*i*) to stand still with feet side-by-side in the start position facing the "8"; (*ii*) to begin walking at their preferred pace when ready; (*iii*) to stop when returning to the start position, placing feet side-by-side again. The test was performed three times for each F8WT direction (clockwise and counterclockwise), alternating the two directions, and the entire trial was considered for further investigations.

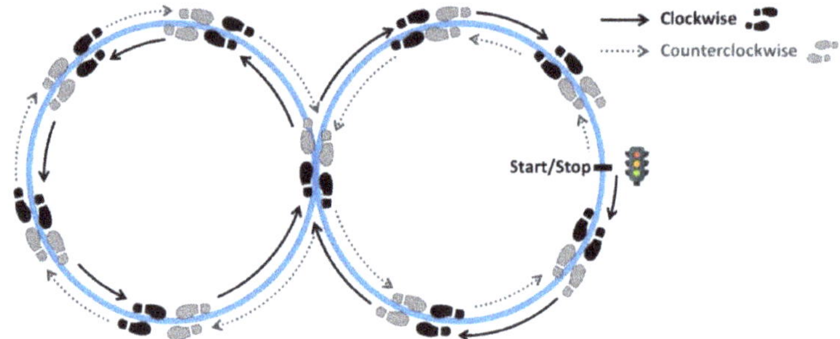

Figure 1. Figure-of-8 shape used for the figure-of-8 walk test (F8WT). Clockwise and counterclockwise directions are indicated with grey and black arrows, respectively.

Fukuda Stepping Test (FST)

The FST is a test used for the diagnosis of vertigo-associated disease [47] and an instrumented version of this test has been recently proposed in the literature [48] and was adopted in this work. Participants were instructed to stand upright blindfolded with both arms frontally outstretched, creating a 90° angle between the arms and the body. Then, they were asked to step on the spot for one minute and to remain still in the final position. Lateral and forward displacements, as well as the amount and side of rotation, were marked on the floor by a piece of tape and subsequently reported as clinical FST parameters. For what concerns the sensor-based parameters, the first and last three strides were discarded in order to evaluate only steady-state stepping.

2.3. Equipment

While performing the three above mentioned motor tasks, each participant was equipped with five synchronized inertial measurement units (IMUs) (128Hz, Opal, APDM, Portland, Oregon, USA): one located on the occipital cranium bone close to the lambdoid suture of the head (H), one on the center of the sternum (S), and one at L4/L5 level, slightly above the pelvis (P), and were used to assess the upper-body stability. The other two IMUs were located on both shanks, slightly above the lateral malleoli, and were used for step and stride segmentation. Each IMU was securely fixed to the participant's body with Velcro straps, except for the head IMU, which was inserted in a tailored pocket of a swim cap worn by each subject.

2.4. Data Processing

All data processing was performed using the Matlab software (The MathWorks Inc., Natick, MA, USA). Each unit embedded three-axial accelerometers and gyroscopes (±6 g with g = 9.81 $m{\cdot}s^{-2}$, and ±1500 °/s of full-range scale, respectively) and provided the quantities with respect to a unit-embedded system of reference. To guarantee a repeatable reference system for the three IMUs located on the upper body, each unit was aligned with the corresponding anatomical axes (antero-posterior: AP, medio-lateral: ML, and cranio-caudal: CC) following the procedure proposed by [49]. The following spatiotemporal parameters were obtained, through a peak detection algorithm, on the ML angular velocity signals measured by the two IMUs on the shanks: average stride duration (SD = time to complete the test/total number of strides) and average stride frequency (SF = total number of strides/time to complete the test). The following gait quality indices were estimated:

- Normalized root mean square (nRMS) values of the accelerations were calculated by dividing the RMS, AP, and ML components by the CC component, at each upper-body level (P, S, H). High RMS values have been associated with higher amount of acceleration, and hence, decreased stability, as reported in [29].
- Attenuation coefficients (AC) [50] between each level pair of the upper-body, for each acceleration component (j), defined as:

$$ACPS_j = \left(1 - \frac{RMS_jS}{RMS_jP}\right),$$

$$ACPH_j = \left(1 - \frac{RMS_jH}{RMS_jP}\right),$$

$$ACSH_j = \left(1 - \frac{RMS_jH}{RMS_jS}\right).$$

  Each coefficient represents the variation of the acceleration from lower to upper-body levels. A positive coefficient indicates an attenuation of the accelerations, while a negative coefficient indicates an amplification of the accelerations from the lower to the upper body level.
- Improved harmonic ratio (iHR), as proposed by [51], was calculated for each acceleration component (j) measured at the pelvis level. This index is based on a spectral analysis of the acceleration signals and is a measure of hemilateral symmetry when stepping (0% = total asymmetry; 100% = total symmetry). It was calculated as follows:

$$\mathbf{iHR_j} = \frac{\sum \textbf{Power of intrinsic harmonics}}{\sum \textbf{Power of intrinsic harmonics} + \sum \textbf{Power of extrinsic harmonics}} \cdot 100.$$

- SPectral ARC length (SPARC), as proposed by [52], calculated for each acceleration component (j) measured at the pelvis level. The calculation of SPARC was performed as follows:

$$-\int_0^{\tilde{\omega}_c} [(\frac{1}{\tilde{\omega}_c})^2 + (\frac{dA(\tilde{\omega})}{d\tilde{\omega}})^2]^{\frac{1}{2}} d\tilde{\omega}; A(\tilde{\omega}) = \frac{A(\tilde{\omega})}{A(0)}$$

$$\tilde{\omega}_c = min\{\tilde{\omega}_c{}^{max} min\{\tilde{\omega} \vee A(r) < \acute{A}, \forall r > \tilde{\omega} \ ,$$

where $A(\tilde{\omega})$ is the Fourier magnitude spectrum of the acceleration signal a(t) and $A(\tilde{\omega})$ is the normalized magnitude spectrum.

2.5. Statistical Analysis

Descriptive and inferential statistical analyses were performed using IBM SPSS Statistics software (v23, IBM Corp., Armonk, NY, USA), and the alpha level of significance was set at 0.05. The normal distribution of each parameter was verified using the Shapiro–Wilk test. As most of the parameters were not normally distributed, the following non-parametric tests were performed:

- Mann–Whitney U test to investigate if significant differences existed between sTBI-1 and sTBI-2 for the clinical scale scores;
- Kruskal–Wallis H-test on the estimated biomechanical parameters, to investigate if significant differences existed among the different levels of walking ability ("group" factor: CG, sTBI-1, or sTBI-2);
- Spearman's rank correlation coefficient (q) between gait quality indices and clinical scale scores, considering the whole sTBI group.

3. Results

3.1. Clinical Scale Score Results

The scores of the administered clinical scales for sTBI-1 and sTBI-2 are reported in Table 2. Results show that sTBI-2 group (defined as very severe TBI according to DGI scores; see methods) presented worse, statistically significant scores in the three clinical scales compared to sTBI-1.

Table 2. Clinical scales results for sTBI. Mean ± standard deviation values are displayed. Statistically significant differences are indicated with *.

	sTBI-1	sTBI-2	p-Value
Dynamic gait index (DGI)	22.1 ± 1.7 *	15.0 ± 3.0 *	0.000
Berg balance scale (BBS)	49.8 ± 2.1 *	42.4 ± 3.9 *	0.000
Community balance and mobility scale (CB&M)	42.0 ± 14.0 *	15.5 ± 8.9 *	0.000

3.2. Spatio-Temporal Parameters and Clinical FST Parameters

Results of temporal (stride frequency and stride duration) and clinical FST parameters (lateral and forward displacements; amount and side of rotation) for the three groups are reported in Table 3. Statistically significant differences were present for all three motor tasks when comparing CG with sTBI-2 and sTBI-1 with sTBI-2. In addition, statistically significant differences between CG and sTBI-1 were found in the spatio-temporal parameters of the FST. Concerning clinical FST parameters, no statistical differences are displayed in terms of lateral and forward displacements, or amount and side of rotation among the three groups. Walking speeds (mean ± standard deviation) obtained during the 10mWT were: 1.48 ± 0.20, 1.10 ± 0.23, and 0.53 ± 0.20, for the CG, sTBI-1, and sTBI-2 groups, respectively. Significant differences were found between CG and both sTBI-1 and sTBI-2, as well as between sTBI-1 and sTBI-2.

Table 3. Temporal and FST parameters. * indicates statistically significant differences between CG and sTBI-2 ($p < 0.001$); § indicates statistically significant differences between sTBI-1 and sTBI-2 ($p < 0.05$); # indicates statistically significant differences between CG and sTBI-1 ($p < 0.001$). Clinical FST parameters: the values of the antero-posterior (AP) and medio-lateral (ML) displacements, the amount of rotation and the side of rotation of the three groups of subjects (CG, sTBI-1, sTBI-2) in the three tasks are reported (mean ± standard deviation).

		Stride Frequency	Stride Duration	Rotation	Side	Displacement	
						AP	ML
		[Strides^{x}s^{-1}]	[s]	[Degrees]	[% Right]	[cm]	[cm]
10mWT	CG	0.9 ± 0.0 *	1.1 ± 0.1 *	-	-	-	-
	sTBI-1	0.8 ± 0.1 §	1.2 ± 0.1 §	-	-	-	-
	sTBI-2	0.7 ± 0.1 *,§	1.4 ± 0.2 *,§	-	-	-	-
F8WT	CG	0.8 ± 0.1 *	1.2 ± 0.1 *	-	-	-	-
	sTBI-1	0.8 ± 0.1 §	1.2 ± 0.2 §	-	-	-	-
	sTBI-2	0.7 ± 0.1 *,§	1.5 ± 0.2 *,§	-	-	-	-
FST	CG	0.8 ± 0.1 *,#	1.2 ± 0.2 #	66 ± 66	30	146 ± 71	44 ± 33
	sTBI-1	0.6 ± 0.2 §,#	1.8 ± 0.9 #	27 ± 17	40	141 ± 38	45 ± 46
	sTBI-2	0.5 ± 0.2 *,§	2.0 ± 1.4	28 ± 23	50	101 ± 60	27 ± 31

3.3. *Root Mean Square, Attenuation Coefficients, Improved Harmonic Ratio, and SPARC*

Significant differences were found for the three motor tasks (10mWT, F8WT, and FST) when comparing both sTBI against CG and sTBI-1 against sTBI-2. Results regarding the 10mWT, the F8WT, and the FST are reported in Figure 2a–c, respectively.

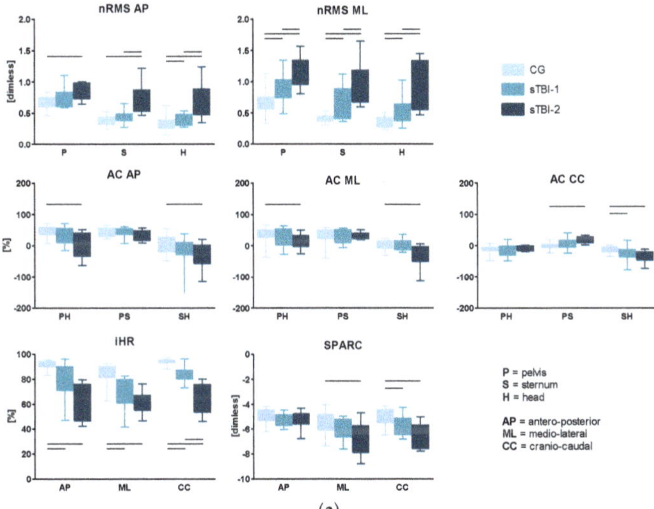

(a)

Figure 2. *Cont.*

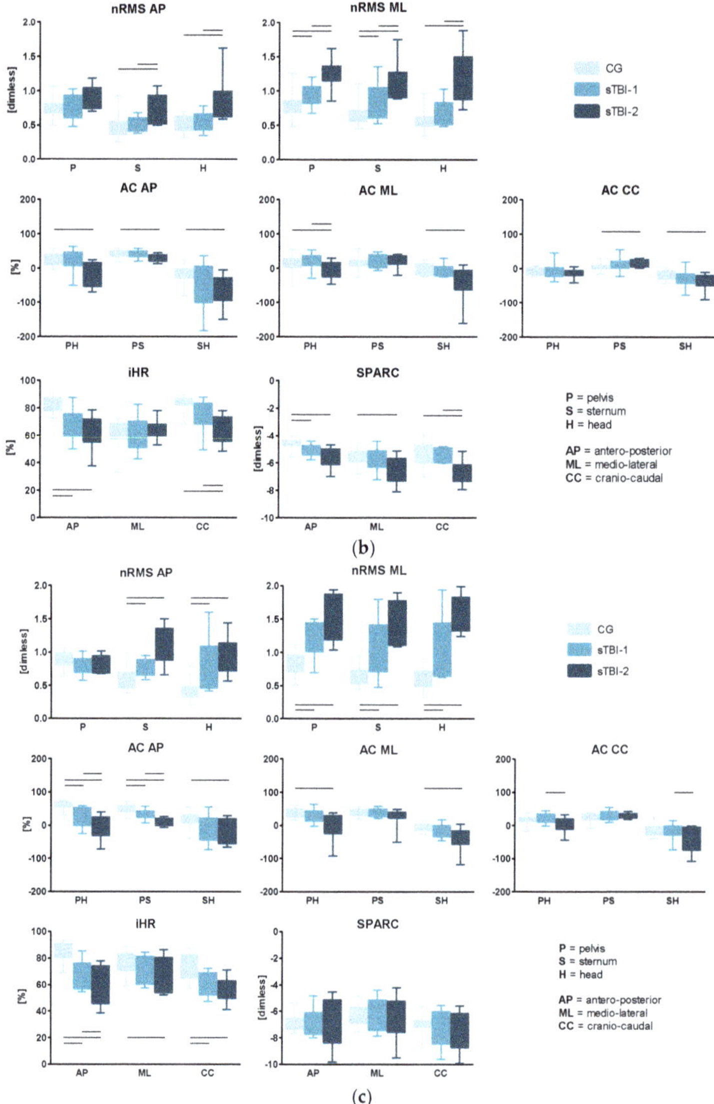

Figure 2. Normalized root mean square (nRMS) values, attenuation coefficients (AC), improved harmonic ratio (iHR), and SPectral ARC length (SPARC) for the sTBI sub-groups and for CG in 10mWT (**a**), F8WT (**b**), and FST (**c**). Medians and interquartile ranges are reported. AP, antero-posterior; ML, medio-lateral; CC, cranio-caudal; P, pelvis; S, sternum; H, head. The horizontal lines indicate statistically significant between-groups differences. (**a**) 10 m walk test. (**b**) Figure-of-8 walk test. (**c**) Fukuda stepping test.

3.4. Association of the Gait Quality Indices with the Clinical Scale Scores

Correlation analysis (Table 4) shows that several indices displayed a significant correlation with the clinical scales scores in the three motor tasks, especially in the 10mWT and F8WT.

Table 4. Spearman's correlation coefficients (ρ) between each estimated parameter and each clinical scale. Statistical significance is indicated by asterisks (* $p < 0.05$; ** $p < 0.001$). Abbreviations: BBS, Berg balance scale; DGI, dynamic gait index; CB&M, community balance and mobility scale, RMS, root mean square; AC, attenuation coefficient; iHR, improved harmonic ratio; SPARC, spectral arc length; AP, antero-posterior; ML, medio-lateral; CC, craniocaudal; P, pelvis; S, sternum; H, head.

		10mWT			F8WT			FST		
		BBS	DGI	CB&M	BBS	DGI	CB&M	BBS	DGI	CB&M
RMS_P	AP	−0.243	−0.309	−0.254	−0.337	−0.335	−0.398	0.043	0.103	0.118
	ML	−0.656 **	−0.467 *	−0.605 **	−0.730 **	−0.666 **	−0.819 **	−0.404	−0.445	−0.500 *
RMS_S	AP	−0.555 *	−0.585 **	−0.679 **	−0.491 *	−0.484 *	−0.600 **	−0.495 *	−0.655 **	−0.500 *
	ML	−0.583 **	−0.503 *	−0.733 **	−0.571 *	−0.463 *	−0.749 **	−0.460 *	−0.516 *	−0.695 **
RMS_H	AP	−0.674 **	−0.641 **	−0.712 **	−0.594 **	−0.611 **	−0.665 **	−0.353	−0.309	−0.246
	ML	−0.781 **	−0.705 **	−0.821 **	−0.796 **	−0.708 **	−0.839 **	−0.618 **	−0.506 *	−0.660 **
ACPH	AP	0.535 *	0.577 **	0.451	0.481 *	0.349	0.418	0.608 **	0.550 *	0.512 *
	ML	0.493 *	0.491 *	0.595 **	0.631 **	0.598 **	0.637 **	0.630 **	0.481 *	0.623 **
	CC	−0.061	−0.076	0.004	0.057	0.059	0.182	0.544 *	0.551 *	0.567 *
ACPS	AP	0.495 *	0.443	0.588 **	0.477 *	0.453	0.454	0.627 **	0.699 **	0.539 *
	ML	0.126	0.159	0.309	0.181	0.090	0.279	0.254	0.122	0.391
	CC	−0.247	−0.286	−0.367	−0.251	−0.190	−0.368	0.093	−0.129	−0.072
ACSH	AP	0.287	0.197	0.242	0.395	0.302	0.402	0.368	0.207	0.330
	ML	0.663 **	0.516 *	0.612 **	0.599 **	0.497 *	0.486 *	0.553 *	0.466 *	0.372
	CC	0.172	0.094	0.337	0.346	0.224	0.451	0.431	0.506 *	0.530 *
iHR	AP	0.423	0.507 *	0.605 **	0.196	0.221	0.361	0.365	0.433	0.391
	ML	0.149	0.319	0.356	−0.143	−0.127	−0.019	0.109	0.188	0.012
	CC	0.734 **	0.733 **	0.677 **	0.693 **	0.667 **	0.658 **	0.016	0.272	0.023
SPARC	AP	0.205	0.051	0.170	0.384	0.308	0.411	−0.056	0.011	−0.061
	ML	0.285	0.390	0.456 *	0.195	0.192	0.160	0.086	0.092	−0.056
	CC	0.390	0.512 *	0.251	0.525 *	0.601 **	0.547 *	0.217	0.211	0.114

4. Discussion

The aims of this study were to quantify gait quality of a sTBI population with different levels of walking ability using a set of wearable inertial sensors and to investigate the association of the estimated gait quality indices with the scores of commonly administered clinical scales. Results show that the instrumented approach allows (*i*) obtainment of quantitative and objective information about patient's motor impairments; (*ii*) discrimination between different levels of walking abilities; (*iii*) exploration of the relationship between the estimated gait quality indices and the clinical scale scores. As expected, clinical scale scores displayed a consistent increasing trend from low to high walking ability levels, showing statistically significant differences between severe (sTBI-1) and very severe (sTBI-2) TBI participants (Table 2). A similar trend was observed when considering the spatio-temporal parameters: for what concerns walking speed, statistically significant differences were found between the control group (CG) and sTBI-2, and between sTBI-1 and sTBI-2 (Table 3). These results are consistent with the existing literature about healthy people [13,53] and TBI participants [26], and confirm the relevance of walking speed as an informative and concise parameter to discriminate between different level of walking ability.

In addition, the values of stride frequency and stride duration obtained in this study are consistent with previously reported results. In particular, in persons with sTBI, a reduced stride frequency, along with an increased stride duration, may be related to post-traumatic Parkinsonism [22,54,55]. Furthermore, as suggested in [56], it can be speculated that people with sTBI increase their stride duration in order to compensate for gait instability and counteract the fear of falling. This significant gait impairment was still observed despite the provision of optimal medication therapy, confirming the very close relationship between altered gait and postural instability in this population [57,58].

Interesting results come from the estimated gait quality indices: almost all parameters in the three motor tasks were able to discriminate between CG and both sTBI groups, especially the sTBI-2, as expected. The actual added value of the proposed approach, however, lies in its ability to detect possible differences between sTBI-1 and sTBI-2, facilitating discriminating between different levels of walking ability. In this respect, in the 10mWT, the two sTBI sub-groups presented differences in gait stability and symmetry. Specifically, considering gait stability, sTBI-2 showed higher nRMS compared to sTBI-1. High nRMS values have been associated with a higher amount of acceleration, and hence, decreased stability [13,29,30,33,34,50]. Both sTBI subgroups, and especially sTBI-2, displayed a decreased stability at the three upper body levels, particularly in the ML direction. This is consistent with previous studies dealing with other neurological populations [13,15,59]. In addition, the attenuation coefficient from pelvis to head in the ML direction discriminates between the two sTBI sub-groups, highlighting that the sTBI-2 sub-group exhibits a limited bottom-up attenuation of upper body accelerations. This result is related to a lack of ability to stabilize the head, impairing the consequent planning of adaptive motor strategies. Concerning gait symmetry, the iHR in the CC component discriminated between sTBI-1 and sTBI-2, showing a reduced gait symmetry, particularly evident in the sTBI-2. Reduced symmetry has been widely associated with an increased fall risk [60,61], thus indicating this parameter as a biomarker for the identification of patients at high risk of falling.

In the F8WT, the nRMS and iHR discriminated between sTBI-1 and sTBI-2, as reported for the 10mWT. In addition, differences were also pointed out considering the smoothness: in fact, the SPARC discriminated sTBI-1 and sTBI-2 well, probably because of greater upper body rigidity in sTBI-2 than sTBI-1 observed in this more difficult task. It is worth mentioning that, being characterized by a curved trajectory, the execution of the F8WT involves the activation of different cortical areas than those required in the planning of straight point-to-point movements. In fact, it is well known that the trajectory planning during curved-path conditions requires additional preparation time [62,63]. The results of the present study show indeed that the F8WT seems to be the walking test that better discriminates among different walking ability levels. This suggests that testing dynamic balance abilities during curved trajectories could be useful for assessing gait in conditions more relevant to cognitive-motor dual tasks, and thus, closer to daily living activities [64,65].

For what concerns the FST, results about clinical FST parameters confirm the previous literature [48,66]: no differences were found in terms of AP–ML displacements, nor side and degree of rotation among groups. The presence of rotation in either direction in the CG shows that turning while stepping on the spot also occurs in healthy people, confirming that clinical FST parameters are not able to distinguish between healthy and pathological subjects, confirming the doubts about its clinical use also for this population. Conversely, when considering gait quality indices obtained from the instrumented FST, the discrimination capability of the test greatly increased: in fact, significant differences were found, not only between CG and both sTBI sub-groups, but also between the sTBI sub-groups. Specifically, for what concerns stability, the nRMS did not discriminate between different levels of walking abilities in pathological subjects, as observed in the 10mWT and the F8WT. On the other hand, attenuation coefficients in the AP and CC directions distinguished between sTBI sub-groups, with sTBI-2 showing less ability in attenuating upper body accelerations from lower to higher levels. Additionally, a reduced symmetry was displayed by sTBI-2 with respect to sTBI-1, with the AP component of the iHR displaying a significant difference between the two sub-groups. The AP direction seems to be the most critical in very severe TBI and this result is in agreement with the existing literature about stroke patients [48], indicating the AP component as the most informative when comparing patients with different walking abilities. In addition, the absence of the visual input during the FST plays an important role on the sensory reweighting, which has been acknowledged as critical in the TBI population [67]. Therefore, these results confirm that the instrumented approach in this test provides valuable information about patients' motor strategies and useful data to tailor rehabilitation protocols [48].

When considering the second aim of the study, several correlations were found between clinical scales and gait quality indices, especially in the 10mWT and the F8WT. nRMS and attenuation coefficients, parameters related to dynamic stability, correlated well with all clinical scales, while worse correlations were present when considering the iHR and the SPARC in both the 10mWT and the F8WT, with no correlations at all for these two parameters in the FST. It should be acknowledged that the proposed clinical scales do not consider tests in which the visual input is removed: this could be one of the reasons why only few correlations have been found when considering the FST. These results highlight the lack of specificity that some clinical scales exhibit [68], while confirming their ability to determine whether or not a patient has a motor impairments. Therefore, the integration of traditional scales and technology-based protocols could assist with improving current clinical routines and with designing rehabilitation treatments, helping to bringing more sensitive, specific, and responsive motor tasks to clinical practice.

Despite the promising results, this study presents some limitations: the main limitation is the heterogeneity of the sample, mainly due to the severities and the locations of the brain injuries. Increasing the sample would likely lead to reduce the heterogeneity of the sample. Furthermore, the relationship between gait characteristics and specific neurological deficits, such as post-traumatic parkinsonism or cerebellar syndromes, and the presence of possible cognitive and behavioral sequelae of sTBI, were not investigated. Although such analyses were beyond the scope of the present study, they could be considered in further studies, in order to obtain more detailed information and better discriminate among people suffering from sTBI.

5. Conclusions

People who suffer a sTBI often complain of balance and gait impairments, but despite the evidence that neuromotor deficits are a common consequence of a sTBI, the existing literature does not adequately describe balance strategies adopted by sTBI survivors. This lack of information depends on various factors: the heterogeneity and severity of the brain damage, the patient's age, and the presence of pre-morbid/co-morbid conditions are the most significant. Furthermore, subtle cognitive functioning deficits, such as executive functions, which are detectable even in persons with good recovery after sTBI [7,63], may interfere with dynamic performances.

The main contribution of the present work is represented by the analysis of gait stability, symmetry, and smoothness indices which objectively describe gait quality in patients with sTBI. Specifically, the lack of ability of both severe and very severe TBI patients to stabilize their head by attenuating body accelerations may have a big impact. In fact, the vestibular system is located at head level; therefore, a high head acceleration could be critical for the planning of adaptive motor strategies.

The data reported herein suggest the appropriateness of an integrated assessment using both clinical scales and wearable sensors to objectively evaluate gait and balance impairments during different dynamic tasks. This integrated approach may be useful to assessing the measures of changes during rehabilitation training aimed at improving patients' gait quality and limiting the risk of falling, supporting rehabilitative staff with designing effective and tailored interventions.

Author Contributions: Conceptualization, R.F., G.V., M.G.B., and M.T.; methodology, V.B., E.B., G.V., and G.A.; software, E.B., formal analysis, V.B. and E.B.; investigation, V.B. and A.O.B.; resources, G.V., M.T., M.G.B., and R.F.; data curation, V.B. and E.B.; writing—original draft preparation, V.B.; writing—review and editing, all authors; visualization, V.B.; supervision, G.V., M.T., and M.G.B.; project administration, G.V. and R.F.

Funding: This research received no external funding.

Conflicts of Interest: The authors declare no conflict of interest.

References

1. Teasdale, G.M. Head Injury. *J. Neurol.* **1995**, 145–147. [CrossRef] [PubMed]
2. Jennett, B. Epidemiology of head injury. *J. Neurol. Psychiatry* **1996**, 1–11. [CrossRef] [PubMed]
3. Kraus, J.; Sullivan, C.; Bowers, S.; Knowlton, S.; Marshall, L. The incidence of acute brain injury and serious impairment in a defined population. *Am. J. Epidemiol.* **1984**, *119*, 186–201. [CrossRef] [PubMed]
4. Popescu, C.; Anghelescu, A.; Daia, C.; Onose, G. Actual data on epidemiological evolution and prevention endeavours regarding traumatic brain injury. *J. Med. Life* **2015**, *8*, 272–277. [PubMed]
5. Ponsford, J.L.; Spitz, G.; Cromarty, F.; Gifford, D.; Attwood, D. Costs of Care after Traumatic Brain Injury. *J. Neurotrauma* **2013**, *30*, 1498–1505. [CrossRef]

6. Teasdale, G.; Maas, A.; Lecky, F.; Manley, G.; Stocchetti, N.; Murray, G. The Glasgow Coma Scale at 40 years: Standing the test of time. *Lancet Neurol.* **2014**, *13*, 844–854. [CrossRef]
7. Ciurli, P.; Bivona, U.; Barba, C.; Onder, G.; Silvestro, D.; Azicnuda, E.; Rigon, J.; Formisano, R. Metacognitive unawareness correlates with executive function impairment after severe traumatic brain injury. *J. Int. Neuropsychol. Soc.* **2010**, *16*, 360–368. [CrossRef]
8. Bivona, U.; Costa, A.; Contrada, M.; Silvestro, D.; Azicnuda, E.; Aloisi, M.; Catania, G.; Ciurli, P.; Guariglia, C.; Caltagirone, C.; et al. Depression, apathy and impaired self-awareness following severe traumatic brain injury: A preliminary investigation. *Brain Inj.* **2019**, *33*, 1245–1256. [CrossRef]
9. Bivona, U.; Formisano, R.; De Laurentiis, S.; Accetta, N.; Rita Di Cosimo, M.; Massicci, R.; Ciurli, P.; Azicnuda, E.; Silvestro, D.; Sabatini, U.; et al. Theory of mind impairment after severe traumatic brain injury and its relationship with caregivers' quality of life. *Restor. Neurol. Neurosci.* **2015**, *33*, 335–345. [CrossRef]
10. Mathias, J.L.; Wheaton, P. Changes in attention and information-processing speed following severe traumatic brain injury: A meta-analytic review. *Neuropsychology* **2007**, *21*, 212–223. [CrossRef]
11. Miotto, E.C.; Cinalli, F.Z.; Serrao, V.T.; Benute, G.G.; Lucia, M.C.S.; Scaff, M. Cognitive deficits in patients with mild to moderate traumatic brain injury. *Arq. Neuropsiquiatr.* **2010**, *68*, 862–868. [CrossRef]
12. Fleminger, S. Long-term psychiatric disorders after traumatic brain injury. *Eur. J. Anaesthesiol.* **2008**, *25*, 123–130. [CrossRef]
13. Bergamini, E.; Iosa, M.; Belluscio, V.; Morone, G.; Tramontano, M.; Vannozzi, G. Multi-sensor assessment of dynamic balance during gait in patients with subacute stroke. *J. Biomech.* **2017**, *61*, 208–215. [CrossRef]
14. Chen, G.; Patten, C.; Kothari, D.H.; Zajac, F.E. Gait differences between individuals with post-stroke hemiparesis and non-disabled controls at matched speeds. *Gait Posture* **2005**, *22*, 51–56. [CrossRef]
15. Iosa, M.; Fusco, A.; Morone, G.; Pratesi, L.; Coiro, P.; Venturiero, V.; De Angelis, D.; Bragoni, M.; Paolucci, S. Assessment of upper-body dynamic stability during walking in patients with subacute stroke. *J. Rehabil. Res. Dev.* **2012**, *49*, 439. [CrossRef]
16. Mancini, M.; King, L.A.; Salarian, A.; Holmstrom, L.; McNames, J.; Horak, F.B. Mobility Lab to Assess Balance and Gait with Synchronized Body-worn Sensors. *J. Bioeng. Biomed. Sci.* **2014**, 1–15. [CrossRef]
17. Patterson, K.K.; Parafianowicz, I.; Danells, C.J.; Closson, V.; Verrier, M.C.; Staines, W.R.; Black, S.E.; McIlroy, W.E. Gait Asymmetry in Community-Ambulating Stroke Survivors. *Arch. Phys. Med. Rehabil.* **2008**, *89*, 304–310. [CrossRef]
18. Schlachetzki, J.C.M.; Aminian, K.; Klucken, J.; Kohl, Z.; Marxreiter, F.; Barth, J.; Reinfelder, S.; Gassner, H.; Eskofier, B.M.; Winkler, J.; et al. Wearable sensors objectively measure gait parameters in Parkinson's disease. *PLoS ONE* **2017**, *12*. [CrossRef]
19. Rinne, M.B.; Pasanen, M.E.; Vartiainen, M.V.; Lehto, T.M.; Sarajuuri, J.M.; Alaranta, H.T. Motor performance in physically well-recovered men with traumatic brain injury. *J. Rehabil. Med.* **2006**, *38*, 224–229. [CrossRef]
20. Williams, G.; Galna, B.; Morris, M.E.; John, O. Spatio-Temporal Deficits and Kinematic Classification of Gait following a Traumatic Brain Injury: A systematic review. *J. Head Trauma Rehabil.* **2010**, *25*, 366–374. [CrossRef]
21. Formisano, R.; Lucia, F.S.; Birbamer, G.; Gerstenbrand, F. Post-Traumatic Cerebellar Syndrome. *New Trends Clin. Neuropharmacol.* **1987**, *1*, 115–118.
22. Formisano, R.; Zasler, N.D. Posttraumatic parkinsonism. *J. Head Trauma Rehabil.* **2014**, *29*, 387–390. [CrossRef]
23. Lehmann, J.; Sherlyn, B.; Price, R.; Burleigh, A.; Hertling, D. Quantitative Evaluation of Sway as an Indicator of Functional Balance in Post-Traumatic Brain Injury. *Arch. Phys. Med. Rehabil.* **1990**, *93*, 2012–2013.
24. Wade, L.D.; Canning, C.G.; Fowler, V.; Felmingham, K.L.; Baguley, I.J. Changes in postural sway and performance of functional tasks during rehabilitation after traumatic brain injury. *Arch. Phys. Med. Rehabil.* **1997**, *78*, 1107–1111. [CrossRef]
25. Geurts, A.C.H.; Ribbers, G.M.; Knoop, J.A.; Van Limbeek, J. Identification of static and dynamic postural instability following traumatic brain injury. *Arch. Phys. Med. Rehabil.* **1996**, *77*, 639–644. [CrossRef]
26. Chou, L.S.; Kaufman, K.R.; Walker-Rabatin, A.E.; Brey, R.H.; Basford, J.R. Dynamic instability during obstacle crossing following traumatic brain injury. *Gait Posture* **2004**, *20*, 245–254. [CrossRef]
27. Ochi, F. Temporal-Spatial Feature of Gait after Traumatic Brain Injury. *J. Head Trauma Rehabil.* **1999**, *14*, 105–115. [CrossRef]

28. Basford, J.R.; Chou, L.S.; Kaufman, K.R.; Brey, R.H.; Walker, A.; Malec, J.F.; Moessner, A.M.; Brown, A.W. An assessment of gait and balance deficits after traumatic brain injury. *Arch. Phys. Med. Rehabil.* **2003**, *84*, 343–349. [CrossRef]
29. Kavanagh, J.J.; Menz, H.B. Accelerometry: A technique for quantifying movement patterns during walking. *Gait Posture* **2008**, *28*, 1–15. [CrossRef]
30. Buckley, C.; Galna, B.; Rochester, L.; Mazzà, C. Upper body accelerations as a biomarker of gait impairment in the early stages of Parkinson's disease. *Gait Posture* **2019**, *71*, 289–295. [CrossRef]
31. Shany, T.; Redmond, S.J.; Marschollek, M.; Lovell, N.H. Assessing fall risk using wearable sensors: A practical discussion. *Z. Gerontol. Geriatr.* **2012**, *45*, 694–706. [CrossRef]
32. Buckley, C.; Brook, G.; Lynn, R.; Claudia, M. Attenuation of Upper Body Accelerations during Gait: Piloting an Innovative Assessment Tool for Parkinson's Disease. *Biomed Res. Int.* **2015**, *2015*. [CrossRef]
33. Belluscio, V.; Bergamini, E.; Salatino, G.; Marro, T.; Gentili, P.; Iosa, M.; Morelli, D.; Vannozzi, G. Dynamic balance assessment during gait in children with Down and Prader-Willi syndromes using inertial sensors. *Hum. Mov. Sci.* **2019**, *63*, 53–61. [CrossRef]
34. Summa, A.; Vannozzi, G.; Bergamini, E.; Iosa, M.; Morelli, D.; Cappozzo, A. Multilevel upper body movement control during gait in children with cerebral palsy. *PLoS ONE* **2016**, *11*, 1–13. [CrossRef]
35. Shumway-Cook, A.; Baldwin, M.; Polissar, N.L.; Gruber, W. Predicting the probability for falls in community-dwelling older adults. *Phys. Ther.* **1997**, *77*, 812–819. [CrossRef]
36. Esser, P.; Dawes, H.; Collett, J.; Feltham, M.G.; Howells, K. Assessment of spatio-temporal gait parameters using inertial measurement units in neurological populations. *Gait Posture* **2011**, *34*, 558–560. [CrossRef]
37. Haggard, P.; Cockburn, J.; Cock, J.; Fordham, C.; Wade, D. Interference between gait and cognitive tasks in a rehabilitating neurological population. *J. Neurol. Neurosurg. Psychiatry* **2000**, *69*, 479–486. [CrossRef]
38. Iosa, M.; De Sanctis, M.; Summa, A.; Bergamini, E.; Morelli, D.; Vannozzi, G. Usefulness of magnetoinertial wearable devices in neurorehabilitation of children with cerebral palsy. *Appl. Bionics Biomech.* **2018**, *2018*. [CrossRef]
39. Cohen, J. Statistical power analysis. *Int. Encycl. Educ.* **1992**, *1*, 98–101. [CrossRef]
40. Levin, H.S.; O'Donnel, V.M.; Grossaman, R.G. The Galveston Orientation and Amnesia Test. *J. Nerv. Ment. Dis.* **1979**, *167*, 675–684. [CrossRef]
41. Herman, T.; Inbar-borovsky, N.; Brozgol, M.; Giladi, N.; Hausdorff, J.M. The Dynamic Gait Index in Healthy Older Adults: The Role of StairClimbing, Fear of Falling and Gender. *Gait Posture* **2009**, *29*, 237–241. [CrossRef]
42. Berg, W.P.; Alessio, H.M.; Mills, E.M.; Tong, C. Circumstances and consequences of falls in independent community-dwelling older adults. *Age Ageing* **1997**, *26*, 261–268. [CrossRef]
43. Howe, J.A.; Inness, E.L.; Venturini, A.; Williams, J.I.; Verrier, M.C. The Community Balance and Mobility Scale—A balance measure for individuals with traumatic brain injury. *Clin. Rehabil.* **2006**, *20*, 885–895. [CrossRef]
44. Van Hedel, H.J.; Wirz, M.; Dietz, V. Assessing walking ability in subjects with spinal cord injury: Validity and reliability of 3 walking tests. *Arch. Phys. Med. Rehabil.* **2005**, *86*, 190–196. [CrossRef]
45. Duncan, P.W.; Studenski, S.; Richards, L.; Gollub, S.; Lai, S.M.; Reker, D.; Perera, S.; Yates, J.; Koch, V.; Rigler, S.; et al. Randomized clinical trial of therapeutic exercise in subacute stroke. *Stroke* **2003**, *34*, 2173–2180. [CrossRef]
46. Turcato, A.M.; Godi, M.; Giardini, M.; Arcolin, I.; Nardone, A.; Giordano, A.; Schieppati, M. Abnormal gait pattern emerges during curved trajectories in high-functioning Parkinsonian patients walking in line at normal speed. *PLoS ONE* **2018**, *13*, 1–26. [CrossRef]
47. Zhang, Y.B.; Wang, W.Q. Reliability of the Fukuda stepping test to determine the side of vestibular dysfunction. *J. Int. Med. Res.* **2011**, *39*, 1432–1437. [CrossRef]
48. Belluscio, V.; Bergamini, E.; Iosa, M.; Tramontano, M.; Morone, G.; Vannozzi, G. The iFST: An instrumented version of the Fukuda Stepping Test for balance assessment. *Gait Posture* **2018**, *60*, 203–208. [CrossRef]
49. Bergamini, E.; Ligorio, G.; Summa, A.; Vannozzi, G.; Cappozzo, A.; Sabatini, A.M. Estimating orientation using magnetic and inertial sensors and different sensor fusion approaches: Accuracy assessment in manual and locomotion tasks. *Sensors (Switzerland)* **2014**, *14*, 18625–18649. [CrossRef]
50. Mazzà, C.; Iosa, M.; Pecoraro, F.; Cappozzo, A. Control of the upper body accelerations in young and elderly women during level walking. *J. Neuroeng. Rehabil.* **2008**, *5*, 30. [CrossRef]

51. Pasciuto, I.; Bergamini, E.; Iosa, M.; Vannozzi, G.; Cappozzo, A. Overcoming the limitations of the Harmonic Ratio for the reliable assessment of gait symmetry. *J. Biomech.* **2017**, *53*, 84–89. [CrossRef]
52. Balasubramanian, S.; Melendez-Calderon, A.; Roby-Brami, A.; Burdet, E. On the analysis of movement smoothness. *J. Neuroeng. Rehabil.* **2015**, *12*, 1–11. [CrossRef]
53. Fang, X.; Liu, C.; Jiang, Z. Reference values of gait using APDM movement monitoring inertial sensor system. *R. Soc. Open Sci.* **2018**, *5*. [CrossRef]
54. Peppe, A.; Stanzione, P.; Pierantozzi, M.; Semprini, R.; Bassi, A.; Santilli, A.M.; Formisano, R.; Piccolino, M.; Bernardi, G. Does pattern electroretinogram spatial tuning alteration in Parkinson's disease depend on motor disturbances or retinal dopaminergic loss? *Electroencephalogr. Clin. Neurophysiol.* **1998**, *106*, 374–382. [CrossRef]
55. Formisano, R.; D'Ippolito, M.; Risetti, M.; Riccio, A.; Caravasso, C.F.; Catani, S.; Rizza, F.; Forcina, A.; Buzzi, M.G. Vegetative state, minimally conscious state, akinetic mutism and parkinsonism as a continuum of recovery from disorders of consciousness: An exploratory and preliminary study. *Funct. Neurol.* **2011**, *26*, 15–24.
56. Tramontano, M.; Morone, G.; Curcio, A.; Temperoni, G.; Medici, A.; Morelli, D.; Caltagirone, C.; Paolucci, S.; Iosa, M. Maintaining gait stability during dual walking task: Effects of age and neurological disorders. *Eur. J. Phys. Rehabil. Med.* **2017**, *53*, 7–13.
57. Galna, B.; Barry, G.; Jackson, D.; Mhiripiri, D.; Olivier, P.; Rochester, L. Accuracy of the Microsoft Kinect sensor for measuring movement in people with Parkinson's disease. *Gait Posture* **2014**, *39*, 1062–1068. [CrossRef]
58. Boonstra, A.M.; Schiphorst, H.R.; Reneman, M.F.; Posthumus, J.B.; Stewart, R.E. Reliability and validity of the visual analogue scale for disability in patients with chronic musculoskeletal pain. *Int. J. Rehabil. Res.* **2008**, *31*, 165–169. [CrossRef]
59. Iosa, M.; Marro, T.; Paolucci, S.; Morelli, D. Stability and harmony of gait in children with cerebral palsy. *Res. Dev. Disabil.* **2012**, *33*, 129–135. [CrossRef]
60. Doi, T.; Hirata, S.; Ono, R.; Tsutsumimoto, K.; Misu, S.; Ando, H. The harmonic ratio of trunk acceleration predicts falling among older people: Results of a 1-year prospective study. *J. Neuroeng. Rehabil.* **2013**, *10*, 7. [CrossRef]
61. Menz, H.B.; Lord, S.R.; Fitzpatrick, R.C. Acceleration patterns of the head and pelvis when walking on level and irregular surfaces. *Gait Posture* **2003**, *18*, 35–46. [CrossRef]
62. Wong, S.S.T.; Yam, M.S.; Ng, S.S.M. The Figure-of-Eight Walk test: Reliability and associations with stroke-specific impairments. *Disabil. Rehabil.* **2013**, *35*, 1896–1902. [CrossRef] [PubMed]
63. Di Russo, F.; Incoccia, C.; Formisano, R.; Sabatini, U.; Zoccolotti, P. Abnormal motor preparation in severe traumatic brain injury with good recovery. *J. Neurotrauma* **2005**, *22*, 297–312. [CrossRef] [PubMed]
64. Foley, J.A.; Cantagallo, A.; Della Sala, S.; Logie, R.H. Dual task performance and post traumatic brain injury. *Brain Inj.* **2010**, *24*, 851–858. [CrossRef] [PubMed]
65. Baker, K.; Rochester, L.; Nieuwboer, A. The Immediate Effect of Attentional, Auditory, and a Combined Cue Strategy on Gait During Single and Dual Tasks in Parkinson's Disease. *Arch. Phys. Med. Rehabil.* **2007**, *88*, 1593–1600. [CrossRef] [PubMed]
66. Paquet, N.; Taillon-Hobson, A.; Lajoie, Y. Fukuda and Babinski-Weil tests: Within-subject variability and test-retest reliability in nondisabled adults. *J. Rehabil. Res. Dev.* **2014**, *51*, 1013–1022. [CrossRef] [PubMed]
67. Akin, F.W.; Murnane, O.D.; Hall, C.D.; Riska, K.M. Vestibular consequences of mild traumatic brain injury and blast exposure: A review. *Brain Inj.* **2017**, *31*, 1188–1194. [CrossRef]
68. Mancini, M.; Horak, F.B. The relevance of clinical balance assessment tools to differentiate balance deficits. *Eur. J. Phys. Rehabil. Med.* **2010**, *46*, 239–248.

Article

Wearable Electronics Assess the Effectiveness of Transcranial Direct Current Stimulation on Balance and Gait in Parkinson's Disease Patients

Mariachiara Ricci [1], Giulia Di Lazzaro [2], Antonio Pisani [2], Simona Scalise [2], Mohammad Alwardat [2], Chiara Salimei [2], Franco Giannini [1] and Giovanni Saggio [1,*]

1 Department of Electronic Engineering, University of Rome "Tor Vergata", 00133 Rome, Italy; rccmch01@uniroma2.it (M.R.); giannini@ing.uniroma2.it (F.G.)
2 Department of Systems Medicine, University of Rome "Tor Vergata", 00133 Rome, Italy; giulia.dilazzaro@students.uniroma2.eu (G.D.L.); pisani@uniroma2.it (A.P.); simona.scalise@alumni.uniroma2.eu (S.S.); mohammadsamimohammad.alwardat.alwardatmohammad01@alumni.uniroma2.eu (M.A.); chiara.salimei@alumni.uniroma2.eu (C.S.)
* Correspondence: saggio@uniroma2.it

Received: 25 October 2019; Accepted: 8 December 2019; Published: 11 December 2019

Abstract: Currently, clinical evaluation represents the primary outcome measure in Parkinson's disease (PD). However, clinical evaluation may underscore some subtle motor impairments, hidden from the visual inspection of examiners. Technology-based objective measures are more frequently utilized to assess motor performance and objectively measure motor dysfunction. Gait and balance impairments, frequent complications in later disease stages, are poorly responsive to classic dopamine-replacement therapy. Although recent findings suggest that transcranial direct current stimulation (tDCS) can have a role in improving motor skills, there is scarce evidence for this, especially considering the difficulty to objectively assess motor function. Therefore, we used wearable electronics to measure motor abilities, and further evaluated the gait and balance features of 10 PD patients, before and (three days and one month) after the tDCS. To assess patients' abilities, we adopted six motor tasks, obtaining 72 meaningful motor features. According to the obtained results, wearable electronics demonstrated to be a valuable tool to measure the treatment response. Meanwhile the improvements from tDCS on gait and balance abilities of PD patients demonstrated to be generally partial and selective.

Keywords: balance; gait; Parkinson's disease; transcranial direct current stimulation; wearable electronics; IMUs

1. Introduction

Wearable electronics are gaining increasing attention and importance as a valid tool for healthcare practitioners in medical treatment [1–3] and patient monitoring [4–6]. In particular, wearable sensors have been applied for assessing the motor performance of patients with neurodegenerative disorders, as it is for Parkinson's disease, in both home and clinical environments [7–12].

Parkinson's disease (PD) can be characterized by motor deficiencies, such as bradykinesia and a combination of rest tremor, rigidity, as well as gait and balance impairment [13]. In routine clinical care, the evaluation of those deficiencies is mainly based on severity-rating standardized scales, such as the Movement Disorder Society Unified Parkinson's disease rating scale (MDS-UPDRS) [14], based on patients' reports and clinicians' vision-based evaluations, and clinical investigators determine the effectiveness of a therapy of a drug by using the MDS-UPDRS score [15]. Inconveniently, patient reports can be affected by mood and unfamiliarity with forms, and clinicians' evaluations can be biased by personal beliefs, experiences, and a priori expectations, resulting in inter- and intra-rater score variability [15,16]. Furthermore, the MDS-UPDRS is quantified according to a discrete scale (0–4, unity step) only, and the human eyes of clinicians hardly detect subtle motor changes during the monitoring of patients. These limitations compel investigators to employ more rigorous, and thus costly, clinical trial designs, with a random assignment of patients, thus blinding investigators to treatment assignment.

The aforementioned limitations can be in some way reduced or overcome through the use of wearable inertial sensors (hereafter wearables), which provide measures of human postures and kinematics, paving the way for objective assessment in clinical trials [17]. In fact, wearables can gather motion parameters in a continuous (analog) or high-step density (digital) scale, and avoid intra- and inter-rater variability, thereby reducing the sample size and simplifying the assessment of the patients, objectively quantifying a possible beneficial effect of a therapeutic intervention. For this reason, even if wearables are still poorly used (only 2.7% of ongoing clinical trials [15]), there is growing attention given to this technological tool, and some pharmaceutical companies are working to develop their own devices [18–20].

Our work approaches the utilization of wearables in the particular case of objectively demonstrating the therapeutic beneficial effects, if any, of transcranial direct current stimulation (tDCS) treatment on the motor impairments of patients affected by Parkinson's disease.

The proven appeal of tDCS is evident as it is a non-invasive, inexpensive, painless brain stimulation technique with many clinical and research applications, ranging from the treatment of depression to neurorehabilitation [21,22]. It consists of applying a direct positive (anodal) or negative (cathodal) 1–2 mA current to the scalp. This stimulation supports the depolarization or hyperpolarization of neurons, thus leading them closer to, or farther away from firing, acting on synaptic transmission or synaptic plasticity [21,23]. Further, tDCS has been used alternatively to (or sometimes concurrently with) dopaminergic drug therapy, because the latter can lose its efficacy during the natural course of the disease, in particular regarding its benefit on postural and gait disorders. Gait is now considered a higher level of cognitive function that involves the integration of attention, planning, memory and other motor, perceptual and cognitive processes. In fact, walking and balance constitute a combination of automatic movement processes, afferent information processing, and intentional adjustments that require a delicate balance between various interacting neuronal systems. In PD, to compensate the loss of motor task, cognitive resources as attention and executive function performed by the dorsolateral pre-frontal cortex (DLPFC) plays a critical role in the relief of gait disorder [24]. In addition, previous studies have shown that anodal tDCS stimulation to either the motor area (M1) or dorsolateral prefrontal cortex (DLPFC) had a significant impact on the motor, non-motor, and balance functional outcomes in PD patients. In fact, brain activation patterns in M1 and DLPFC are extremely involved in successful locomotion performance in patients with PD [21,25–27]. Further, the effectiveness of tDCS for alleviating gait and postural instability seems promising [28–31], however, evidence of its benefit remains unclear and controversial [23,32] because different tDCS protocols and target areas of scalp have been considered, leading to conflicting evidence on MDS-UPDRS scores [23,28].

Our work aims to objectively quantify the motor performance improvements, if any, due to tDCS treatment in a population of patients with PD and gait disturbances. To this aim, we used wearables to measure specific motor tasks, and analyzed the related results by means of the standardized response mean (SRM) index, comparing them with those obtained by the clinical evaluation.

2. Materials and Methods

2.1. Subjects

Ten PD patients (Table 1) with postural and gait disturbances were recruited at Tor Vergata University Hospital, Rome, Italy. Idiopathic PD was diagnosed according to the MDS clinical diagnostic criteria for PD [13], and patients were enrolled at Hoehn & Yahr disease stages between 1.5

and 4, and with MDS-UPDRS III scores related to a gait higher than 1. Exclusion criteria were age (younger than 30 or older than 85), dementia (mini mental status evaluation, MMSE, score < 24 [33]), therapy changes in the last three months, orthopedic comorbidities, other neurological disorders, and therapy with drugs possibly interfering with motor function (e.g., antipsychotics).

Table 1. Patients' information.

Age	77.2 ± 6.3 y
Gender	7 M, 3 F
Disease duration	10.37 ± 3.8 y
MDS-UPDRS II	15.6 ± 3.66
MDS-UPDRS III	35.2 ± 5.63
Hoehn & Yahr	2.9 ± 0.16
Levodopa equivalent daily dose	771.7 ± 213.58 mg

This study was conducted in agreement with the ethical principles of the Helsinki declaration. Informed consent was obtained from each participant and ethical approval was obtained by the local committee (RS 190/18). Patients consented to participate and did not change the therapy during the study, from T0 to T2 (Figure 2), in order to minimize any alteration of motor performance due to dopaminergic therapy variations.

2.2. Motor Tests

We requested each participant to perform six motor tasks which, according to clinical standards, are relevant for a comprehensive evaluation of balance and gait. Tasks included stance feet together (SFT), tandem stance (TS), the pull test (PT), timed up and go test (TUG), stop and go test (S&G), and narrow walking test (NW). In particular, SFT and TS are useful to test balance; PT corresponds to the item 3.12 of MDS-UPDRS III to test postural response; TUG, S&G and NW are used to assess mobility and gait. Wearables were placed by means of Velcro strips on segments of the body, according to the particular test, as schematized in Figure 1. The descriptions of the tests and corresponding placements of the wearable sensors are specified in the following.

2.2.1. Stance Feet Together (SFT) and Tandem Stance (TS)

In SFT and TS tests, the patient has to stand and maintain the posture for 30 s. More particularly, in the SFT with feet side-by-side and close together, in TS with feet in tandem position (i.e., one ahead, aligned and close to the other). The wearables were placed on the posterior trunk at the level of T5 and on the external parts of the calf segments of both legs.

2.2.2. Pull Test (PT)

The subject, comfortably standing upright with shoulders to the examiner, is rapidly and vigorously pushed backward on his/her shoulders so as to be forced to make one, or more, steps backwards, recovering his/her balance. The sensors were placed as for SFT and TS.

2.2.3. Timed Up and Go (TUG)

The subject starts seated on a straight-backed chair with arms across the chest, then gets up, walks straight 6 m, turns around, walks straight back and, turning on his/her-self, sits down returning to the initial condition. The sensors were placed on the patient's pelvis at the level of L5, posterior trunk at the level of T5, on the external parts of thighs and calf segments of both lower limbs, arms, and forearms.

2.2.4. Stop and Go (S&G)

The subject walks for six meters in a straight line, turns around, walks six meters back while the examiner tells him/her to stop and go for 6 times. The sensors were placed on the patient's pelvis at L5 level, posterior trunk at T5 level, on the external parts of thighs and calf segments of both lower limbs. The time, when the examiner tells the patient to stop was recorded.

2.2.5. Narrow Walking (NW)

The subject walks 6 m straight, but passing through a 70 cm narrow door in the middle of the path. The sensors were placed on the patient's pelvis at L5 level, posterior trunk at T5 level, on the external parts of thighs, and calf segments of both lower limbs. The time, the time when the patient passes through the door was recorded.

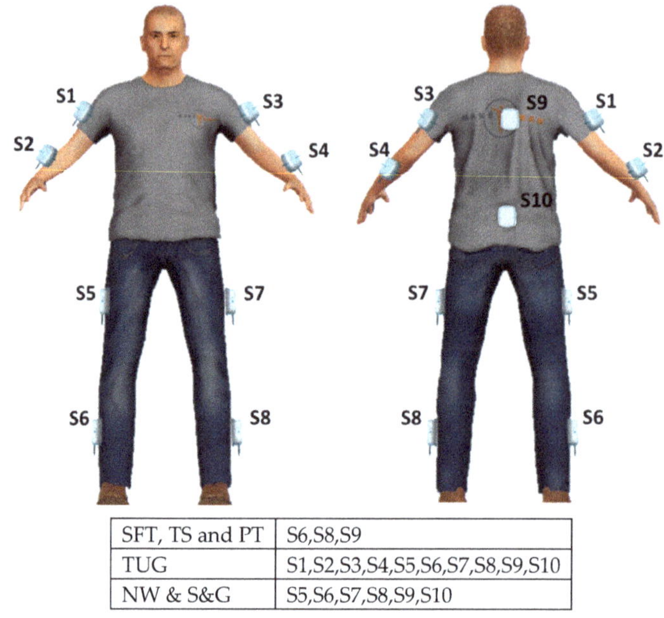

SFT, TS and PT	S6,S8,S9
TUG	S1,S2,S3,S4,S5,S6,S7,S8,S9,S10
NW & S&G	S5,S6,S7,S8,S9,S10

Figure 1. Sensors, labeled from S1 to S10, as located on the body of the patients. Different motor tests resulted with a different number of used sensors.

2.3. *tDCS Stimulation*

Direct current (DC) was delivered to stimulate the left dorsolateral-prefrontal cortex (DLPFC) by means of a tDCS low-intensity stimulator (BrainStim, EMS Srl, Bologna, Italy). Two saline-soaked electrodes (35 cm^2) were placed on F4 (according to the 10–20 international EEG nomenclature) and on the right forearm, respectively. The stimulation was of 2mA DC (0.057 mA/cm^2 in density) delivered for 20 min (30 s step-up ramp, 30 s step-down ramp), repeated ten times, obtaining one session/day, for five consecutive days. Such a stimulation session was followed by two non-stimulation days, and again by another five days of long stimulation (Figure 2). During each tDCS application, patients were at rest without any concurrent motor tasks.

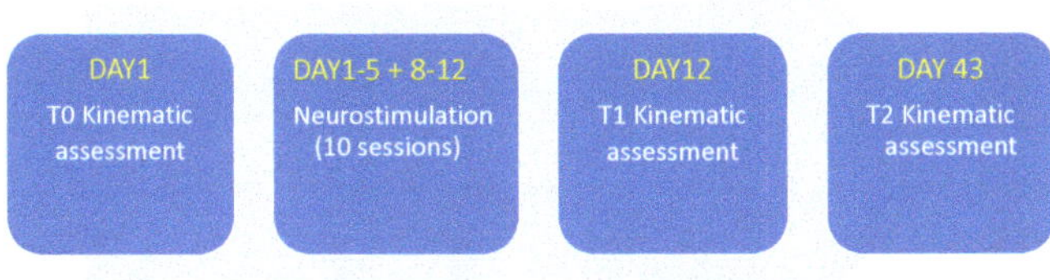

Figure 2. Flow diagram showing the study design and stimulation protocol.

2.4. Wearable Electronics

Different technologies can furnish data in terms of gait and balance performances. We can refer, for instance, to pressure sensors embedded into the floor and electro-goniometers, etc., with the optical-based systems considered as the gold standard because of their high accuracy. However, optical-based systems have some important drawbacks, such as the necessities of a free line of sight, time-consuming calibration procedures, necessity of skilled personnel and, above all, a very high cost. Wearable electronics have none of those drawbacks, and have been demonstrated to perform with the appropriate accuracy for our purposes [34,35].

Wearable electronics constitute a network of validated inertial measurement units (IMUs) termed Movit (by Captiks Srl, Rome Italy) [7,34,35], each housing a 3-axis accelerometer (±8 g) and a 3-axis gyroscope (±2000°/s), synchronized to a personal computer receiver, with a 50 Hz data transfer rate. A proprietary application, termed Motion Studio, processes and stores data.

The number of used IMUs and the position of patients' bodies (by means of elastic bands) varied according to the particular motor tasks performed. Measured data consist of accelerations, angular velocities, and joint angles, computed from the related quaternions via Euler decomposition. In turn, the quaternions are generated using a Kalman filter on data coming from the accelerometers and the gyroscopes, sampled at 200 Hz. By means of a patented calibration procedure, the spatial orientations of the dressed IMUs are represented on a computer screen as a human avatar, which replicates patient movements, with his/her joint angles gathered with a forward kinematic procedure in a parent-child hierarchy

2.5. Features

For each task, we obtained several features, as reported in Table 2 and described in the following paragraphs.

2.5.1. Stance Feet Together (SFT) and Tandem Stance (TS)

Eleven features from the sensor located on the trunk were taken into consideration: range of accelerations, angular velocities and angles of the trunk in the medial-lateral (ML), anterior-posterior (AP) and vertical (V) directions; Jerk and Sway Area. In particular, Jerk, gathered from the accelerometers, represents the time derivative of acceleration [36], and is used as an empirical measure of the smoothness of the movements [37,38]. The Sway Area is the area of the ellipse that encompasses 95% of the values of medial lateral and anterior posterior accelerations around their mean values.

2.5.2. Pull Test (PT)

The PT test is useful to evaluate the postural responses to an unexpected external perturbation. We extracted the 11 features as for the SFT, plus the number of steps following the pushing as resulted from data gathered by the sensors placed on the ankles.

2.5.3. Time Up and Go (TUG)

TUG is one of the most widely used clinical tests and allows for the assessment of several aspects of gait. Parkinsonian gait is characterized by a slowed speed, decreased arm swing, shuffling steps, and difficulty to turn [39]. TUG is composed by four phases: the sit-to-stand phase (patient gets up from the sitting position with arms across the chest), the walking phase (patient walks for 6 m forth and back), the turning phase (the patient turns 180°), and the turn-to-sit phase (the patient turns and sit back on the chair). Each phase is segmented considering data gathered by the IMU on the trunk. We detected the sit-to-stand and turn-to-sit phases considering the interval between the two local minimum values before and after a local maximum of the accelerometer data, in the AP direction, corresponding to the flexion/extension movement of trunk. The turning phase is identified using thresholds on the trunk angle in the vertical direction (the turning component looks as a positive or negative ramp, depending on the direction of the turn). Further details on the segmentation of TUG test are reported in [7].

From these segmentations, 24 features were computer, as described in Table 2, including:

1. Temporal gait characteristics, such as number of steps, step duration, stance duration and swing duration;
2. Features related to upper and lower limb movements, such as the range of motion of arms and legs (Flex Arm, Flex Leg), the average angular velocity (Average Vel) of arms, forearms, legs and thighs, and the asymmetry between right and left limbs (Asym Arm, Asym Leg);
3. Turning parameters, such as the angular velocity of the trunk (Peak Turning Vel), the turning velocity (Turning Vel) and the number of steps (Steps Turning).

2.5.4. Stop and Go (S&G) & Narrow Walking (NW)

Parkinsonian gait problems are often triggered by some circumstances such as spaces with a narrow passage (e.g., a door), unexpected visual or auditory stimuli, stressful situations, cognitive load anxiety and difficulty in starting and stopping [39]. The results are a decreasing step length and step time, decreasing velocity, and increasing variability of step length and time [40,41]. The S&G and NW tests are used to provide evidence for these symptoms. We computed seven features for each task.

For the S&G test, we computed the duration of steps, stance and swing, as well as the angular velocity of the leg of the first steps at the beginning of gait, thus, after each stop signal of the examiner and the variability of the temporal step variables (CV Step, CV Stance, CV Swing).

For the NW test, we computed the same features but extracted them during the 3 s when the patient was passing through the door.

Table 2. Extracted Features from each motor test.

Task	Feature	Description
SFT, TS, PT	Jerk	Time derivative of acceleration in ML and AP directions [42]
	Sway Area	The ellipse that encompasses 95% of the values of ML and AP acceleration around their mean values [42]
	Range	The range of acceleration and angular velocity signals in all the three directions (6 features in total)
PT	# of Steps	The number of steps performed by the subject following the push
TUG	TUG phases duration	Include TUG time (duration of the entire test), sit-to-stand time, walk time, turning time and turn-to-sit time
	# of Steps	Number of steps during the walking phase.
	Gait metrics	Include mean and coefficient of variation of step duration, stance duration, and swing duration
	Flex Arm, Flex Leg	The angular flexion range of arms and legs
	Asym Arm, Asym Leg	Difference in angular flexion range between the faster and slower arm/leg divided by the larger value (lv%)
	Average Vel	The average angular velocity of arm, forearm and thigh along the medial lateral axis during the walking phase
	Turning Vel	The range of turning (180°) divided by turning time
	Peak Turning Vel	The maximum achieved angular velocity of the trunk rotation in the vertical axis during the turning phase
	Steps Turning	The number of steps during the turning phase
	Average Vel SitStand	The average angular velocity of trunk during sit-to-stand in in the anterior posterior plane
S&G	Gait metrics	Mean and coefficient of variation of duration of step, stance and swing computed on first four steps at the beginning of gait, after each stop signal of the examiner
	Step velocity	The angular velocity of legs computed on first four steps at the beginning of gait, after each stop signal of the examiner
NW	Gait metrics	Mean and coefficient of variation of duration of step, stance and swing computed on the 3 s time with patient passing through the door.
	Step velocity	The angular velocity of legs computed on the 3 s time with the patient passing through the door

2.6. Clinical and Wearables-Based Evaluations

Motor test performances of each of the ten PD patients just before the stimulation protocol (T0 time), just soon after the protocol (T1 time), and 1 month after (T2 time) were evaluated in order to quantify the effect of the tDCS and its persistence, if any.

The evaluations were performed both as standard clinical ones and by the analysis of data gathered through the wearable electronics.

All patients were evaluated by a movement disorder specialist, with general neurological examination, clinical tests, and questionnaires. Clinical tests consisted in the administration of MDS unified Parkinson's disease rating scale (MDS-UPDRS) and the Berg balance scale (BBS) [43], a clinical five-point ordinal scale that assess balance. Each patient was also evaluated with the freezing of gait questionnaire (FOG-Q) [44], a 6-item questionnaire used to assess gait disturbance severity in patients with PD, and the Hoehn and Yahr scale (H&Y) [45], a commonly used system for describing the progress of symptoms.

To evaluate the responsiveness of a treatment, we considered two aspects. First, we assessed the ability of wearable features to detect change over a particular time frame. Then, we evaluated the relationship between a change in the feature values and the external measure (e.g., the clinical score).

The standardized response mean (SRM) [46] was used to assess the responsiveness to the tDCS therapy. A reason for choosing SRM is because, differently from the paired t-test, it has no dependence on sample size [47]. The SRM expresses the ratio of TT:SDC, where TT is the mean change between T1 and T0 and between T2 and T1, and SDC the standard deviation of the change. Empirically, an SRM value of 0.20 represents a small, 0.50 a moderate, and 0.80 a large responsiveness, respectively.

We used Spearman's rank correlation coefficient to investigate the relation between the clinical scores and the features. Stance feet together (SFT) and tandem stance (TS) tasks were used to evaluate the static balance, assessed by the clinicians using the BBS scale. Features extracted from SFT and TS are compared with the BBS score. PT features were correlated to the corresponding UPDRS III item 3.12 score (PT is part of UPDRS III tasks). Features extracted from gait related tasks (TUG; ST and NW) were correlated with the UPDRS III gait item score (3.10). The significance level was set at 0.05.

3. Results

Table 3 shows the mean, standard deviation values, and SRM of the clinical evaluation results. Tables 4–9 report the motor features of SFT, TS, PT, TUG, S&G and NW tests, and correlation analysis between the features and the corresponding clinical evaluation.

Table 3. Clinical evaluation.

Clinical Evaluation	T0 Mean ± SD	T1 Mean ± SD	T2 Mean ± SD	SRM (T0 vs. T1)	SRM (T0 vs. T2)	SRM (T1 vs. T2)
MDS-UPDRS II	15.6 ± 3.67	13.9 ± 3.21	14.3 ± 3.23	−0.53	−0.42	0.23
MDS-UPDRS III	35.2 ± 5.64	30.5 ± 6.8	30.4 ± 3.47	−0.67	−1.15	−0.01
Gait item (3.10)	2.20 ± 0.60	1.60 ± 0.49	1.50 ± 0.50	−0.90	−0.90	−0.33
PT item (3.12)	1.80 ± 0.75	1.20 ± 0.75	1.60 ± 0.49	−0.65	−0.23	0.82
FOGQ	13.4 ± 3.69	12.5 ± 3.47	12.4 ± 2.11	−0.62	−0.33	−0.04
BBS	42.3 ± 12.35	47.2 ± 7.97	49.3 ± 6.96	0.79	0.82	0.50

Table 4. Stance feet together (SFT): feature values at T0, T1, T2; values of SRM comparing times; correlation with BBS score.

Feature (SFT)	T0 Mean ± SD	T1 Mean ± SD	T2 Mean ± SD	SRM (T0 vs. T1)	SRM (T0 vs. T2)	SRM (T1 vs. T2)	Correlation with BBS
Jerk	0.08 ± 0.03	0.07 ± 0.03	0.07 ± 0.03	−0.22	−0.72	−0.15	−0.38 *
Sway Area	0.32 ± 0.22	0.32 ± 0.3	0.32 ± 0.26	−0.01	−0.01	0.01	−0.22
Range Acc V	0.66 ± 0.53	0.61 ± 0.42	0.5 ± 0.34	−0.13	−0.53	−0.39	−0.60 *
Range Acc ML	0.56 ± 0.16	0.59 ± 0.28	0.59 ± 0.24	0.12	0.18	0.05	−0.46 *
Range Acc AP	0.99 ± 0.35	0.92 ± 0.38	0.9 ± 0.37	−0.14	−0.22	−0.07	−0.17
Range Gyr V	7.76 ± 3.45	10.71 ± 5.77	9.04 ± 5.13	0.53	0.24	−0.28	−0.48 *
Range Gyr ML	11.66 ± 5.78	10.95 ± 6.81	9.88 ± 4.87	−0.08	−0.32	−0.20	−0.55 *
Range Gyr AP	4.55 ± 2.35	5.05 ± 4.05	4.32 ± 2.49	0.13	−0.10	−0.27	−0.44 *

* p value < 0.05.

Table 5. Tandem stance (TS): features values at T0, T1, T2; values of SRM comparing times; correlation with BBS score.

Feature (TS)	T0 Mean ± SD	T1 Mean ± SD	T2 Mean ± SD	SRM (T0 vs. T1)	SRM (T0 vs. T2)	SRM (T1 vs. T2)	Correlation with BBS
Jerk	0.78 ± 1.32	0.21 ± 0.14	0.42 ± 0.45	−0.41	−0.33	0.44	−0.43 *
Sway Area	3.05 ± 4.44	1 ± 0.75	1.14 ± 1.28	−0.43	−0.41	0.13	−0.37 *
Range Acc V	2.79 ± 2.4	1.42 ± 1.36	1.32 ± 1.23	−0.47	−0.62	−0.11	−0.45 *
Range Acc ML	2.73 ± 2.45	1.73 ± 1.48	2.52 ± 2.45	−0.33	−0.08	0.29	−0.51 *
Range Acc AP	3.04 ± 2.88	1.87 ± 0.82	1.77 ± 1.19	−0.39	−0.43	−0.08	−0.35 *
Range Gyr V	40.01 ± 27.54	26.24 ± 12.28	40.06 ± 41.53	−0.42	0.00	0.35	−0.54 *
Range Gyr ML	58.97 ± 73.46	20.48 ± 13.44	29.28 ± 30.11	−0.47	−0.37	0.29	−0.53 *
Range Gyr AP	27.67 ± 26.98	14.68 ± 10.79	15.18 ± 10.01	−0.42	−0.46	0.04	−0.52 *

* p value < 0.05.

Table 6. Pull test (PT): feature values at T0, T1, T2; values of SRM comparing times; correlation with UPDRS item 3.12 (PT) score.

Feature (PT)	T0 Mean ± SD	T1 Mean ± SD	T2 Mean ± SD	SRM (T0 vs. T1)	SRM (T0 vs. T2)	SRM (T1 vs. T2)	Correlation with PT Item
Number of Steps	4.5 ± 1.8	4.2 ± 1.25	4.2 ± 2.27	−0.15	−0.09	0.00	−0.10
Jerk	11.03 ± 13.43	13.87 ± 18.82	13.64 ± 13.88	0.28	0.40	−0.02	−0.22
Sway Area	99.28 ± 140.03	83.99 ± 91.55	66.66 ± 63.16	−0.22	−0.38	−0.38	−0.27
Range Acc V	15.05 ± 6.23	14.92 ± 5.95	15.71 ± 5.95	−0.02	0.10	0.16	−0.45 *
Range Acc ML	16.33 ± 8.07	15.46 ± 6.17	15.89 ± 8.59	−0.11	−0.06	0.05	−0.28
Range Acc AP	11.52 ± 6.59	13.85 ± 8.27	12.33 ± 6.14	0.34	0.14	−0.23	−0.23
Range Gyr V	238.32 ± 209.22	246.18 ± 166.59	207.18 ± 104.41	0.08	−0.16	−0.27	−0.32
Range Gyr ML	456.96 ± 241.32	340.59 ± 230.85	402.52 ± 228.77	−0.34	−0.22	0.31	−0.47 *
Range Gyr AP	114.13 ± 111.22	91.72 ± 30.45	80.3 ± 31.42	−0.19	−0.33	−0.27	−0.16

* p-value < 0.05.

Table 7. TUG: feature values at T0, T1, T2; values of SRM comparing times; correlation with UPDRS item 3.10 (Gait) score.

Feature (TUG)	T0 Mean ± SD	T1 Mean ± SD	T2 Mean ± SD	SRM (T0 vs. T1)	SRM (T0 vs. T2)	SRM (T1 vs. T2)	Correlation with Gait Item
Tug Time	32.19 ± 10.24	28.27 ± 9.7	26.8 ± 6.49	−0.48	−0.93	−0.24	0.55 *
Sit-to-Stand Time	3.03 ± 2.64	1.94 ± 0.99	2.12 ± 0.89	−0.41	−0.39	0.13	0.28
Walk Time	20.58 ± 6.96	17.92 ± 6.02	17.81 ± 4.62	−0.52	−0.78	−0.04	0.56 *
Turning Time	3.9 ± 1.71	3.92 ± 2.24	3.09 ± 0.89	0.01	−0.68	−0.44	0.53 *
Turn-to-Sit Time	4.67 ± 1.66	4.48 ± 1.16	3.78 ± 1.2	−0.11	−0.48	−0.67	0.23
Number of Steps	40.13 ± 10.3	37.78 ± 11.76	38.3 ± 9.38	−0.05	−0.57	−0.21	0.49 *
Step duration	1.17 ± 0.08	1.14 ± 0.11	1.15 ± 0.1	−0.34	−0.24	0.21	0.24
Stance	57.94 ± 3.53	55.9 ± 7.88	57.56 ± 3.31	−0.23	−0.30	0.18	0.20
Swing	42.26 ± 3.62	43.97 ± 7.52	42.44 ± 3.31	0.20	0.20	−0.17	−0.22
CV step	0.07 ± 0.03	0.06 ± 0.03	0.09 ± 0.08	−0.24	0.17	0.54	0.47 *
CV Stance	0.08 ± 0.04	0.36 ± 0.89	0.08 ± 0.03	0.30	−0.09	−0.29	0.12
CV Swing	0.13 ± 0.1	0.11 ± 0.08	0.1 ± 0.04	−0.17	−0.27	−0.08	0.21
Flex Leg	23.41 ± 4.43	23.37 ± 7.39	23.91 ± 6.32	−0.01	0.09	0.10	−0.54 *
Flex Arm	30.38 ± 13.57	28.53 ± 18.85	28.1 ± 14.63	−0.13	−0.22	−0.05	0.28
Asym Leg	12.68 ± 6.62	17.03 ± 20.31	16.22 ± 13.22	0.19	0.22	−0.05	0.20
Asym Arm	40.06 ± 27.2	43.99 ± 22.93	40.81 ± 22.05	0.15	0.04	−0.22	−0.07
Average Vel Thigh	00.00 ± 6.56	43.01 ± 8.07	41.88 ± 7.55	0.62	0.49	−0.31	−0.60 *
Average Vel Leg	72.72 ± 17.24	89.6 ± 17.35	88.24 ± 16.19	0.83	0.83	0.15	−0.52 *
Average Vel Arm	24.42 ± 12.82	25.41 ± 11.03	23.57 ± 9.24	0.14	−0.10	−0.27	0.11
Average Vel Forearm	38.82 ± 17.6	40.26 ± 21.41	34.79 ± 10.74	0.11	−0.28	−0.34	0.06
Turning Vel	51.82 ± 14.09	58.92 ± 23.8	62.47 ± 14.75	0.32	0.93	0.22	−0.53 *
Peak Turning Vel	91.12 ± 18.8	105.04 ± 30.13	101.76 ± 26.65	0.61	0.60	−0.19	−0.25
Steps Turning	5 ± 1	6.5 ± 3.67	5.6 ± 2.65	0.39	0.31	−0.22	0.48 *
Average Vel Sit Stand	27.27 ± 7.96	34.42 ± 11.1	33.37 ± 10.46	0.93	0.61	−0.09	−0.43 *

* p-value < 0.05.

Table 8. Stop and go (S&G): feature values at T0, T1, T2; values of SRM comparing times; correlation with UPDRS item 3.10 (gait) score.

Feature (S&G)	T0 Mean ± SD	T1 Mean ± SD	T2 Mean ± SD	SRM (T0 vs. T1)	SRM (T0 vs. T2)	SRM (T1 vs. T2)	Correlation with Gait Item
Step duration	1.44 ± 0.38	1.33 ± 0.29	1.34 ± 0.17	−0.21	−0.31	0.04	−0.42 *
Stance	0.99 ± 0.39	0.91 ± 0.33	0.86 ± 0.21	−0.14	−0.35	−0.13	−0.18
Swing	0.45 ± 0.06	0.42 ± 0.07	0.48 ± 0.08	−0.44	0.34	0.86	−0.22
Step velocity	179.34 ± 46.87	184.93 ± 60.03	174.42 ± 47.18	0.08	−0.10	−0.24	−0.08
CV step	0.15 ± 0.12	0.1 ± 0.06	0.1 ± 0.06	−0.35	−0.34	0.05	0.02
CV Stance	0.31 ± 0.22	0.24 ± 0.18	0.33 ± 0.16	−0.21	0.06	0.48	−0.13
CV Swing	0.17 ± 0.04	0.18 ± 0.05	0.17 ± 0.05	0.13	0.04	−0.14	0.07

* p-value < 0.05.

Table 9. Narrow walking (NW): feature values at T0, T1, T2; values of SRM comparing times; correlation with UPDRS item 3.10 (gait) score.

Feature (NW)	T0 Mean ± SD	T1 Mean ± SD	T2 Mean ± SD	SRM (T0 vs. T1)	SRM (T0 vs. T2)	SRM (T1 vs. T2)	Correlation with Gait Item
Step duration	1.18 ± 0.09	1.09 ± 0.08	1.11 ± 0.09	−1.60	−0.91	0.77	0.25
Stance	0.65 ± 0.12	0.63 ± 0.08	0.65 ± 0.07	−0.17	0.02	0.49	0.05
Swing	0.5 ± 0.03	0.46 ± 0.05	0.47 ± 0.04	−1.58	−0.90	0.50	−0.01
Step velocity	266.98 ± 40.93	297.96 ± 50.31	284.76 ± 39.79	1.56	0.92	−0.66	−0.44*
CV step	0.1 ± 0.05	0.07 ± 0.03	0.08 ± 0.04	−0.59	−0.23	0.27	−0.02
CV Stance	0.14 ± 0.07	0.12 ± 0.04	0.14 ± 0.08	−0.27	−0.01	0.21	0.04
CV Swing	0.13 ± 0.06	0.09 ± 0.03	0.11 ± 0.04	−0.74	−0.37	0.60	0.35*

* p-value < 0.05.

3.1. Clinical Evaluation

MDS-UPDRS sections two and three, BBS, and FOG-Q (Table 3) demonstrated moderate responsiveness to tDCS at the end of the treatment. The effect appears stable after one month with some improvement in BBS and MDS-UPDRS Section 2 score.

3.2. Stance Feet Together (SFT) and Tandem Stance (TS)

Jerk demonstrated a decrement, but only in a small percentage, in SFT (Table 4) and TS (Table 5) in both T1 and T2. During TS, Sway Area, range of the accelerations and angular velocities in the three directions decreased in T1 with a responsiveness around 0.4. The effect is stable at T2 compared to T1 with low improvements in some features.

The BBS score correlates significantly with almost all the features extracted from SFT and TS such as Jerk, Sway area (only TS, r = −0.37) and range of the accelerations and angular velocities. So, features highly reflect the clinical evaluation in this case.

3.3. Pull Test (PT)

During the PT, the obtained results (Table 6) showed an unchanged number of steps after tDCS treatment, a small increment of Jerk, and a small reduction of Sway Area at the end of the treatment and one month after.

Regarding the clinical evaluation, only few features (Range Acc V, r = −0.45; Range Gyr ML, r = −0.47) correlated with the UPDRS PT sub score.

3.4. Time Up and Go (TUG)

It was found that tDCS showed a moderate effect on the duration of sit-to-stand and walking phase in T1 and T2, as compared to the baseline (Table 7). A lower duration of the Turning phase is present only at T2. In correlation with a lower duration of the walking phase, our results show a reduction of the number of steps and stance duration. No changes were found in features related to the upper limbs. Conversely, the velocity of the lower extremities meaningfully increased. Finally, patients increased the velocity to turn and sit at T1 and T2, with comparison to the baseline values.

The UPDRS gait item score correlates significantly with several features extracted from TUG. Significant correlations regard the features representing the duration of the TUG phases (namely tug time, walk time and turning time). So, patients that take time to complete TUG have higher score on gait item. Weak correlation was for the temporal gait characteristics with the exception of number of steps and CV step. Gait item correlates significantly with features related to lower limb movements (Flex Leg, Average Vel Thigh, and Average Vel Leg) and the turning phase (Turning Vel, Steps Turning).

3.5. Stop and Go (S&G) & Narrow Walking (NW)

Both S&G (Table 8) and NW (Table 9) tests show a shorter duration of the step and swing phase and decreased variability of step duration in both T1 and T2 with respect to the baseline. The velocity remained unchanged in S&G but increased in NW. Large responsiveness is found in NW related to step duration, swing duration, velocity, and all the temporal step variability features.

One feature from S&G (step duration, $r = -0.42$) and two features from NW (Step Velocity, $r = -0.44$; CV Swing, $r = 0.35$) are significantly related to the UPDRS gait item.

4. Discussion

The response to dopaminergic drug replacement therapy in PD may lose its effectiveness during the course of the disease. Postural and gait disturbances, in particular, are symptoms that are difficult to treat with currently available pharmacological therapies.

Recent studies suggest a potential positive impact of tDCS on gait and balance in PD patients, symptoms of the late stage of PD, poorly responding to the classic dopaminergic treatment.

Our work focused on objectively quantifying the effect of tDCS on gait and postural stability from measured data gathered by wearable electronics used during motor tests of Parkinson's disease patients.

Within this context, the obtained results demonstrate the impact of wearable electronics with respect to standard clinical evaluation, allowing for interesting insights on the range of change on motor performance following the therapy. In fact, wearable electronics can evidence key elements of postural instability or gait abnormalities, both for evaluating the progression in PD and even to identify the disease at early stages [7,48–50]. Accordingly, in this study, specific motor tests were considered to assess the effects of tDCS therapy on balance and gait disturbances, taking into account the effects on measured motor features, soon after the delivery and one month later.

For balance assessment, three different motor tests were adopted to evaluate the equilibrium in three different conditions: SFT for static balance, TS to assess the balance when a low perturbation is introduced, and PT to assess postural responses to an unexpected perturbation. According to the kinematic assessment, Jerk is the only feature that presents a significant variation in SFT, TS and PT, suggesting that it is a highly sensitive measure of balance. This confirms the finding reported in previous studies, wherein Jerk was suggested as a valid biomarker of PD [7,49].

For gait assessment, the TUG test was useful to evaluate the slower speed, decreased arm swing, shuffling steps and difficulty to turn. Further S&G and NW tests were useful to evaluate step time, velocity, and variability of steps, due to the difficulty to start/stop and pass through a narrow door.

Our results show a reduction of step and stance duration and an increment of lower limb velocity during TUG, S&G and NW tests. These achievements confirm the findings reported in other works, which evidenced some improvement of hypokinetic gait in PD after tDCS treatment [29,30,51].

The effect is more evident in NW test, where we observed a large responsiveness to tDCS. The reason why PD patients tend to decrease step time and velocity when approaching a narrowed space is not completely understood [39], however tDCS in some way improves this aspect. We evidenced an improvement of gait in turning and standing tasks during TUG test too, when patients increased the velocity to turn and sit after the stimulation protocol. In particular, changes in turning are one of the early motor deficiencies in PD, as previously reported [50]. The wearable impact in analyzing this complex motor task is relevant. In fact, clinical evaluation alone demonstrated an amelioration in gait and pull test items but was not able to disclose which features of these two motor functions improved. Being able to thoroughly phenotype patients' motor performances is crucial to understanding the effect of a therapeutic intervention and to allow for speculation with respect to its dynamics.

In order to provide clinical validity for our approach, we investigated the relation between the clinical scores, given by the examiners, and the measured features. Clinical vs. wearables outcomes demonstrated general significant results (Tables 4–9). In particular, a higher correlation was found between features extracted from static balance tasks (SFT and TS) and BBS scores and between TUG features and UPDRS gait item scores.

Not all of the features presented a perfect correlation with clinical rating, and this is also expected since these measures should be more sensitive than clinical scales, mostly due to the fact that clinical examination is based on a rating scale with only a few steps, while wearables produce a density scale with a high number of steps [52]. For example, in the TUG test, the duration of the performance is a significant parameter for both the classical clinical exam and "technology-based assessment". Conversely, the average velocity of lower limbs was significantly and accurately measured only by the wearable sensors. The same consideration applies for the other features extracted from the balance and gait tests. These results are in accordance with a recent work [7], evidencing that several features extracted by sensors were able to detect subtle abnormalities in early stage PD patients where the corresponding clinical score, obtained by visual examination, was considered normal for the majority of subjects.

It could be argued that a better sensitivity can be clinically irrelevant, detecting differences too small to have a real impact on a patient's life and functioning. Alternatively, it allows investigators to better phenotype motion alterations and their changes after a therapy, and to objectively measure the benefit from a standard intervention, in view of its customization and relevant optimization.

We are aware of some limitations of the present study. First, tDCS was adopted for patients under other medical treatments that had already been adjusted for the optimal dose. We did not use a test-retest design, thus we cannot exclude variability due to participants' physical or mental conditions, or to drug response fluctuations. To minimize the effects of the aforementioned limitations, we performed the study at the same time of the day for every patient, and no modification to the therapy was allowed in the three months preceding the study and during its course. The study cannot exclude a placebo effect. Moreover, we performed the experiment on a small sample size. Indeed, further studies, on larger cohorts, are mandatory in order to confirm our findings.

5. Conclusions

Our study aimed to demonstrate the advantages of outcomes from technology-based measures in clinical trials. These advantages are particularly important for revealing the effectiveness of tDCS protocols in late stage PD patients. This is because the benefit of tDCS remains unclear and controversial, thus the outcomes from electronic wearables can help the clinical rating of the tDCS effectiveness. In particular, our results provide evidence of the wearable electronic impact, as a complementary tool to the standard clinical evaluation.

The adoption of wearables furnished a number of motor features, some of them with a good correlation with standard clinical assessment, others adding information not evident to human eyes.

Nonetheless, even if wearables can provide motor features for an insight of each patient's motor performances, they remain rarely adopted in clinical trials. We believe that relevant reasons for this can be ascribed to the lack of an integrated platform that can be easily used by nurses and clinicians, and a lack of regulatory approval and appropriate cost–benefit ratios [15,52]. However, the idea to develop and integrate technologies into the assessment of therapy effectiveness has become so evident that several academic centers and companies have started to bring them to the market.

Author Contributions: conceptualization, writing—review and editing, M.R., G.D.L., A.P., F.G. and G.S.; methodology and investigation, M.R., G.D.L., A.P., S.S., M.A., C.S. and G.S.; software, M.R.; formal analysis and data curation, M.R. and G.D.L.; validation, M.R., G.D.L., A.P., F.G. and G.S.; writing—original draft preparation, M.R., and G.D.L.; supervision, A.P., F.G. and G.S.

Funding: This research received no external funding.

Acknowledgments: We acknowledge the support given by Luca Pietrosanti in measurements.

Conflicts of Interest: F.G. and G.S. owe 6% each of Captiks srl. The other authors declare no conflicts of interests. Except from the authors, no others had role in the design of the study; in the collection, analyses, or interpretation of data; in the writing of the manuscript, and in the decision to publish the results.

Ethical Statements: All subjects gave their informed consent for inclusion before they participated in the study. The study was conducted in accordance with the Declaration of Helsinki, and the protocol was approved by the Ethics Committee of Fondazione PTV Policlinico Tor Vergata.

References

1. Saggio, G.; Lazzaro, A.; Sbernini, L.; Carrano, F.M.; Passi, D.; Corona, A.; Panetta, V.; Gaspari, A.L.; Di Lorenzo, N. Objective surgical skill assessment: An initial experience by means of a sensory glove paving the way to open surgery simulation? *J. Surg. Educ.* **2015**, *72*, 910–917. [CrossRef] [PubMed]
2. Saggio, G.; Santosuosso, G.L.; Cavallo, P.; Pinto, C.A.; Petrella, M.; Giannini, F.; Di Lorenzo, N.; Lazzaro, A.; Corona, A.; D'Auria, F.; et al. Gesture recognition and classification for surgical skill assessment. In Proceedings of the IEEE International Symposium on Medical Measurements and Applications, Bari, Italy, 30–31 May 2011.
3. Sbernini, L.; Quitadamo, L.R.; Riillo, F.; Di Lorenzo, N.; Gaspari, A.L.; Saggio, G. Sensory-Glove-Based Open Surgery Skill Evaluation. *IEEE Trans. Hum.-Mach. Syst.* **2018**, *48*, 213–218. [CrossRef]
4. Lukowicz, P.; Kirstein, T.; Tröster, G. Wearable systems for health care applications. *Methods Inf. Med.* **2004**, *43*, 232–238. [PubMed]
5. Park, S.; Jayaraman, S. Enhancing the Quality of Life Through Wearable Technology. *IEEE Eng. Med. Biol. Mag.* **2003**, *22*, 41–48. [CrossRef] [PubMed]
6. Darwish, A.; Hassanien, A.E. Wearable and implantable wireless sensor network solutions for healthcare monitoring. *Sensors* **2011**, *11*, 5561–5595. [CrossRef]
7. Ricci, M.; Di Lazzaro, G.; Pisani, A.; Mercuri, N.B.; Giannini, F.; Saggio, G. Assessment of motor impairments in early untreated Parkinson's disease patients: The wearable electronics impact. *IEEE J. Biomed. Health. Inform.* **2019**. [CrossRef]
8. Piro, N.E.; Baumann, L.; Tengler, M.; Piro, L. Telemonitoring of patients with Par-kin son's disease using inertia sensors. *Appl. Clin. Inform.* **2014**, *5*, 503–511.
9. Giuberti, M.; Ferrari, G.; Contin, L.; Cimolin, V.; Azzaro, C.; Albani, G.; Mauro, A. Assigning UPDRS scores in the leg agility task of parkinsonians: Can it be done through BSN-based kinematic variables? *IEEE Internet Things J.* **2015**, *2*, 41–51. [CrossRef]
10. Dai, H.; Zhang, P.; Lueth, T.C. Quantitative assessment of parkinsonian tremor based on an inertial measurement unit. *Sensors* **2015**, *15*, 25055–25071. [CrossRef]
11. Weiss, A.; Sharifi, S.; Plotnik, M.; van Vugt, J.P.P.; Giladi, N.; Hausdorff, J.M. Toward automated, at-home assessment of mobility among patients with Parkinson disease, using a body-worn accelerometer. *Neurorehabil. Neural Repair* **2011**, *25*, 810–818. [CrossRef]
12. Rovini, E.; Maremmani, C.; Cavallo, F. How wearable sensors can support parkinson's disease diagnosis and treatment: A systematic review. *Front. Neurosci.* **2017**, *11*, 555. [CrossRef] [PubMed]
13. Postuma, R.B.; Berg, D.; Stern, M.; Poewe, W.; Olanow, C.W.; Oertel, W.; Obeso, J.; Marek, K.; Litvan, I.; Lang, A.E.; et al. MDS clinical diagnostic criteria for Parkinson's disease. *Mov. Disord.* **2015**, *30*, 1591–1601. [CrossRef] [PubMed]
14. Christopher, G.G. MDS-UPDRS. *Off. Work. Doc.* **2008**, *23*, 2129–2170. [CrossRef]

15. Artusi, C.A.; Mishra, M.; Latimer, P.; Vizcarra, J.A.; Lopiano, L.; Maetzler, W.; Merola, A.; Espay, A.J. Integration of technology-based outcome measures in clinical trials of Parkinson and other neurodegenerative diseases. *Park. Relat. Disord.* **2018**, *46*, S53–S56. [CrossRef] [PubMed]
16. Goetz, C.G.; Tilley, B.C.; Shaftman, S.R.; Stebbins, G.T.; Fahn, S.; Martinez-Martin, P.; Poewe, W.; Sampaio, C.; Stern, M.B.; Dodel, R.; et al. Movement Disorder Society-sponsored revision of the Unified Parkinson's Disease Rating Scale (MDS-UPDRS): Scale presentation and clinimetric testing results. *Mov. Disord.* **2008**, *23*, 2129–2170. [CrossRef]
17. Lipsmeier, F.; Taylor, K.I.; Kilchenmann, T.; Wolf, D.; Scotland, A.; Schjodt-Eriksen, J.; Cheng, W.Y.; Fernandez-Garcia, I.; Siebourg-Polster, J.; Jin, L.; et al. Evaluation of smartphone-based testing to generate exploratory outcome measures in a phase 1 Parkinson's disease clinical trial. *Mov. Disord.* **2018**, *33*, 1287–1297. [CrossRef] [PubMed]
18. Parkinson's KinetiGraph®system. Available online: https://www.globalkineticscorporation.com/the-pkg-system/ (accessed on 15 September 2019).
19. What is Wearable Technology and How Can It Help People with Parkinson's Disease? Available online: https://www.apdaparkinson.org/article/wearable-technology-in-parkinsons/ (accessed on 15 September 2019).
20. Roche Technology Measures Parkinson's Disease Fluctuations. Available online: https://www.roche.com/media/store/roche_stories/roche-stories-2015-08-10.htm (accessed on 15 September 2019).
21. Fregni, F.; Boggio, P.S.; Santos, M.C.; Lima, M.; Vieira, A.L.; Rigonatti, S.P.; Silva, M.T.A.; Barbosa, E.R.; Nitsche, M.A.; Pascual-Leone, A. Noninvasive cortical stimulation with transcranial direct current stimulation in Parkinson's disease. *Mov. Disord.* **2006**, *21*, 1693–1702. [CrossRef]
22. Brunoni, A.R.; Nitsche, M.A.; Bolognini, N.; Bikson, M.; Wagner, T.; Merabet, L.; Edwards, D.J.; Valero-Cabre, A.; Rotenberg, A.; Pascual-Leone, A.; et al. Clinical research with transcranial direct current stimulation (tDCS): Challenges and future directions. *Brain Stimul.* **2012**, *5*, 175–195. [CrossRef]
23. Elsner, B.; Kugler, J.; Pohl, M.; Mehrholz, J. Transcranial direct current stimulation (tDCS) for idiopathic Parkinson's disease. *Cochrane Database Syst. Rev.* **2016**, *2016*. [CrossRef]
24. Snijders, A.H.; Takakusaki, K.; Debu, B.; Lozano, A.M.; Krishna, V.; Fasano, A.; Aziz, T.Z.; Papa, S.M.; Factor, S.A.; Hallett, M. Physiology of freezing of gait. *Ann. Neurol.* **2016**, *80*, 644–659. [CrossRef]
25. Dagan, M.; Herman, T.; Harrison, R.; Zhou, J.; Giladi, N.; Ruffini, G.; Manor, B.; Hausdorff, J.M. Multitarget transcranial direct current stimulation for freezing of gait in Parkinson's disease. *Mov. Disord.* **2018**, *33*, 642–646. [CrossRef] [PubMed]
26. Vitorio, R.; Stuart, S.; Rochester, L.; Alcock, L.; Pantall, A. fNIRS response during walking—Artefact or cortical activity? A systematic review. *Neurosci. Biobehav. Rev.* **2017**, *83*, 160–172. [CrossRef] [PubMed]
27. Stuart, S.; Vitorio, R.; Morris, R.; Martini, D.N.; Fino, P.C.; Mancini, M. Cortical activity during walking and balance tasks in older adults and in people with Parkinson's disease: A structured review. *Maturitas* **2018**, *113*, 53–72. [CrossRef] [PubMed]
28. Hadoush, H.; Al-Jarrah, M.; Khalil, H.; Al-Sharman, A.; Al-Ghazawi, S. Bilateral anodal transcranial direct current stimulation effect on balance and fearing of fall in patient with Parkinson's disease. *NeuroRehabilitation* **2018**, *42*, 63–68. [CrossRef] [PubMed]
29. Manenti, R.; Brambilla, M.; Rosini, S.; Orizio, I.; Ferrari, C.; Borroni, B.; Cotelli, M. Time up and go task performance improves after transcranial direct current stimulation in patient affected by Parkinson's disease. *Neurosci. Lett.* **2014**, *580*, 74–77. [CrossRef] [PubMed]
30. Yotnuengnit, P.; Bhidayasiri, R.; Donkhan, R.; Chaluaysrimuang, J.; Piravej, K. Effects of Transcranial Direct Current Stimulation Plus Physical Therapy on Gait in Patients with Parkinson Disease: A Randomized Controlled Trial. *Am. J. Phys. Med. Rehabil.* **2018**, *97*, 7–15. [CrossRef]
31. Lattari, E.; Costa, S.S.; Campos, C.; de Oliveira, A.J.; Machado, S.; Maranhao Neto, G.A. Can transcranial direct current stimulation on the dorsolateral prefrontal cortex improves balance and functional mobility in Parkinson's disease? *Neurosci. Lett.* **2017**, *636*, 165–169. [CrossRef]
32. Lefaucheur, J.P.; Antal, A.; Ayache, S.S.; Benninger, D.H.; Brunelin, J.; Cogiamanian, F.; Cotelli, M.; De Ridder, D.; Ferrucci, R.; Langguth, B.; et al. Evidence-based guidelines on the therapeutic use of transcranial direct current stimulation (tDCS). *Clin. Neurophysiol.* **2017**, *128*, 56–92. [CrossRef]
33. O'Neill, D. The Mini—Mental Status Examination. *J. Am. Geriatr. Soc.* **1991**, *39*, 733. [CrossRef]

34. Ricci, M.; Terribili, M.; Giannini, F.; Errico, V.; Pallotti, A.; Galasso, C.; Tomasello, L.; Sias, S.; Saggio, G. Wearable-based electronics to objectively support diagnosis of motor impairments in school-aged children. *J. Biomech.* **2019**, *83*, 243–252. [CrossRef]
35. Alessandrini, M.; Micarelli, A.; Viziano, A.; Pavone, I.; Costantini, G.; Casali, D.; Paolizzo, F.; Saggio, G. Body-worn triaxial accelerometer coherence and reliability related to static posturography in unilateral vestibular failure. *Acta Otorhinolaryngol. Ital.* **2017**, *37*, 231–236. [PubMed]
36. Flash, T.; Hogan, N. The coordination of arm movements: An experimentally confirmed mathematical model. *J. Neurosci.* **1985**, *5*, 1688–1703. [CrossRef] [PubMed]
37. Chen, T.Z.; Xu, G.J.; Zhou, G.A.; Wang, J.R.; Chan, P.; Du, Y.F. Postural sway in idiopathic rapid eye movement sleep behavior disorder: A potential marker of prodromal Parkinsons disease. *Brain Res.* **2014**, *1559*, 26–32. [CrossRef] [PubMed]
38. Mancini, M.; Horak, F.B.; Zampieri, C.; Carlson-Kuhta, P.; Nutt, J.G.; Chiari, L. Trunk accelerometry reveals postural instability in untreated Parkinson's disease. *Park Relat. Disord.* **2011**, *17*, 557–562. [CrossRef] [PubMed]
39. Beck, E.N.; Martens, K.A.E.; Almeida, Q.J. Freezing of gait in Parkinson's disease: An overload problem? *PLoS ONE* **2015**, *10*, e0144986. [CrossRef] [PubMed]
40. Morris, M.E.; Iansek, R.; Matyas, T.A.; Summers, J.J. Stride length regulation in Parkinson's disease: Normalization strategies and underlying mechanisms. *Brain* **1996**, *119*, 551–568. [CrossRef]
41. Hausdorff, J.M.; Cudkowicz, M.E.; Firtion, R.; Wei, J.Y.; Goldberger, A.L. Gait variability and basal ganglia disorders: Stride-to-stride variations of gait cycle timing in Parkinson's disease and Huntington's disease. *Mov. Disord.* **1998**, *13*, 428–437. [CrossRef]
42. Mancini, M.; Salarian, A.; Carlson-Kuhta, P.; Zampieri, C.; King, L.; Chiari, L.; Horak, F.B. ISway: A sensitive, valid and reliable measure of postural control. *J. Neuroeng. Rehabil.* **2012**, *9*, 1–8. [CrossRef]
43. Li, S. The balance scale. *Nature* **2010**, *464*, 804. [CrossRef]
44. Giladi, N.; Shabtai, H.; Simon, E.S.; Biran, S.; Tal, J.; Korczyn, A.D. Construction of freezing of gait questionnaire for patients with Parkinsonism. *Park Relat. Disord.* **2000**, *6*, 165–170. [CrossRef]
45. Martinez-Martin, P. Hoehn and Yahr Staging Scale. *Encycl. Mov. Disord.* **2010**, *1*, 23–25.
46. Nahler, G.; Nahler, G. standardized response mean (SRM). In *Dictionary of Pharmaceutical Medicine*; Springer: Vienna, Austria, 2009; p. 173.
47. Husted, J.A.; Cook, R.J.; Farewell, V.T.; Gladman, D.D. Methods for assessing responsiveness: A critical review and recommendations. *J. Clin. Epidemiol.* **2000**, *53*, 459–468. [CrossRef]
48. Horak, F.B.; Mancini, M. Objective biomarkers of balance and gait for Parkinson's disease using body-worn sensors. *Mov. Disord.* **2013**, *28*, 1544–1551. [CrossRef] [PubMed]
49. Mancini, M.; Carlson-Kuhta, P.; Zampieri, C.; Nutt, J.G.; Chiari, L.; Horak, F.B. Postural sway as a marker of progression in Parkinson's disease: A pilot longitudinal study. *Gait Posture* **2012**, *36*, 471–476. [CrossRef] [PubMed]
50. Salarian, A.; Zampieri, C.; Horak, F.B.; Carlson-Kuhta, P.; Nutt, J.G.; Aminian, K. Analyzing 180° turns using an inertial system reveals early signs of progression of Parkinson's disease. In Proceedings of the 31st Annual International Conference of the IEEE Engineering in Medicine and Biology Society: Engineering the Future of Biomedicine, Minneapolis, MN, USA, 3–6 September 2009; pp. 224–227.
51. Von Papen, M.; Fisse, M.; Sarfeld, A.S.; Fink, G.R.; Nowak, D.A. The effects of 1 Hz rTMS preconditioned by tDCS on gait kinematics in Parkinson's disease. *J. Neural Transm.* **2014**, *121*, 743–754. [CrossRef] [PubMed]
52. Espay, A.J.; Hausdorff, J.M.; Sánchez-Ferro, Á.; Klucken, J.; Merola, A.; Bonato, P.; Paul, S.S.; Horak, F.B.; Vizcarra, J.A.; Mestre, T.A.; et al. A roadmap for implementation of patient-centered digital outcome measures in Parkinson's disease obtained using mobile health technologies. *Mov. Disord.* **2019**, *34*, 657–663. [CrossRef]

Review

Cueing Paradigms to Improve Gait and Posture in Parkinson's Disease: A Narrative Review

Niveditha Muthukrishnan [1], James J. Abbas [1], Holly A. Shill [2] and Narayanan Krishnamurthi [1,3,*]

[1] Center for Adaptive Neural Systems, School of Biological and Health Systems Engineering, Arizona State University, Tempe, AZ 85287, USA; Niveditha.Muthukrishnan@asu.edu (N.M.); James.Abbas@asu.edu (J.J.A.)

[2] Muhammad Ali Parkinson Center, Barrow Neurological Institute, St. Joseph's Hospital and Medical Center, Phoenix, AZ 85013, USA; Holly.Shill@DignityHealth.org

[3] Edson College of Nursing and Health Innovation, Arizona State University, Phoenix, AZ 85004, USA

* Correspondence: Narayanan.Krishnamurthi@asu.edu; Tel.: +1-(602)-496-0912

Received: 11 November 2019; Accepted: 9 December 2019; Published: 11 December 2019

Abstract: Progressive gait dysfunction is one of the primary motor symptoms in people with Parkinson's disease (PD). It is generally expressed as reduced step length and gait speed and as increased variability in step time and step length. People with PD also exhibit stooped posture which disrupts gait and impedes social interaction. The gait and posture impairments are usually resistant to the pharmacological treatment, worsen as the disease progresses, increase the likelihood of falls, and result in higher rates of hospitalization and mortality. These impairments may be caused by perceptual deficiencies (poor spatial awareness and loss of temporal rhythmicity) due to the disruptions in processing intrinsic information related to movement initiation and execution which can result in misperceptions of the actual effort required to perform a desired movement and maintain a stable posture. Consequently, people with PD often depend on external cues during execution of motor tasks. Numerous studies involving open-loop cues have shown improvements in gait and freezing of gait (FoG) in people with PD. However, the benefits of cueing may be limited, since cues are provided in a consistent/rhythmic manner irrespective of how well a person follows them. This limitation can be addressed by providing feedback in real-time to the user about performance (closed-loop cueing) which may help to improve movement patterns. Some studies that used closed-loop cueing observed improvements in gait and posture in PD, but the treadmill-based setup in a laboratory would not be accessible outside of a research setting, and the skills learned may not readily and completely transfer to overground locomotion in the community. Technologies suitable for cueing outside of laboratory environments could facilitate movement practice during daily activities at home or in the community and could strongly reinforce movement patterns and improve clinical outcomes. This narrative review presents an overview of cueing paradigms that have been utilized to improve gait and posture in people with PD and recommends development of closed-loop wearable systems that can be used at home or in the community to improve gait and posture in PD.

Keywords: Parkinson's disease; cueing; gait; posture; rehabilitation; wearable sensors

1. Introduction

Parkinson's disease (PD), which is the second most common progressive neurodegenerative disease, results in motor and non-motor dysfunctions caused by the degeneration of dopamine-producing cells of the substantia nigra and other brain regions [1,2]. Clinical motor symptoms include bradykinesia, tremor, rigidity, freezing of gait, and instability of posture and gait [3–6]. Some of the common manifestations of PD that affect gait and posture are stooped posture and shuffling of gait, increases in gait asymmetry and double support time, reductions in step length and walking speed, impairments in postural responses to perturbations, and increases in variability of step/stride time as well as step/stride length [7]. Considerable efforts are being taken to improve options for treating mobility deficits in persons with PD because of the associated risk of falls and loss of independence.

Pharmacological and deep brain stimulation (DBS) surgical treatments have been demonstrated to be partially effective in managing some of the manifestations of gait impairments and postural instability. As the primary pharmacological treatment in PD, the dopamine replacement therapy (i.e., levodopa) improves stride length, gait speed, and double support time variability, whereas it does not have any significant benefits on cadence and other temporal characteristics of gait [8]. The effects of levodopa on postural sway is controversial [9,10]. Regarding inadequate postural responses (compensatory stepping) leading to falls in PD, levodopa seems to offer no benefit [11,12]. Thus, the effects of levodopa on gait and posture in PD is inconsistent.

Concerning the effects of the DBS, stimulation of subthalamic nucleus (STN-DBS) consistently improved stride length but no effects on stride time and its variability were found. Stimulation of globus pallidum internus (GPi-DBS) significantly improved gait velocity but without any significant improvements in stride length. Also, many people with PD reported postoperative worsening of gait and increased risk of falls [13]. In the case of the stimulation of pedunculopontine nucleus (PPN-DBS) at 15–70 Hz, improvements in postural instability and gait disorder, including freezing of gait and falls, have been noticed. However, the improvements varied depending on the duration of follow-up and types of outcome measures obtained [14]. Low-frequency STN-DBS and GPi-DBS (below 100 Hz) have shown encouraging beneficial effects on axial symptoms in PD; however, higher levels of evidence with randomized and blinded studies are needed to confirm the benefits [15]. Also, the overall benefits of low-frequency STN-DBS decrease with long-term use [16].

2. Pathophysiology of Motor Dysfunction in PD

The loss of dopaminergic neurons in the substantia nigra pars compacta within the basal ganglia leads to classical parkinsonian motor symptoms. The basal ganglia play significant roles in the production and control of automatic and well-learned motor movements. First, the basal ganglia generate internal cues or trigger to facilitate the initiation of movement sequences without attention. Second, they contribute to the cortical "motor set", i.e., they aid in the preparation and maintenance of motor schemes in a state of action readiness thereby enabling appropriate motor function execution. The widely accepted model of basal ganglia consists of two circuits, the direct and indirect pathways, which originate from striatal neurons and project to various output structures. The direct pathway is postulated to promote movement by direct inhibitory projections to the globus pallidus internus/substantia nigra reticulata (GPi/SNr), whereas the indirect pathway is hypothesized to inhibit movement projecting to the GPi/SNr through globus pallidus externus (GPe) and subthalamic nucleus (STN). In PD, striatal dopaminergic depletion results in the reduced inhibitory direct pathway and increased indirect pathway output onto the GPi/SNr and, subsequently, increased GPi/SNr inhibition to the output structures. This consequently leads to deficiencies in the execution of a movement [6,17–19] (Figure 1). This deficiency in execution results in hypokinesia, a central feature in PD, or lack of movement together with muscular rigidity.

Evidence indicates that the basal ganglia are also important for sensorimotor integration. Striatal cells are robustly activated when a sensory event functions as a cue for a movement. In addition, the caudate nucleus and substantia nigra contain a large proportion of cells that are multisensory; such cells could be used to integrate sensory inputs and form a multimodal representation of the environment in the basal ganglia. Disruption of basal ganglia processes enhances the response of pallidal neurons to passive limb movement, suggesting an impaired gain mechanism because of dopamine depletion [2,4,20]. A common consequence of striatal dopamine loss is attenuation of the transfer of critical information to the basal ganglia which leads to a decrease in the ability to detect relevant internal sensory and or movement cues [1,21,22]. Such a disruption of information flow to the basal ganglia may worsen impaired movement selection and sequencing in striatum with dopamine loss thereby resulting in gait impairments [23]. The pattern of deficits in people with PD is consistent with a disruption of this integration mechanism. Persons with PD may become increasingly dependent on external stimuli to initiate and shape motor output and may be unable to effectively execute movements because of the lack of critical proprioceptive information [24–27].

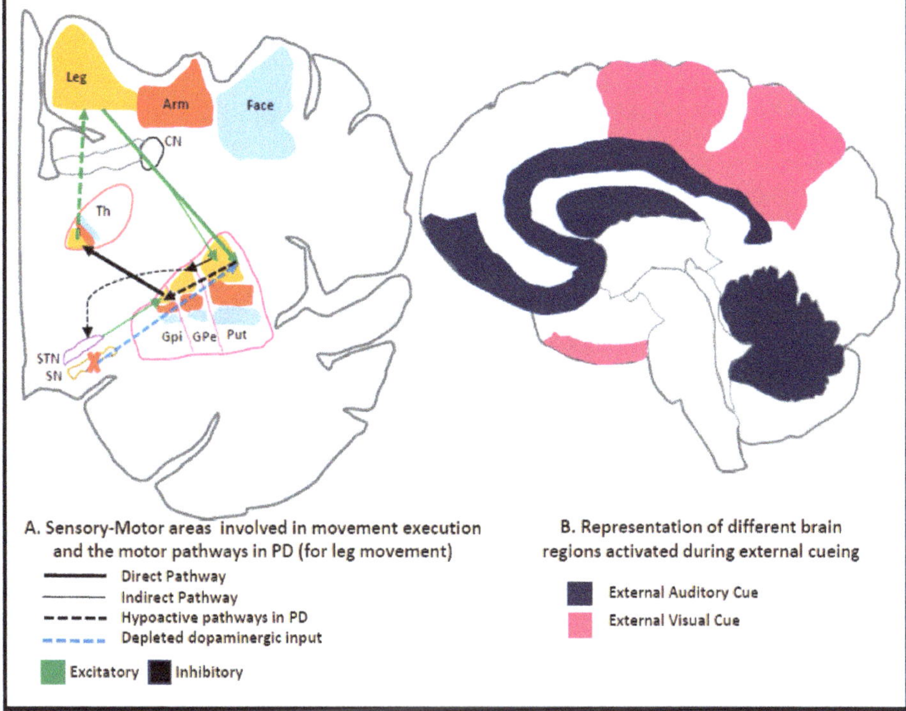

Figure 1. (**A**) Sensory-motor areas for movement execution in the basal ganglia and the impaired motor pathways in Parkinson's disease (PD) with the prevalence of the indirect pathway over the direct pathway and the affected SN's input to the circuit. SN—Substantia nigra, GPi—globus pallidus internus, GPe—globus pallidus externus, Put—putamen, Th—thalamus, CN—caudate nucleus, STN—sub-thalamic nucleus. This results in increased neuronal firing activity in the output nuclei of the basal ganglia that leads to excessive inhibition of thalamo-cortical and brainstem motor systems which, in turn, interferes with movement onset and execution [28,29]. (**B**) Representation of brain areas activated during external cueing reported from findings of image analysis studies conducted on people with PD during cueing experiments [17,30–32].

The presentation of cues in PD is hypothesized to compensate for the pathology by increasing cortical activation which diminishes pathological activity (10–30 Hz) in the basal ganglia [33], mainly by suppressing the subthalamic nucleus through direct pathways [34]. In the case of visual cues, the unaffected visual-motor pathways are believed to play a major role in facilitating movements bypassing the basal ganglia [35].

3. Methodology

In this review, the current state of scientific knowledge associated with cueing to improve gait and posture in PD is presented. The search for research articles involving cueing/feedback to improve gait/posture in PD used combinations of the following keywords: Parkinson's disease, cueing/cues/cue, real-time feedback, gait, and posture. From the set of 304 articles returned by the search, only studies that used quantitative gait and posture outcome measures (e.g., step length, stride length, walking speed, cadence, and posture) were included. The set of studies were then categorized by type of feedback implemented (visual, auditory, somatosensory), by wearability/non-wearability of the cueing device/mechanism, and by study duration (single-session or long-term training). References cited in the selected publications were also examined for other relevant studies to be considered. Studies were excluded if they were not directed for people with PD, did not measure spatiotemporal parameters of gait and/or posture, or used non-cue-based gait and posture rehabilitation strategies.

4. Cueing for Rehabilitation in PD

Given the limited ability of pharmacological and surgical treatments to address gait and postural impairments in PD, various forms of external cueing (visual, auditory, or somatosensory) are being investigated for inclusion in neuromotor rehabilitation programs. Cueing can be defined as a mechanism of applying a spatial or a temporal stimulus to facilitate initiating or maintaining motor activity [32]. Numerous studies have shown that external cueing can improve the amplitude and timing of the intended movement by increasing body position/movement awareness, making it a suitable modality for gait and posture rehabilitation [25,26,36–40]. In addition, cueing has also been increasingly used in helping with the initiation of a movement [41].

Cueing studies could be classified as open-loop cueing or closed-loop cueing based on how the cue is presented. In open-loop cueing, the user is presented a series of cues in a periodic or preset manner that is independent of the user's performance. Metronome beats and a set of lines on the floor separated by a preset distance are examples of open-loop temporal and spatial cues, respectively. Open-loop studies have most widely utilized auditory or visual forms of cues to improve gait in people with PD. While auditory cues have most often been delivered as rhythmic auditory stimulation (RAS) or metronome beats in accordance to the user's preferred gait speed or cadence [36,42–46], other types of cues such as highly rhythmic music or verbal instructions have also been investigated in some of the studies [43,44,47]. Most forms of visual cueing present lines or markers on the floor as targets for foot placement. Markers such as stripes/tapes on the floor, projections from laser pointers, and lights mounted on the user or embedded on a walking stick or walker [48–52] have been utilized. Visual cues were spaced at distances based on the subject's average step/stride length measured at baseline trials. A few studies have investigated the use of somatosensory cues [53–55] using vibrating wrist-worn devices and a combination of audio/visual and or/somatosensory cues for rehabilitation [40,56,57].

Studies of open-loop cueing used as a therapeutic modality have demonstrated short-term and long-term gait improvements [41–43,48,58–61]. Short-term studies investigated immediate effects with and without different cue interventions [62–64]. Laboratory-based long-term training studies compared walking with cues to without cues [50]. One long-term auditory cueing study investigated differences between ecological-based footstep cues (sound recorded while walking on gravel) to artificially synthesized RAS [43] and compared walking with auditory cues to walking with visual cues [40]. In the studies that presented cues as training, cues were provided progressively [65] or in combination with physical therapy improved step time variability [59], posture and bradykinesia [66], stride length, gait speed, and cadence [42].

In contrast with open-loop cues, closed-loop cueing provides feedback on the user's performance in real-time which can facilitate modifying one's performance to achieve the desired movements. Real-time feedback of step length [67–70] and uprightness of posture [69] have been investigated for targeting PD-specific gait and posture deficits. However, these studies used treadmill-based cueing systems and, therefore, are not suitable for overground locomotion during free-living conditions.

Many studies have been performed using virtual reality (VR) which provides visual stimuli that can help in motor and cognitive training [60,62,71–76]. These studies have used augmented visual/auditory- or somatosensory-based feedback for training, but a meta-analysis [77] indicated that there is only limited evidence of improvements in gait and balance due to the use of VR compared to an active intervention without the VR component. Importantly, most of these VR systems require a very sophisticated and expensive setup and may not be suitable for use at home.

5. Benefits of Open-Loop Cueing on Gait in PD

Evaluation of the acute/immediate effects of cues demonstrated that gait variables, such as cadence [48,52,56,59,78,79], speed [48,52,57,59,80], and step length [42,44,45,48,63,80], increased during walking with rhythmic auditory stimulation (RAS) when compared to walking without cues. In some instances, the improvements in step length were reported to be a consequence of using a cadence that was higher than the baseline. In addition to improving stride length and temporal measures, RAS also reduced stride-time variability [81] and helped persons without freezing of gait (FoG) more than those with FoG [79]. It was suggested that RAS might provide an external rhythm that can compensate for the defective internal rhythm of the basal ganglia in PD [45,50,80].

Use of visual cues, on the other hand, consistently improved step/stride length [49–51,70,71,73] with or without increasing walking speed or cadence. Plausible explanations for these acute effects could be that visual cues may help fill in for the motor set deficiency by providing visual-spatial data [17,82] and help in focusing attention on gait [57,73]. However, in studies that involved visual cueing during treadmill walking, it is not clear whether the gait benefits were due to the visual cueing or to the external pacemaker effect of the treadmill. Also, treadmill walking at speeds greater than the comfortable speed may demand more attention to the task of walking itself, which may result in worsening gait automaticity (ability to perform upper and lower limbs movements automatically during gait with little attention) which is already reduced in PD compared to age-matched controls [52,83]. An investigation of a visual cueing strategy that used a subject-mounted light device to present step length cues at a preset distance in front of the user reported improvements in stride length and gait speed [49]. Cueing studies that combine auditory, visual, or somatosensory cues [40,56] also reported improvements in cadence, gait speed, and stride length. Moreover, studies that focused on attention strategy by asking people with PD to think about taking larger strides were found to be effective in normalizing gait deficits observed in PD [47,57].

Notably, studies that have investigated the impact of long-term training demonstrated that RAS was effective in improving both temporal and spatial gait measures, such as walking speed, cadence, and step/stride length, regardless of the type of sound stimulation (ecological, synthetic auditory cue) that was provided. A follow-up evaluation conducted after three months revealed that the effects of the training were still largely maintained. When RAS was used for one-week training in PD people with FoG [58], walking speed was increased, but no change in freezing episodes was noted, whereas, in another study that used RAS for a three-week training, stride length, walking speed, and cadence were significantly increased [42]. Effects of long-term gait training with and without visual cues showed increases in step length and gait speed [50]. An open-loop cueing study that demonstrated improvements in both temporal gait parameters and stride length attributed temporal improvements with the use of auditory cues and improved stride length to the visual cues [40]. Results from a similar study [66] showed improvements in postural stability and bradykinesia (as measured using Unified Parkinson's Disease Rating Scale (UPDRS)-Part III items) that were retained six weeks after the training period was completed.

6. Benefits of Closed-Loop Cueing on Gait in PD

Closed-loop cueing provides feedback based on the user's movements in real-time so that the user can be aware of their performance and modulate it to achieve the desired/target performance. Studies that investigated closed-loop cueing are fewer in number and are more recent as compared to open-loop strategies. Both single-session and long-term training studies using closed-loop cueing were conducted using auditory, visual, somatosensory, and combined cueing strategies to evaluate their effects on gait and posture. Studies that used closed-loop feedback systems have demonstrated a higher degree of gait and posture improvement as well as residual carry-over effects in comparison with open-loop, feed-forward systems [39,81]. This could be because performance-based cues have been shown to help the user understand the delivered cue.

A single-session closed-loop study provided visual feedback based on the patient's own motion using eye-glasses and observed acute improvements in walking speed and stride length [84]. Two studies used treadmill walking with closed-loop visual cueing to demonstrate that people with PD could successfully follow the cues and improve the targeted gait parameters; one involved projection of target step length and uprightness cues (only one type of feedback was used at a given time) on the monitor in front of the treadmill [69] and the other projected visual cues (transverse lines) on the treadmill belt [70]. A few studies developed smartphone applications and utilized data from inertial measurement units to measure surrogates of current gait performance, which were obtained by calculating an average of the parameter over several steps, and provided feedback when the gait parameter was not in the target zone [85,86]. The feedback was provided to the user only when the gait pattern was insufficient and was referred to as "on-demand" feedback [86]. Of the closed-loop cueing studies listed in Table 1, two of them examined the immediate/acute effects of auditory cues in a single session using a wearable sensor system. The Ambulosono sensor system and the StepPlus system [87,88] were developed to provide auditory feedback to inform users when their current spatiotemporal gait parameters are out of a specified target range. Both systems [87,88] were designed for use by people with PD but have not yet been tested in people with PD. Preliminary results on a control population (a group of individuals without PD) showed improvements in stride length, stride length CoV, and cadence. The Armsense device, a portable device to measure arm-swing and provide tactile feedback, was tested in a single-session study on individuals with PD and demonstrated improvements in spatiotemporal gait parameters [89].

With mounting evidence suggesting greater gait and posture improvements as a result of closed-loop cueing training, a pilot study [69] was extended to assess the performance of cues on improving gait and posture in PD in a six-week training study [90,91].

Other long-term training studies using closed-loop visual and auditory cueing evaluated the effects of closed-loop cueing on a variety of gait parameters: gait speed [67,68,70,86,92,93], cadence [70,86], stride length [67,68,70,86], fall incidences [94], and other gait and dynamic balance measures [74,84,95] at follow-up and post-training. Only two of the long-term training studies used a wearable sensor-based, closed-loop system [86,95].

Some closed-loop training studies used augmented reality devices and game-based motion therapy for combinational cueing [72,75,77,84,96,97]. Results from these studies suggested that the closed-loop sensory feedback with or without long-term training was an effective non-pharmacologic intervention for gait and balance improvement in PD. The abovementioned studies involving virtual reality and game-based visual cueing have provided feedback to the user using monitors placed at the eye-level which may help people with PD to be upright at least while following the feedback.

The regular practice of being upright during the training and any sustained benefits may reduce the issue of stoopness experienced by people with PD. To date, only a few closed-loop studies [70,74,86,94,95] included a randomized control trial (RCT) research design to confirm that the gait and posture improvements observed are mainly due to the presentation of cues.

Table 1. Summary of key findings of closed-loop cueing strategies that were included in this review.

Study	Intervention Type	Sensors; Feedback Mode	Outcome Measures	Study Protocol	Results	Limitations
Badarny et al. 2014 [84]	Visual	Wearable motion sensors; virtual reality based eye-glasses	Walking speed, stride length	Single-session study with cue and a follow-up evaluation (1 week later)	Increases in both walking speed and stride length, immediate effects and at follow-up	No control group; only assessed short-term effects
Jellish et al. 2015 [69]	Visual	Treadmill-based, video-based motion capture system and a feedback monitor	Step length, postural (back) angle measured during treadmill walking	Single-session study using multiple trials with and without cues	Increases in uprightness and step length	Utilized technology that is only available in research labs
Chomiak et al. 2019 [87]	Auditory	IMU sensors with a smartphone application-(Ambulosono sensor system)	Step length, walking distance, velocity, and cadence	Single-session study multiple trials	Evaluation of the sensor's performance on healthy controls	Use of iPod Touch for feedback is not cost-effective and the system has not been evaluated on PD population
Bartels et al. 2018 [88]	Auditory	IMU sensors with a smartphone application	Stride length	Single-session study with multiple trials	Evaluation of the sensor's performance on healthy controls	The system has not been evaluated on PD population
Young et al. 2014 [98]	Auditory-sonification of gait–swing phase	Video-based motion capture system with a smart phone application	Step length CoV	Single-session study in the lab with multiple trials	Reduction in step length variability	Utilized technology that is only available in research labs
Thompson et.al. 2017 [89]	Somatosensory	IMU sensors with a software application on the laptop and a vibratory device	Step length, lateral trunk sway, cadence, gait velocity, arm swing	Single-session study in the lab with multiple trials.	Increases in step length, arm swing magnitude, reduced cadence	Though somatosensory cues have been successful in helping with the rhythm of the movement, they are less effective in increasing the amplitude of the desired movement
Schlick et al. 2016 [70]	Visual	Treadmill-based pressure platform and video feedback monitor	Gait speed, stride length and cadence	Long-term training (5 weeks) at lab, RCT	Both the training and control group showed increases in gait speed and stride length post training, but sustained effects after 2 months were observed only in the case of feedback-based training	Small sample at follow-up because of attrition
Mirelman et al. 2016 [94]	Visual	Video-based motion capture system with virtual reality feedback	Fall incident rates	Long-term training (3 times/week for 6 weeks) at lab	Reduction in the rate of falls during the 6 month follow-up evaluation	No control group
Baskaran. 2017 [90]	Visual	Treadmill-based video-based motion capture system and a feedback monitor	Step length, postural (back) angle measured during treadmill walking	Long-term training (3 times/week for 6 weeks) at lab	Increases in uprightness and step length	No control group and a small sample size
Yang et al. 2016 [75]	Visual	Video-based motion capture system with virtual reality feedback	BBS, DGI, TUG test	Long-term training (2 times/week for 6 weeks) at lab, RCT	Increase in clinical score, BBS performance which was retained at 2 week follow-up	Small sample size
Van den Heuvel et al. 2014 [60]	Visual	Video-based augmented feedback system with treadmill and IMU sensors	FRT, BBS, UPDRS	Long-term training (2 times/week for 5 weeks) at lab, RCT	Improvements in balance scores in favor of the feedback system	Changes in scores were not statistically significant
Ginis et al. 2015 [86]	Auditory	IMU sensors with a smartphone application (CuPiD system)	Gait speed, cadence, stride length and stride length asymmetry	Long-term training (3 times/week for 6 weeks) at home, RCT	Increase in gait speed at post-training	Assessors were not blinded

Table 1. *Cont.*

Study	Intervention Type	Sensors; Feedback Mode	Outcome Measures	Study Protocol	Results	Limitations
Carpinella et al. 2016 [95]	Auditory and visual	IMU sensors and monitor for exercise therapy with a Gamepad (Gaming Experience in Parkinson's Disease)	BBS and gait speed	Long-term training (3 times/week for 6 weeks) at lab, RCT	Increase in clinical score, BBS performance and retained effects at 1 month follow-up	Lack of online computation of gait measures and the use of technology that is only available in research labs
Frazzitta et al. 2009 [68]	Auditory and visual	Treadmill-based strain gauge and a visual feedback monitor	Stride length, gait speed	Long-term training (4 weeks) at lab	Greater increase in gait speed and stride length following treadmill-based cue training than with overground-based cue training	No control group and the study did not evaluate residual effect at follow-up
Rochester et al. 2010 [81]	Auditory, visual, and somatosensory	IMU-based rhythmical feedback system	Walking speed, step length, step frequency	Long-term training study for 6 weeks at lab	Increase in walking speed and step length with all cue types in both single and dual-tasking after training.	No control group
Espay et al. 2010 [96]	Auditory and visual	IMU sensors and a head-mounted display and headphones	Gait velocity, stride length and cadence	Home-based training for 2 weeks	Increase in gait velocity and stride length after training	No control group
Pompeu et al. 2012 [74]	Auditory and visual	Wii Fit games	UPDRS	Long-term training (2 times/week for 7 weeks) with exercise therapy at lab	Decrease (improvement) in UPDRS post-training and at 2 month follow-up evaluation	No control group

Inertial measurement unit (IMU), Coefficient of variation (CoV)), Unified Parkinson's Disease Rating Scale (UPDRS), Berg balance score (BBS), Dynamic gait index (DGI), Functional reach test (FRT), Randomized controlled trial (RCT). The table includes only experimental studies and does not list the review articles mentioned in the text.

7. Discussion

7.1. Different Cueing Types May Engage Different Mechanisms

Findings from the literature indicate that both types of cueing (i.e., auditory and visual) result in improved gait and posture in individuals with PD. The hypothesized neural mechanism for external cueing, suggested by Morris et al. [92], bypasses the hypoactive basal ganglia-supplementary motor cortex (SMA) circuit by slightly altering the way the neural circuits control movement in individuals with PD [31,35,99]. In general, sensory cues are known to enable the dorsolateral pre-motor control system [30,32,63] which bypasses the SMA that is deficient in PD. Specifically, it has been suggested that auditory cues help in improving the temporal parameters, such as cadence and gait speed, and that external cues help because they are able to bypass the internal rhythm deficit associated with PD. Visual cues, on the other hand, are believed to enable the visual–cerebellar motor circuit that influences the spatial aspects of gait, such as step/stride length [71,82,92,100].

7.2. Effect of Disease Stage on Cueing Strategy

The effect of cueing in PD rehabilitation may depend on the stage of the disease and the type of dominant symptoms being experienced. The studies included in this review focused predominantly on individuals classified as Hoehn and Yahr stages II, III, and IV. For people in the early stage of disease severity, external cues can compensate for small deviations from their normal gait pattern thereby maintaining optimal gait quality and preventing deconditioning through training. Severely affected individuals with PD rely on external cues to compensate for deficits in the automatic control mechanisms (i.e., the ability to automatically generate normal stride length in a timely manner) thus improving gait and reducing the incidence of falls and freezing of gait [30,32,37,55].

7.3. Open-loop Cueing: Challenges and Limitations

The primary challenge in cueing, whether open-loop or closed-loop, is to present the cue in a manner that is informative, but does not have detrimental side effects on gait or balance. Despite numerous studies that demonstrated the benefits of cues on gait in PD, most of them did not investigate cues effects on balance control. Also, studies that specifically used cues to improve balance control in PD are very limited, and they were focused on improving posture during quiet stance, sit-to-stand, and dynamic balance maneuvers [101–103]. It is possible that the presentation of visual cues in the form of markers on the floor or on the treadmill belt [38,70,81] may further degrade posture and stability because it requires people with PD, who may already experience stooped posture, to look down.

Similarly, the use of auditory cues provided via earphones may reduce the awareness of environmental sounds which may make it unsuitable for use outside of a laboratory environment. This could be particularly problematic if sounds are provided continuously, i.e., with every step. Another major limitation associated with open-loop systems is that the user is required to detect any mismatch between the cue and their performance and decide how to respond in a manner that will get them entrained (in sync) with the cues. For auditory cues, the user might have to make a quick or a long-duration step in order to get in phase with the cues; for visual cues, the user might have to make a short or a long step in order to get in phase. Finally, although the literature on open-loop cueing in PD includes several studies that observed considerable improvements in spatiotemporal parameters, future studies along these lines could help to move the field forward by documenting how well users are able to follow the cues and by utilizing an RCT research design. Documentation of performance in following the cues could provide insight into the limitations of the cue presentation technique and could help to document the progression of learning throughout an intervention; the use of an RCT design would provide more reliable and actionable evidence for a decision to use a technique in the clinic.

7.4. Closed-loop Cueing: Challenges and Limitations

As with open-loop cueing, closed-loop strategies must also present information in a manner that is informative and does not have detrimental side-effects on gait. In addition, closed-loop paradigms must also measure/calculate the feedback parameter in real-time and, if it is to be useful outside of the laboratory, the entire system should be wearable and affordable. Setups that use motion capture systems in the laboratory or clinic are expensive and require travel and staff time. Treadmill-based systems pose limitations because some people with PD do not feel comfortable walking on a treadmill, whether that be at home or a facility with supervision. For these reasons, a low-cost wearable system that could readily be used on a daily basis during overground walking might be more widely accepted. However, there are technical challenges in measuring gait parameters from wearable sensors in real-time and conveying feedback in a manner that is safe and easy-to-use. Our group and others are working to develop low-cost, wearable systems for real-time feedback in home or community environments [87,104,105]. Once the technical development challenges are overcome, these systems will be evaluated for accuracy and safety and then clinical efficacy will have to be assessed in an RCT. These types of technologies have potential for widespread use, but they would require regulatory approval before commercialization and marketing.

8. Conclusions

Based on the review of the literature presented here, it is clear that cueing can be an effective component of locomotor therapy for people with PD who experience gait deficits. Rhythmic auditory cueing has been the most widely used technique, but it is most effective only in influencing the temporal parameters of gait. Visual cueing techniques have been used to increase spatial parameters, such as step/stride length, and to reduce step/stride length variability and asymmetry. Such improvements could have a high clinical impact, as they are important factors in gait and posture rehabilitation for people with PD. However, the usefulness of visual cueing techniques has been limited by challenges in presenting cues in a manner that is practical outside the laboratory and in a manner that encourages upright walking. To overcome the limitations of currently available techniques, several groups are developing unobtrusive wearable systems for closed-loop cueing to provide feedback of performance on a step-by-step or on-demand basis. These systems seek to improve locomotion during activities of daily living by providing feedback of gait and posture parameters that are often deficient in PD and by providing it in a way that can be readily used on a regular basis in the home or the community. Recent engineering developments have produced technology that is suitable for applications that require wearable sensors. Current challenges are to develop algorithms to interpret information from the sensors in real-time and to present it to the user in a manner that is intuitive, non-distracting, and actionable. Such advances that lead to technology for cueing that is effective, affordable, and wearable may enable adoption of these techniques by individuals with PD for use on a regular basis at home and in the community.

Author Contributions: N.K., J.J.A., and N.M. conceptualized the article, wrote the text and performed critical revision of the manuscript; H.A.S. provided support for interpretation of important clinical content and for review and editing of the entire manuscript; Funding was obtained by N.K., J.J.A., and H.A.S.

Funding: The preparation of the manuscript was supported by funds from the National Institutes of Health (1R21NR017484).

Conflicts of Interest: The authors have no conflict of interest/financial disclosures to report.

References

1. Rogers, M.W. Disorders of posture, balance, and gait in Parkinson's disease. *Clin. Geriatr. Med.* **1996**, *12*, 825–845. [CrossRef]
2. Kalia, L.V.; Lang, A.E. Parkinson disease in 2015: Evolving basic, pathological and clinical concepts in PD. *Nat. Rev. Neurol.* **2016**, *12*, 65–66. [CrossRef] [PubMed]

3. Miller-Patterson, C.; Buesa, R.; McLaughlin, N.; Jones, R.; Akbar, U.; Friedman, J.H. Motor asymmetry over time in Parkinson's disease. *J. Neurol. Sci.* **2018**, *393*, 14–17. [CrossRef] [PubMed]
4. Takakusaki, K.; Tomita, N.; Yano, M. Substrates for normal gait and pathophysiology of gait disturbances with respect to the basal ganglia dysfunction. *J. Neurol.* **2008**, *255* (Suppl. 4), 19–29. [CrossRef] [PubMed]
5. Plotnik, M.; Hausdorff, J.M. The role of gait rhythmicity and bilateral coordination of stepping in the pathophysiology of freezing of gait in Parkinson's disease. *Mov. Disord.* **2008**, *23* (Suppl. 2), S444–S450. [CrossRef] [PubMed]
6. Boonstra, T.A.; van der Kooij, H.; Munneke, M.; Bloem, B.R. Gait disorders and balance disturbances in Parkinson's disease: Clinical update and pathophysiology. *Curr. Opin. Neurol.* **2008**, *21*, 461–471. [CrossRef]
7. Hobert, M.A.; Nussbaum, S.; Heger, T.; Berg, D.; Maetzler, W.; Heinzel, S. Progressive Gait Deficits in Parkinson's Disease: A Wearable-Based Biannual 5-Year Prospective Study. *Front. Aging Neurosci.* **2019**, *11*, 22. [CrossRef]
8. Rochester, L.; Baker, K.; Nieuwboer, A.; Burn, D. Targeting dopa-sensitive and dopa-resistant gait dysfunction in Parkinson's disease: Selective responses to internal and external cues. *Mov. Disord.* **2011**, *26*, 430–435. [CrossRef]
9. Rocchi, L.; Chiari, L.; Horak, F.B. Effects of deep brain stimulation and levodopa on postural sway in Parkinson's disease. *J. Neurol. Neurosurg. Psychiatry* **2002**, *73*, 267–274. [CrossRef]
10. Beuter, A.; Hernandez, R.; Rigal, R.; Modolo, J.; Blanchet, P.J. Postural sway and effect of levodopa in early Parkinson's disease. *Can. J. Neurol. Sci.* **2008**, *35*, 65–68. [CrossRef]
11. King, L.A.; Horak, F.B. Lateral stepping for postural correction in Parkinson's disease. *Arch. Phys. Med. Rehabil.* **2008**, *89*, 492–499. [CrossRef] [PubMed]
12. King, L.A.; St George, R.J.; Carlson-Kuhta, P.; Nutt, J.G.; Horak, F.B. Preparation for compensatory forward stepping in Parkinson's disease. *Arch. Phys. Med. Rehabil.* **2010**, *91*, 1332–1338. [CrossRef] [PubMed]
13. Potter-Nerger, M.; Volkmann, J. Deep brain stimulation for gait and postural symptoms in Parkinson's disease. *Mov. Disord. Off. J. Mov. Disord. Soc.* **2013**, *28*, 1609–1615. [CrossRef] [PubMed]
14. Wang, J.W.; Zhang, Y.Q.; Zhang, X.H.; Wang, Y.P.; Li, J.P.; Li, Y.J. Deep Brain Stimulation of Pedunculopontine Nucleus for Postural Instability and Gait Disorder After Parkinson Disease: A Meta-Analysis of Individual Patient Data. *World Neurosurg.* **2017**, *102*, 72–78. [CrossRef]
15. Baizabal-Carvallo, J.F.; Alonso-Juarez, M. Low-frequency deep brain stimulation for movement disorders. *Parkinsonism Relat. Disord.* **2016**, *31*, 14–22. [CrossRef]
16. Xie, T.; Bloom, L.; Padmanaban, M.; Bertacchi, B.; Kang, W.; MacCracken, E.; Dachman, A.; Vigil, J.; Satzer, D.; Zadikoff, C.; et al. Long-term effect of low frequency stimulation of STN on dysphagia, freezing of gait and other motor symptoms in PD. *J. Neurol. Neurosurg. Psychiatry* **2018**, *89*, 989–994. [CrossRef]
17. Debaere, F.; Wenderoth, N.; Sunaert, S.; Van Hecke, P.; Swinnen, S.P. Internal vs external generation of movements: Differential neural pathways involved in bimanual coordination performed in the presence or absence of augmented visual feedback. *NeuroImage* **2003**, *19*, 764–776. [CrossRef]
18. Cunnington, R.; Iansek, R.; Bradshaw, J.L.; Phillips, J.G. Movement-related potentials in Parkinson's disease. Presence and predictability of temporal and spatial cues. *Brain* **1995**, *118 Pt 4*, 935–950. [CrossRef]
19. Georgiou, N.; Iansek, R.; Bradshaw, J.L.; Phillips, J.G.; Mattingley, J.B.; Bradshaw, J.A. An evaluation of the role of internal cues in the pathogenesis of parkinsonian hypokinesia. *Brain* **1993**, *116 Pt 6*, 1575–1587. [CrossRef]
20. Ferrarin, M.; Rizzone, M.; Lopiano, L.; Recalcati, M.; Pedotti, A. Effects of subthalamic nucleus stimulation and L-dopa in trunk kinematics of patients with Parkinson's disease. *Gait Posture* **2004**, *19*, 164–171. [CrossRef]
21. Redgrave, P.; Rodriguez, M.; Smith, Y.; Rodriguez-Oroz, M.C.; Lehericy, S.; Bergman, H.; Agid, Y.; DeLong, M.R.; Obeso, J.A. Goal-directed and habitual control in the basal ganglia: Implications for Parkinson's disease. *Nat. Rev. Neurosci.* **2010**, *11*, 760–772. [CrossRef] [PubMed]
22. Benninger, F.; Khlebtovsky, A.; Roditi, Y.; Keret, O.; Steiner, I.; Melamed, E.; Djaldetti, R. Beneficial effect of levodopa therapy on stooped posture in Parkinson's disease. *Gait Posture* **2015**, *42*, 263–268. [CrossRef] [PubMed]
23. Weiss, D.; Schoellmann, A.; Fox, M.D.; Bohnen, N.I.; Factor, S.A.; Nieuwboer, A.; Hallett, M.; Lewis, S.J.G. Freezing of gait: Understanding the complexity of an enigmatic phenomenon. *Brain* **2019**. [CrossRef] [PubMed]

24. Abbruzzese, G.; Berardelli, A. Sensorimotor integration in movement disorders. *Mov. Disord.* **2003**, *18*, 231–240. [CrossRef]
25. Zia, S.; Cody, F.; O'Boyle, D. Joint position sense is impaired by Parkinson's disease. *Ann. Neurol.* **2000**, *47*, 218–228. [CrossRef]
26. Schubert, M.; Prokop, T.; Brocke, F.; Berger, W. Visual kinesthesia and locomotion in Parkinson's disease. *Mov. Disord. Off. J. Mov. Disord. Soc.* **2005**, *20*, 141–150. [CrossRef]
27. Baroni, A.; Benvenuti, F.; Fantini, L.; Pantaleo, T.; Urbani, F. Human ballistic arm abduction movements: Effects of L-dopa treatment in Parkinson's disease. *Neurology* **1984**, *34*, 868–876. [CrossRef]
28. Obeso, J.A.; Rodriguez-Oroz, M.C.; Benitez-Temino, B.; Blesa, F.J.; Guridi, J.; Marin, C.; Rodriguez, M. Functional organization of the basal ganglia: Therapeutic implications for Parkinson's disease. *Mov. Disord.* **2008**, *23* (Suppl. 3), S548–S559. [CrossRef]
29. Magrinelli, F.; Picelli, A.; Tocco, P.; Federico, A.; Roncari, L.; Smania, N.; Zanette, G.; Tamburin, S. Pathophysiology of Motor Dysfunction in Parkinson's Disease as the Rationale for Drug Treatment and Rehabilitation. *Parkinsons Dis.* **2016**, *2016*, 9832839. [CrossRef]
30. Sweeney, D.; Quinlan, L.R.; Browne, P.; Richardson, M.; Meskell, P.; ÓLaighin, G. A Technological Review of Wearable Cueing Devices Addressing Freezing of Gait in Parkinson's Disease. *Sensors* **2019**, *19*, 1277. [CrossRef]
31. Nieuwboer, A. Cueing for freezing of gait in patients with Parkinson's disease: A rehabilitation perspective. *Mov. Disord.* **2008**, *23* (Suppl. 2), S475–S481. [CrossRef] [PubMed]
32. Nieuwboer, A.; Kwakkel, G.; Rochester, L.; Jones, D.; van Wegen, E.; Willems, A.M.; Chavret, F.; Hetherington, V.; Baker, K.; Lim, I. Cueing training in the home improves gait-related mobility in Parkinson's disease: The RESCUE trial. *J. Neurol. Neurosurg. Psychiatry* **2007**, *78*, 134–140. [CrossRef] [PubMed]
33. Amirnovin, R.; Williams, Z.M.; Cosgrove, G.R.; Eskandar, E.N. Visually guided movements suppress subthalamic oscillations in Parkinson's disease patients. *J. Neurosci.* **2004**, *24*, 11302–11306. [CrossRef] [PubMed]
34. Sarma, S.V.; Cheng, M.L.; Eden, U.; Williams, Z.; Brown, E.N.; Eskandar, E. The effects of cues on neurons in the basal ganglia in Parkinson's disease. *Front. Integr. Neurosci.* **2012**, *6*, 40. [CrossRef]
35. Glickstein, M.; Stein, J. Paradoxical movement in Parkinson's disease. *Trends Neurosci.* **1991**, *14*, 480–482. [CrossRef]
36. Ashoori, A.; Eagleman, D.M.; Jankovic, J. Effects of Auditory Rhythm and Music on Gait Disturbances in Parkinson's Disease. *Front. Neurol.* **2015**, *6*, 234. [CrossRef]
37. Ginis, P.; Nackaerts, E.; Nieuwboer, A.; Heremans, E. Cueing for people with Parkinson's disease with freezing of gait: A narrative review of the state-of-the-art and novel perspectives. *Ann. Phys. Rehabil. Med.* **2018**, *61*, 407–413. [CrossRef]
38. Griffin, H.J.; Greenlaw, R.; Limousin, P.; Bhatia, K.; Quinn, N.P.; Jahanshahi, M. The effect of real and virtual visual cues on walking in Parkinson's disease. *J. Neurol.* **2011**, *258*, 991–1000. [CrossRef]
39. Mancini, M.; Smulders, K.; Harker, G.; Stuart, S.; Nutt, J.G. Assessment of the ability of open-and closed-loop cueing to improve turning and freezing in people with Parkinson's disease. *Sci. Rep.* **2018**, *8*, 12773. [CrossRef]
40. Suteerawattananon, M.; Morris, G.S.; Etnyre, B.R.; Jankovic, J.; Protas, E.J. Effects of visual and auditory cues on gait in individuals with Parkinson's disease. *J. Neurol. Sci.* **2004**, *219*, 63–69. [CrossRef]
41. Lu, C.; Amundsen Huffmaster, S.L.; Tuite, P.J.; Vachon, J.M.; MacKinnon, C.D. Effect of Cue Timing and Modality on Gait Initiation in Parkinson Disease With Freezing of Gait. *Arch. Phys. Med. Rehabil.* **2017**, *98*, 1291–1299. [CrossRef] [PubMed]
42. Thaut, M.H.; McIntosh, G.C.; Rice, R.R.; Miller, R.A.; Rathbun, J.; Brault, J.M. Rhythmic auditory stimulation in gait training for Parkinson's disease patients. *Mov. Disord.* **1996**, *11*, 193–200. [CrossRef] [PubMed]
43. Murgia, M.; Pili, R.; Corona, F.; Sors, F.; Agostini, T.A.; Bernardis, P.; Casula, C.; Cossu, G.; Guicciardi, M.; Pau, M. The Use of Footstep Sounds as Rhythmic Auditory Stimulation for Gait Rehabilitation in Parkinson's Disease: A Randomized Controlled Trial. *Front. Neurol.* **2018**, *9*, 348. [CrossRef] [PubMed]
44. Hausdorff, J.M.; Lowenthal, J.; Herman, T.; Gruendlinger, L.; Peretz, C.; Giladi, N. Rhythmic auditory stimulation modulates gait variability in Parkinson's disease. *Eur. J. Neurosci.* **2007**, *26*, 2369–2375. [CrossRef] [PubMed]

45. McIntosh, G.C.; Brown, S.H.; Rice, R.R.; Thaut, M.H. Rhythmic auditory-motor facilitation of gait patterns in patients with Parkinson's disease. *J. Neurol. Neurosurg. Psychiatry* **1997**, *62*, 22–26. [CrossRef] [PubMed]
46. Howe, T.E.; Lovgreen, B.; Cody, F.W.; Ashton, V.J.; Oldham, J.A. Auditory cues can modify the gait of persons with early-stage Parkinson's disease: A method for enhancing parkinsonian walking performance? *Clin. Rehabil.* **2003**, *17*, 363–367. [CrossRef] [PubMed]
47. Behrman, A.L.; Teitelbaum, P.; Cauraugh, J.H. Verbal instructional sets to normalise the temporal and spatial gait variables in Parkinson's disease. *J. Neurol. Neurosurg. Psychiatry* **1998**, *65*, 580–582. [CrossRef] [PubMed]
48. Bagley, S.; Kelly, B.; Tunnicliffe, N.; Turnbull, G.I.; Walker, J.M. The effect of visual cues on the gait of independently mobile Parkinson's patients. *Phyiotherapy* **1991**, *77*, 415–420. [CrossRef]
49. Lewis, G.N.; Byblow, W.D.; Walt, S.E. Stride length regulation in Parkinson's disease: The use of extrinsic, visual cues. *Brain* **2000**, *123 Pt 10*, 2077–2090. [CrossRef]
50. Sidaway, B.; Anderson, J.; Danielson, G.; Martin, L.; Smith, G. Effects of long-term gait training using visual cues in an individual with Parkinson disease. *Phys. Ther.* **2006**, *86*, 186–194.
51. Azulay, J.P.; Mesure, S.; Amblard, B.; Blin, O.; Sangla, I.; Pouget, J. Visual control of locomotion in Parkinson's disease. *Brain* **1999**, *122 Pt 1*, 111–120. [CrossRef]
52. Luessi, F.; Mueller, L.K.; Breimhorst, M.; Vogt, T. Influence of visual cues on gait in Parkinson's disease during treadmill walking at multiple velocities. *J. Neurol. Sci.* **2012**, *314*, 78–82. [CrossRef] [PubMed]
53. Xu, J.; Bao, T.; Lee, U.H.; Kinnaird, C.; Carender, W.; Huang, Y.; Sienko, K.H.; Shull, P.B. Configurable, wearable sensing and vibrotactile feedback system for real-time postural balance and gait training: Proof-of-concept. *J. Neuroeng. Rehabil.* **2017**, *14*, 102. [CrossRef] [PubMed]
54. Gopalai, A.A.; Senanayake, S.M.; Kiong, L.C.; Gouwanda, D. Real-time stability measurement system for postural control. *J. Bodyw. Mov. Ther.* **2011**, *15*, 453–464. [CrossRef] [PubMed]
55. Van Wegen, E.; de Goede, C.; Lim, I.; Rietberg, M.; Nieuwboer, A.; Willems, A.; Jones, D.; Rochester, L.; Hetherington, V.; Berendse, H.; et al. The effect of rhythmic somatosensory cueing on gait in patients with Parkinson's disease. *J. Neurol. Sci.* **2006**, *248*, 210–214. [CrossRef] [PubMed]
56. De Oliveira Souza, C.; Callil Voos, M.; Fen Chien, H.; Ferreira Barbosa, A.; Brant Rodrigues, R.; Colucci Fonoff, F.; Caromano, F.A.; de Abreu, L.C.; Reis Barbosa, E.; Talamoni Fonoff, E. Combined auditory and visual cueing provided by eyeglasses influence gait performance in Parkinson Disease patients submitted to deep brain stimulation: A pilot study. *Int. Arch. Med.* **2015**. [CrossRef]
57. Lohnes, C.A.; Earhart, G.M. The impact of attentional, auditory, and combined cues on walking during single and cognitive dual tasks in Parkinson disease. *Gait Posture* **2011**, *33*, 478–483. [CrossRef] [PubMed]
58. Cubo, E.; Leurgans, S.; Goetz, C.G. Short-term and practice effects of metronome pacing in Parkinson's disease patients with gait freezing while in the 'on' state: Randomized single blind evaluation. *Parkinsonism Relat. Disord.* **2004**, *10*, 507–510. [CrossRef]
59. Del Olmo, M.F.; Cudeiro, J. Temporal variability of gait in Parkinson disease: Effects of a rehabilitation programme based on rhythmic sound cues. *Parkinsonism Relat. Disord.* **2005**, *11*, 25–33. [CrossRef]
60. Van den Heuvel, M.R.; Kwakkel, G.; Beek, P.J.; Berendse, H.W.; Daffertshofer, A.; van Wegen, E.E. Effects of augmented visual feedback during balance training in Parkinson's disease: A pilot randomized clinical trial. *Parkinsonism Relat. Disord.* **2014**, *20*, 1352–1358. [CrossRef]
61. Morris, M.E.; Iansek, R.; Matyas, T.A.; Summers, J.J. Ability to modulate walking cadence remains intact in Parkinson's disease. *J. Neurol. Neurosurg. Psychiatry* **1994**, *57*, 1532–1534. [CrossRef] [PubMed]
62. Shen, X.; Mak, M.K. Balance and Gait Training with Augmented Feedback Improves Balance Confidence in People With Parkinson's Disease: A Randomized Controlled Trial. *Neurorehabil. Neural Repair* **2014**, *28*, 524–535. [CrossRef] [PubMed]
63. Lim, I.; van Wegen, E.; de Goede, C.; Deutekom, M.; Nieuwboer, A.; Willems, A.; Jones, D.; Rochester, L.; Kwakkel, G. Effects of external rhythmical cueing on gait in patients with Parkinson's disease: A systematic review. *Clin. Rehabil.* **2005**, *19*, 695–713. [CrossRef] [PubMed]
64. Nieuwboer, A.; Baker, K.; Willems, A.M.; Jones, D.; Spildooren, J.; Lim, I.; Kwakkel, G.; Van Wegen, E.; Rochester, L. The short-term effects of different cueing modalities on turn speed in people with Parkinson's disease. *Neurorehabil. Neural Repair* **2009**, *23*, 831–836. [CrossRef]
65. Ford, M.P.; Malone, L.A.; Nyikos, I.; Yelisetty, R.; Bickel, C.S. Gait training with progressive external auditory cueing in persons with Parkinson's disease. *Arch. Phys. Med. Rehabil.* **2010**, *91*, 1255–1261. [CrossRef]

66. Marchese, R.; Diverio, M.; Zucchi, F.; Lentino, C.; Abbruzzese, G. The role of sensory cues in the rehabilitation of parkinsonian patients: A comparison of two physical therapy protocols. *Mov. Disord.* **2000**, *15*, 879–883. [CrossRef]
67. Frazzitta, G.; Bertotti, G.; Ucellini, D.; Maestri, R. Parkinson's Disease Rehabilitation-a pilot study with 1 year follow-up. *Mov. Disord.* **2010**, *25*, 1762–1763.
68. Frazzitta, G.; Maestri, R.; Uccellini, D.; Bertotti, G.; Abelli, P. Rehabilitation treatment of gait in patients with Parkinson's disease with freezing: A comparison between two physical therapy protocols using visual and auditory cues with or without treadmill training. *Mov. Disord.* **2009**, *24*, 1139–1143. [CrossRef]
69. Jellish, J.; Abbas, J.J.; Ingalls, T.; Mahant, P.; Samanta, J.; Ospina, M.; Krishnamurthi, N. A System for Real-Time Feedback to Improve Gait and Posture in Parkinson's Disease. *IEEE J. Biomed. Health Inform.* **2015**, *19*, 1809–1819. [CrossRef]
70. Schlick, C.; Ernst, A.; Botzel, K.; Plate, A.; Pelykh, O.; Ilmberger, J. Visual cues combined with treadmill training to improve gait performance in Parkinson's disease: A pilot randomized controlled trial. *Clin. Rehabil.* **2016**, *30*, 463–471. [CrossRef]
71. Lee, N.Y.; Lee, D.K.; Song, H.S. Effect of virtual reality dance exercise on the balance, activities of daily living, and depressive disorder status of Parkinson's disease patients. *J. Phys. Ther. Sci.* **2015**, *27*, 145–147. [CrossRef] [PubMed]
72. Liao, Y.Y.; Yang, Y.R.; Cheng, S.J.; Wu, Y.R.; Fuh, J.L.; Wang, R.Y. Virtual Reality-Based Training to Improve Obstacle-Crossing Performance and Dynamic Balance in Patients With Parkinson's Disease. *Neurorehabil. Neural Repair* **2015**, *29*, 658–667. [CrossRef] [PubMed]
73. Pedreira, G.; Prazeres, A.; Cruz, D.; Gomes, I.; Monteiro, L.; Melos, A. Virtual games and quality of life in PD-a randomized controlled trial. *Adv. Parkinson's Dis.* **2013**, *2*, 97–101. [CrossRef]
74. Pompeu, J.E.; Mendes, F.A.; Silva, K.G.; Lobo, A.M.; Oliveira Tde, P.; Zomignani, A.P.; Piemonte, M.E. Effect of Nintendo Wii-based motor and cognitive training on activities of daily living in patients with Parkinson's disease: A randomised clinical trial. *Physiotherapy* **2012**, *98*, 196–204. [CrossRef] [PubMed]
75. Yang, W.C.; Wang, H.K.; Wu, R.M.; Lo, C.S.; Lin, K.H. Home-based virtual reality balance training and conventional balance training in Parkinson's disease: A randomized controlled trial. *J. Formos. Med. Assoc.* **2016**, *115*, 734–743. [CrossRef]
76. Yen, C.Y.; Lin, K.H.; Hu, M.H.; Wu, R.M.; Lu, T.W.; Lin, C.H. Effects of virtual reality-augmented balance training on sensory organization and attentional demand for postural control in people with Parkinson disease: A randomized controlled trial. *Phys. Ther.* **2011**, *91*, 862–874. [CrossRef]
77. Dockx, K.; Bekkers, E.M.; Van den Bergh, V.; Ginis, P.; Rochester, L.; Hausdorff, J.M.; Mirelman, A.; Nieuwboer, A. Virtual reality for rehabilitation in Parkinson's disease. *Cochrane Database Syst. Rev.* **2016**, *12*, CD010760. [CrossRef]
78. Zijlstra, W.; Rutgers, A.W.; Van Weerden, T.W. Voluntary and involuntary adaptation of gait in Parkinson's disease. *Gait Posture* **1998**, *7*, 53–63. [CrossRef]
79. Willems, A.M.; Nieuwboer, A.; Chavret, F.; Desloovere, K.; Dom, R.; Rochester, L.; Jones, D.; Kwakkel, G.; Van Wegen, E. The use of rhythmic auditory cues to influence gait in patients with Parkinson's disease, the differential effect for freezers and non-freezers, an explorative study. *Disabil. Rehabil.* **2006**, *28*, 721–728. [CrossRef]
80. McCoy, R.W.; Kohl, R.M.; Elliott, S.M.; Joyce, A.S. The impact of auditory cues on gait control of individuals with Parkinson's disease. *J. Hum. Mov. Stud.* **2002**, *42*, 229–236.
81. Rochester, L.; Baker, K.; Hetherington, V.; Jones, D.; Willems, A.M.; Kwakkel, G.; Van Wegen, E.; Lim, I.; Nieuwboer, A. Evidence for motor learning in Parkinson's disease: Acquisition, automaticity and retention of cued gait performance after training with external rhythmical cues. *Brain Res.* **2010**, *1319*, 103–111. [CrossRef] [PubMed]
82. Spaulding, S.J.; Barber, B.; Colby, M.; Cormack, B.; Mick, T.; Jenkins, M.E. Cueing and gait improvement among people with Parkinson's disease: A meta-analysis. *Arch. Phys. Med. Rehabil.* **2013**, *94*, 562–570. [CrossRef] [PubMed]
83. Wu, T.; Hallett, M. Neural correlates of dual task performance in patients with Parkinson's disease. *J. Neurol. Neurosurg. Psychiatry* **2008**, *79*, 760–766. [CrossRef] [PubMed]
84. Badarny, S.; Aharon-Peretz, J.; Susel, Z.; Habib, G.; Baram, Y. Virtual reality feedback cues for improvement of gait in patients with Parkinson's disease. *Tremor Other Hyperkinetic Mov.* **2014**, *4*, 225. [CrossRef]

85. Machado, J.P.F. Smartphone Based Closed-Loop Auditory Cueing System. 2014. Available online: https://repositorio-aberto.up.pt/bitstream/10216/75447/2/31866.pdf (accessed on 4 October 2019).
86. Ginis, P.; Nieuwboer, A.; Dorfman, M.; Ferrari, A.; Gazit, E.; Canning, C.G.; Rocchi, L.; Chiari, L.; Hausdorff, J.M.; Mirelman, A. Feasibility and effects of home-based smartphone-delivered automated feedback training for gait in people with Parkinson's disease: A pilot randomized controlled trial. *Parkinsonism Relat. Disord.* **2016**, *22*, 28–34. [CrossRef] [PubMed]
87. Chomiak, T.; Sidhu, A.S.; Watts, A.; Su, L.; Graham, B.; Wu, J.; Classen, S.; Falter, B.; Hu, B. Development and Validation of Ambulosono: A Wearable Sensor for Bio-Feedback Rehabilitation Training. *Sensors* **2019**, *19*, 686. [CrossRef]
88. Bartels, B.M.; Moreno, A.; Quezada, M.J.; Sivertson, H.; Abbas, J.; Krishnamurthi, N. Real-Time Feedback Derived from Wearable Sensors to Improve Gait in Parkinson's Disease. *Technol. Innov.* **2018**, *20*, 37–46. [CrossRef]
89. Thompson, E.; Agada, P.; Wright, W.G.; Reimann, H.; Jeka, J. Spatiotemporal gait changes with use of an arm swing cueing device in people with Parkinson's disease. *Gait Posture* **2017**, *58*, 46–51. [CrossRef]
90. Baskaran, D. *Real-Time Feedback Training to Improve Gait and Posture in Parkinson's Disease*; Arizona State University: Tempe, AZ, USA, 2017.
91. Krishnamurthi, N.; Baskaran, D.; Parikh, S.; Venugopal, V.; Muthukrishnan, N.; Driver-Dunckley, E.; Mahant, P.; Ospina, M.C.; Abbas, J.J. *Real-Time Feedback during Treadmill Training for Individuals with Parkinson's Disease*; Society for Neuroscience: Chicago, IL, USA, 2019; p. 1.
92. Morris, M.E.; Iansek, R.; Matyas, T.A.; Summers, J.J. Stride length regulation in Parkinson's disease. Normalization strategies and underlying mechanisms. *Brain* **1996**, *119 Pt 2*, 551–568. [CrossRef]
93. Schlick, C.; Struppler, A.; Boetzel, K.; Plate, A.; Ilmberger, J. Dynamic visual cueing in combination with treadmill training for gait rehabilitation in Parkinson disease. *Am. J. Phys. Med. Rehabil.* **2012**, *91*, 75–79. [CrossRef]
94. Mirelman, A.; Rochester, L.; Maidan, I.; Del Din, S.; Alcock, L.; Nieuwhof, F.; Rikkert, M.O.; Bloem, B.R.; Pelosin, E.; Avanzino, L.; et al. Addition of a non-immersive virtual reality component to treadmill training to reduce fall risk in older adults (V-TIME): A randomised controlled trial. *Lancet* **2016**, *388*, 1170–1182. [CrossRef]
95. Carpinella, I.; Cattaneo, D.; Bonora, G.; Bowman, T.; Martina, L.; Montesano, A.; Ferrarin, M. Wearable Sensor-Based Biofeedback Training for Balance and Gait in Parkinson Disease: A Pilot Randomized Controlled Trial. *Arch. Phys. Med. Rehabil.* **2017**, *98*, 622–630. [CrossRef]
96. Espay, A.J.; Baram, Y.; Dwivedi, A.K.; Shukla, R.; Gartner, M.; Gaines, L.; Duker, A.P.; Revilla, F.J. At-home training with closed-loop augmented-reality cueing device for improving gait in patients with Parkinson disease. *J. Rehabil. Res. Dev.* **2010**, *47*, 573. [CrossRef]
97. Karatsidis, A.; Richards, R.E.; Konrath, J.M.; van den Noort, J.C.; Schepers, H.M.; Bellusci, G.; Harlaar, J.; Veltink, P.H. Validation of wearable visual feedback for retraining foot progression angle using inertial sensors and an augmented reality headset. *J. Neuroeng. Rehabil.* **2018**, *15*, 78. [CrossRef]
98. Young, W.R.; Shreve, L.; Quinn, E.J.; Craig, C.; Bronte-Stewart, H. Auditory cueing in Parkinson's patients with freezing of gait. What matters most: Action-relevance or cue-continuity? *Neuropsychologia* **2016**, *87*, 54–62. [CrossRef]
99. Pereira, M.P.; Gobbi, L.T.; Almeida, Q.J. Freezing of gait in Parkinson's disease: Evidence of sensory rather than attentional mechanisms through muscle vibration. *Parkinsonism Relat. Disord.* **2016**, *29*, 78–82. [CrossRef]
100. Rocha, P.A.; Porfirio, G.M.; Ferraz, H.B.; Trevisani, V.F. Effects of external cues on gait parameters of Parkinson's disease patients: A systematic review. *Clin. Neurol. Neurosurg.* **2014**, *124*, 127–134. [CrossRef]
101. Bhatt, T.; Yang, F.; Mak, M.K.; Hui-Chan, C.W.; Pai, Y.C. Effect of externally cued training on dynamic stability control during the sit-to-stand task in people with Parkinson disease. *Phys. Ther.* **2013**, *93*, 492–503. [CrossRef]
102. Mak, M.K.; Hui-Chan, C.W. Audiovisual cues can enhance sit-to-stand in patients with Parkinson's disease. *Mov. Disord.* **2004**, *19*, 1012–1019. [CrossRef]
103. Schlenstedt, C.; Mancini, M.; Horak, F.; Peterson, D. Anticipatory Postural Adjustment during Self-Initiated, Cued, and Compensatory Stepping in Healthy Older Adults and Patients With Parkinson Disease. *Arch. Phys. Med. Rehabil.* **2017**, *98*, 1316–1324. [CrossRef]

104. Lopez, W.O.; Higuera, C.A.; Fonoff, E.T.; Souza Cde, O.; Albicker, U.; Martinez, J.A. Listenmee and Listenmee smartphone application: Synchronizing walking to rhythmic auditory cues to improve gait in Parkinson's disease. *Hum. Mov. Sci.* **2014**, *37*, 147–156. [CrossRef]
105. Muthukrishnan, N.; Turaga, P.; Abbas, J.J.; Ingalls, T.; Krishnamurthi, N. *Gait and Balance Monitoring Using Wearable Technology for Real-Time Feedback in Parkinson's Disease*; Society for Neuroscience: Chicago, IL, USA, 2019; p. 1.

Review

Gait Analysis in Parkinson's Disease: An Overview of the Most Accurate Markers for Diagnosis and Symptoms Monitoring

Lazzaro di Biase [1,*], Alessandro Di Santo [1], Maria Letizia Caminiti [1], Alfredo De Liso [1], Syed Ahmar Shah [2], Lorenzo Ricci [1] and Vincenzo Di Lazzaro [1]

1 Unit of Neurology, Neurophysiology, Neurobiology, Department of Medicine, Università Campus Bio-Medico di Roma, Via Álvaro del Portillo 21, 00128 Rome, Italy; a.disanto@unicampus.it (A.D.S.); m.caminiti@unicampus.it (M.L.C.); a.deliso@unicampus.it (A.D.L.); lorenzo.ricci@unicampus.it (L.R.); v.dilazzaro@unicampus.it (V.D.L.)

2 Usher Institute, Edinburgh Medical School: Molecular, Genetic and Population Health Sciences, The University of Edinburgh, EH16 4UX Edinburgh, UK; ahmar.shah@ed.ac.uk

* Correspondence: l.dibiase@unicampus.it

Received: 10 June 2020; Accepted: 17 June 2020; Published: 22 June 2020

Abstract: The aim of this review is to summarize that most relevant technologies used to evaluate gait features and the associated algorithms that have shown promise to aid diagnosis and symptom monitoring in Parkinson's disease (PD) patients. We searched PubMed for studies published between 1 January 2005, and 30 August 2019 on gait analysis in PD. We selected studies that have either used technologies to distinguish PD patients from healthy subjects or stratified PD patients according to motor status or disease stages. Only those studies that reported at least 80% sensitivity and specificity were included. Gait analysis algorithms used for diagnosis showed a balanced accuracy range of 83.5–100%, sensitivity of 83.3–100% and specificity of 82–100%. For motor status discrimination the gait analysis algorithms showed a balanced accuracy range of 90.8–100%, sensitivity of 92.5–100% and specificity of 88–100%. Despite a large number of studies on the topic of objective gait analysis in PD, only a limited number of studies reported algorithms that were accurate enough deemed to be useful for diagnosis and symptoms monitoring. In addition, none of the reported algorithms and technologies has been validated in large scale, independent studies.

Keywords: Parkinson's disease; gait analysis; diagnosis; symptoms monitoring; wearable; home-monitoring; machine learning

1. Introduction

Parkinson's' disease (PD) gold standard for diagnosis and symptoms monitoring is based on clinical evaluation, which includes several subjective components. The lack of objective and quantitative biomarkers for diagnosis and symptoms monitoring leads to significant direct and indirect healthcare cost. Based on the current diagnostic criteria [1], the diagnostic error rate is around 20% [2]. In addition, PD is a dynamic disease (i.e., symptoms changes during the disease course) that requires continuous adjustment of therapy.

In the early stages of PD, the most effective treatment to alleviate motor symptoms is oral L-DOPA [3]. However, during moderate and advanced stages, in addition to cardinal motor symptoms, the patient may show motor fluctuations and dyskinesia. During this stage, the brain becomes very sensitive to dopamine level fluctuations, and a continuous stimulation (instead of pulsatile drugs administration) may help in controlling motor fluctuations, dyskinesias and cardinal motor symptoms. This stimulation may be pharmacological with levodopa [4–6] or dopamine agonists [7], or provided by DBS (deep brain stimulation) [8–12]. During moderate and advanced stages, gait problems, like freezing of gait and reduced balance and postural control, become more evident and unlike cardinal motor symptoms, PD patients respond less to conventional therapy (i.e., oral L-DOPA).

In line with cardinal motor symptoms, to date, gait problems are evaluated with semiquantitative rating scales like the unified Parkinson's disease rating scale (UPDRS) [13] or the movement disorders society unified Parkinson's disease rating scale (MDS-UPDRS) [14]. In an effort to improve PD management and move towards a quantitative and home-oriented assessment and recognition of PD motor symptoms, different technologies have been used to evaluate bradykinesia [15–17], rigidity [17–20], tremor [21–23] and axial symptoms [24–27].

Gait impairment is an evolving condition and different patterns of gait disturbances can be detected throughout the progression of the disease [28]: reduced amplitude of arm swing, reduced smoothness of locomotion, increased interlimb asymmetry [29], low speed, reduced step length [29], shuffling steps, increased double-limb support, increased cadence [28], defragmentation of turns (i.e., turning en block), problems with gait initiation [30], freezing of gait and reduced balance and postural control [28].

Some gait features in PD are specific, and get worse during the disease course. An objective and quantitative gait analysis system could, therefore, potentially improve the current practice (semiquantitative gait evaluation) that may aid in diagnosis, symptom monitoring, therapy management, rehabilitation and fall risk assessment and prevention in Parkinson's disease patients. Among all these promising applications of gait analysis in Parkinson's disease, we have confined the scope of our review to two main unmet needs in this disorder: the diagnostic error, and the lack of objective biomarkers for motor status discrimination. Therefore, the main aim of this overview is to summarize the most important technologies used to evaluate gait features and the associated algorithms that have shown promise in using gait analysis to aid diagnosis and symptom monitoring in Parkinson's disease (PD) patients. The scope of the review was confined to studies that showed any promise of being clinically useful (i.e., are both highly sensitive and specific defined as those with at least 80% sensitivity and 80% specificity) for diagnosis or motor status discrimination.

1.1. Gait Features

Human gait is a sequence of involuntary movements, cyclically repeated and triggered by voluntary movement. Several components could be used to objectively measure and analyze gait cycle. These components are typically categorized into spatiotemporal, kinematics and kinetics features [31,32].

1.2. Spatiotemporal Features

Several spatiotemporal features can be used for gait analysis (Table 1). These features are the more commonly used types of features to objectively describe the gait pattern in healthy subjects and patients with several diseases. Spatiotemporal features could refer to the global gait cycle or to the stride cycle.

Table 1. Spatiotemporal gait and stride features.

Gait Cycle	The time from initial contact to initial contact on the same foot including both the stance phase and swing phase.
Stance Phase	The period during which the foot is in contact with the support surface during one gait cycle.
Swing Phase	The period during which the foot is airborne during one gait cycle.
Double Limb Support	The period during which both feet are in contact with the support surface during one gait cycle.
Single Limb Support	The period during which only one foot is in contact with the support surface during one gait cycle.
Step Duration	The period between 2 successive events of the same type on opposite limbs.
Stride Length	The linear distance between 2 successive events (initial contact) on the same limb.
Step Length	The linear distance between 2 successive events of same type on opposite limbs.
Step Width	The horizontal distance between 2 points on opposite limbs.
Foot Progression Angle	The angle between the longitudinal axis of the foot and the line of gait progression.

Each gait cycle starts with the initial contact of one foot and ends with a new initial contact of the same foot (Figure 1). One single cycle is composed of a stance and a swing phase: the stance phase is the period during which the foot is in contact with a support surface, and the swing phase is the period during which the same foot is airborne in preparation for the next gait cycle. During the gait cycle, the legs can be individually or simultaneously placed on the ground, so it is possible to identify a single limb support stage, that is the phase during which only one foot is on the surface, and a double limbs support stage during which both legs are on the support surface during a one step cycle (Table 1).

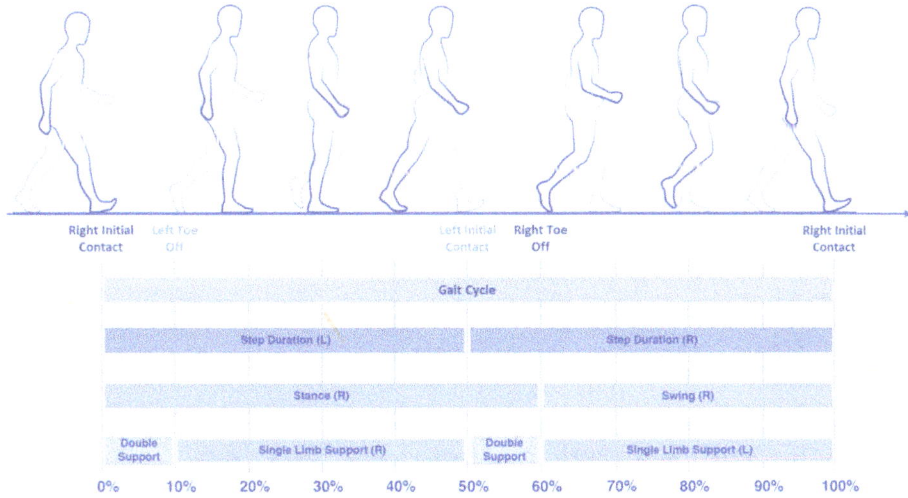

Figure 1. Human gait cycle.

The step and the stride are defined as the length/duration between 2 successive events of same type on opposite limbs, and on the same limb, respectively (Table 1; Figure 2).

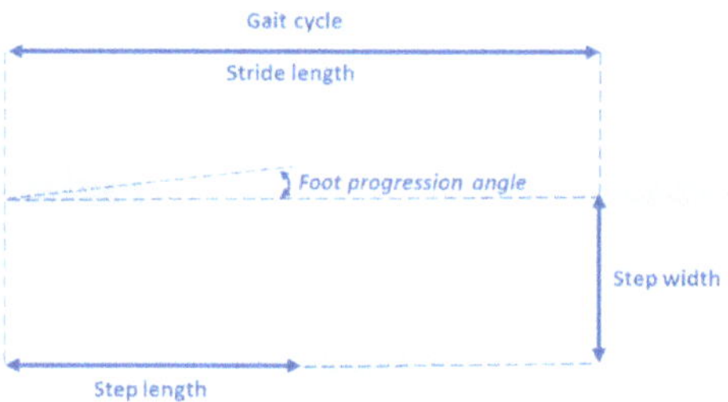

Figure 2. Stride analysis of a single gait cycle.

The step width represents the horizontal distance measured between the position of the feet on the same event. Finally, foot progression angle can be measured, this represents the angle between the longitudinal axis of the foot and the line of gait progression (Table 1).

1.3. Kinetics Features

The kinetics analysis (or dynamics of gait) is the study of the forces and their effect on motion. The dynamics forces are the causes of the motion that result in the kinematic movements. Commonly, these forces are represented by the ground reaction force (GRF) on the hip, knee and ankle joints calculated on the sagittal plane. GRF is described only when feet are in contact with the ground (stance phase) and represents the effect of gravity on a body area counterbalanced by the contact with ground and the limb muscular activation [31,32] (Figure 3). GRF refers to a center of pressure (CoP) that is the point of force application.

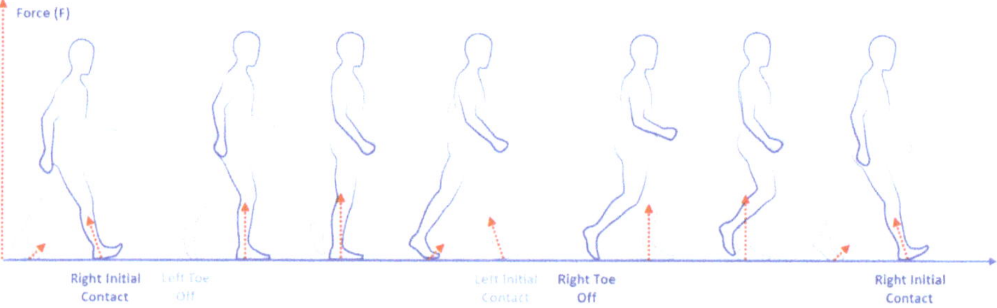

Figure 3. Gait kinetics (dynamics) features.

1.4. Kinematics Features

Kinematics features describe the movements without taking the forces causing the movements into account [31,32]. Kinematics analysis could describe gait features based on the sagittal, horizontal or frontal plane for several body areas and joints such as the ankle, knee, hip and pelvis. Kinematics features could be extrapolated both from the stance and swing phases. The kinematic analysis of the position, velocity and acceleration of a body part can be determined. Angular kinematics objectively quantify (as degrees) the joint's motion around axes in different gait phases (Figure 4).

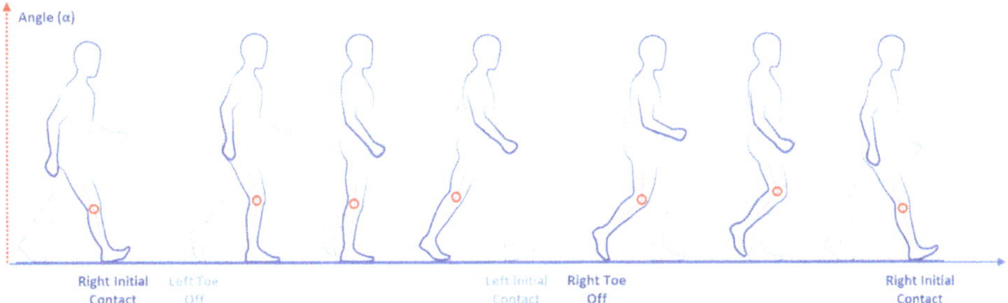

Figure 4. Gait kinematics features.

1.5. Gait Analysis Technologies

Several technologies can be used for quantitative gait analysis. Technologies can be divided into two main subtypes: wearables and non-wearables [33]. Wearable sensors used for gait analysis are inertial sensors [34], goniometer [35], pressure and force sensors [36], electromyography (EMG) [37,38], IR-UWB (impulse radio ultra-wideband) [39] and ultrasound [40]. Among non-wearable sensors, the most common types are floor sensors [41,42] and image processing-based technologies (such as a single or multiple cameras [43–45], time of flight [46–48], stereoscopic vision [49,50], structured light [51] and IR thermography [52]).

Accelerometers, gyroscopes and magnetometers can be the component of the same inertial measurement unit (IMU) device (Figure 5), one of the most widely used type of sensors in gait analysis especially in PD [33,53]. It can measure velocity, acceleration, orientation and gravitational forces and can be used to study gait initiation [54], assess standing balance [55] and quantify bradykinesia [56].

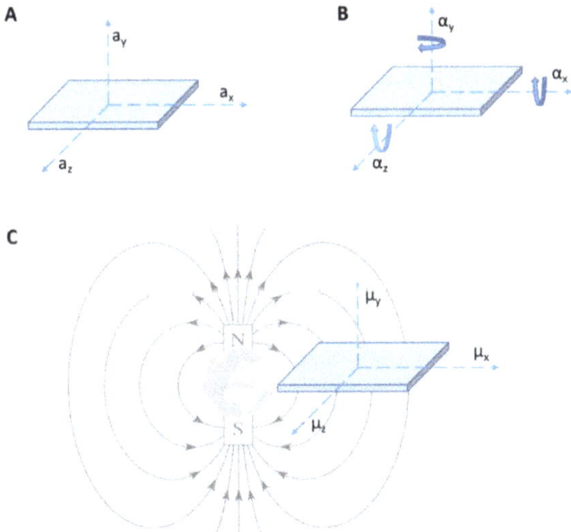

Figure 5. The same inertial measurement unit (IMU) device composed by accelerometers (**A**), a gyroscope (**B**) and a magnetometer (**C**). Legend: (**A**) a_x, a_y and a_z = linear acceleration on the three axis x, y and z; (**B**) α_x, α_y and α_z = angular acceleration on the three axis x, y and z and (**C**) μ_x, μ_y and μ_z = magnetic moment on the three axis x, y and z.

Accelerometers are composed of a mechanical sensing element with a proof mass attached to a mechanical suspension system, with respect to a reference frame, that can be forced to deflect by the inertial force according to the acceleration of gravity. The acceleration can be measured electrically using the physical changes in the displacement of the proof mass with respect to the reference frame [34]. Accelerometer can be attached to the feet, legs or waist [33]. Gyroscope is based on the property that all body that revolves around an axis develop rotational inertia determined by the body's moment of inertia [33]. Basically, a gyroscope is an angular velocity sensor. Magnetometer is based on the magneto resistive effect and can estimate changes in the orientation of a body segment in relation to the magnetic north. It can provide information that cannot be determined by both an accelerometer and gyroscope [34].

Goniometers work with resistance that changes depending on how flexed the sensors is. When flexed, the resistance increases proportionally to the flex angle. Goniometers are easy to set up and use a simple algorithm [33]. Goniometers are commonly used to study the angles for ankles, knees, hips and metatarsals [35].

Pressure and force sensors measure the force applied on the sensor without considering the components of this force on all other axes [33].

Force sensors measure the ground reaction force under the foot and return a current or voltage proportional to the pressure measured. Usually this kind of sensor is easily integrated into instrumented shoes [36]. Pressure and force sensors have been used to study stride length variability in PD patients with freezing of gait (FOG) [57]. Electromyography is a neurophysiologic technique that registers the electrical signals associated with motor unit activity, both voluntary and involuntary. The electrical signal can be recorded with surface electrodes (non-invasively) or needle electrodes (invasively) [58]. Some EMG electrodes are commercialized in combination with wireless technology and play an important role in evaluating walking performance during gait [38]. EMG can be used to study postural disorders in Parkinson's disease patients, like exploring muscular activity in the Pisa syndrome [59].

The impulse radio ultra-wideband (IR-UWB) technique can detect and track movements non-invasively with high resolution and accuracy through emitting impulse radio waves of very short duration, and receiving the reflected waves from the target body [60]. It has been used to quantify activity measurement in movement disorders [61]. This technology also shows a good penetrating power able to detect the motions of internal organs. This technique can also be used for a wearable healthcare system to continuously estimate foot clearance due to its high temporal resolution, low power consumption and multipath immunity [62]. This technology can be used for step and gait phase detection [39].

Ultrasonic sensors can measure the time a sound takes to send and receive the wave produced as it is reflected from an object. Knowing the time and the speed, we can estimate the distance between two points [33]. This kind of technology is useful for step length measurement and gait phase detection [40] to analyze bilateral gait symmetry and coordination [63].

Among non-wearable sensors, the single camera image processing system is composed of single or multiple cameras that can be used to obtain information about gait in selected individuals. This technique allows individual recognition and segment position localization. Image processing has been used to identify people by the way they walk [44] and has several medical applications such as gait recognition considering changes in the subject path [64], and study of the gait kinematic [65].

Time of flight (TOF) systems are based on cameras using signal modulation that measure distances between the camera and the subject based on the phase-shift principle [46]. The TOF system can detect the segment position, gait phase, foot plantar pressure distribution and are useful for individual recognition [47]. TOF systems have been used to assess medication adherence in patients with movement disorders [66].

Stereoscopic vision is used to determine the depth of points in the scene by using a model through the calculation of similar triangles between the optical sensor, the light-emitter and the object in the scene. This could be useful in gait phase detection, segment position and individual recognition [49].

Structured light is the projection of a light pattern under geometric calibration on an object whose shape is to be recovered [33]. This technology is used for segment position study and gait phase detection. Kinetic sensor is one of the most common devices using this technology to create a marker-based real-time biofeedback system for gait retraining [51].

Infrared thermography (IR) creates visual images based on surface temperatures. For studying human gait, its functioning is based on skin emissivity. This method has been applied to recognize the human gait pattern [52].

1.6. Machine Learning Algorithms Application for Gait Analysis

There has been increasing use of machine learning (ML) in medicine including neurology to aid diagnosis, and patient management using risk stratification [67,68]. ML algorithms learn from data (past experiences) by identifying underlying patterns and relationships. The field of ML can broadly be categorized into supervised, unsupervised and reinforcement learning.

Supervised learning (SL) begins with the aim of predicting a known output or target. Indeed, an SL algorithm takes a known set of input data (the training set) and known responses to the data (output), and trains a model to generate reasonable predictions for the response to new input data. In such algorithms, the artificial intelligence (AI) is approximating what a trained physician is already able to perform with high accuracy. This approach means that the learning algorithm generalized the training data to previously unobserved situations in a "reasonable" way.

All forms of SL algorithms can be classified as either classification or regression. Classification techniques predict discrete responses. Regression techniques, instead, are used to predict continuous responses. They can also be used for modeling the risk, meaning that the computer is doing more than merely reproducing the physician skills. These algorithms are also capable of discovering new associations not apparently evident to human's preliminary interpretation.

Differently from SL, in the unsupervised learning (UL) algorithm, we were no longer interested in predicting outputs. Instead, we aimed to discover naturally occurring patterns or groupings within the data. It is important to emphasize that the examples given to learners were unlabeled; thus, there was no error or reward signal to evaluate a potential solution. Common UL clustering algorithms could broadly divided into three groups: hard clustering, where each data point belongs to only one cluster, and soft clustering, where each data point can belong to more than one cluster; and dimensionality reduction techniques.

Reinforcement learning (RL) is an approach in ML that states what actions an agent should take in an environment to capitalize on the idea of an increasing reward. RL is different form standard SL in that correct input/output pairs are never presented, nor are suboptimal actions explicitly corrected. The primary goal is the direct performance, which involves finding a balance between exploration of unknown datasets and exploitation of current knowledge [69].

The most widely used ML technique in gait analysis is SL, with varying levels of complexity and interpretability. Table 2 describes the most used algorithms in this field.

Table 2. Machine learning algorithm used for gait analysis.

Algorithm	How It Works	Interpretability (+): Min (+++++): Max
k Nearest Neighbor (kNN):	Categorizes objects based on the classes of the nearest neighbors in the dataset. The function is estimated only locally and all of the calculations are delayed up to the prediction or classification. The kNN method is sensitive to the dataset [70].	+++
Linear Support Vector Machine (SVM):	Classifies data by finding the linear decision boundary (hyperplane) that separates all data points of one class from those of the other class [71]. The best hyperplane for an SVM is the one with the largest margin between the two classes, when the data is linearly separable [72,73].	+++
Kernel Support Vector Machine (Kernel SVM):	Similar to SVM but additionally uses the "kernel trick" to transform the input data (not linearly separable) into a new feature space (linearly separable)	++
Artificial Neural Networks (ANNs)	Inspired by the connectivity of neurons in the human brain, a neural network consists of highly connected networks of neurons that relate the inputs to the desired outputs [74]. Each nonlinear function in the network can be used for the mapping from the training inputs to the training outputs.	+
Naïve Bayes (NB)	A naïve Bayes classifier assumes that the presence of a particular feature in a class is unrelated to the presence of any other feature and uses the Bayes theorem to determine the posterior probability	+++
Linear Discriminant Analysis (LDA)	It classifies data by finding linear combinations of features. Discriminant Analysis (DA) assumes that different classes generate data based on Gaussian distributions. The distributions parameters are used to calculate boundaries, which can be linear or quadratic functions.	++++
Decision Tree (DT)	It predicts responses to data by following the decisions in the tree-algorithm from the root (beginning) down to a leaf node. DTs can solve a classification problem by continuously dividing the input space to build a tree on which the nodes are as pure as possible and contain points of a single class. DTs are considered naïve algorithms; however, they have great performances in prediction and classification applications.	+++++
Random forest	An ensemble technique that uses a very large number of decision trees, often resulting in improved accuracy over DTs at the expense of reduced interpretability	+

2. Materials and Methods

In line with the study of Sánchez-Ferro, et al. [75] we used a similar search string for axial symptoms in Parkinson's disease patients, in PubMed for articles published between 1 January 2005, and 30 August 2019 (Table 3). We identified studies that used technologies to distinguish PD patients from healthy subjects or to differentiate PD motor status or different disease stages. We only selected studies that declared a sensitivity and specificity of at least 80% when using gait analysis for either diagnosis or motor status discrimination. Additionally, further relevant articles based on the author's knowledge of the state of the art in this field were also added.

Table 3. Search strategy.

Domain	Search String
Disease	("Parkinsonian Disorders" OR "Parkinson disease" OR "Parkinson Disease, Secondary" OR "Basal Ganglia Diseases" OR "Parkinsonism" OR "Parkinson's Disease") AND
Technology	("Technology" OR "Technologies" OR "Diagnostic Techniques, Neurological" OR "Assessment" OR "Patient Outcome Assessment" OR "Symptom Assessment" OR "Evaluation" OR "Diagnostic Self Evaluation" OR "Investigative Techniques" OR "Wireless Technology" OR "Remote Sensing Technology" OR "Biomedical Technology" OR "Technology Assessment, Biomedical" OR "Medical Informatics" OR "Cloud Computing" OR "Point of Care systems" OR "Biomedical Engineering" OR "Machine Learning" OR "Artificial Intelligence" OR "Kinesis" OR "Mobile Applications" OR "Cell Phones" OR "Smartphones" OR "Software" OR "Software Validation" OR "Platform" OR "Accelerometer" OR "Gyroscope" OR "Magnetometer" OR "Actigraph" OR "Wearable" OR "Device" OR "Big Data" OR "Sensor" OR "Internet of Things" OR "Closed-loop System" OR "Hybrid" OR "Home monitoring" OR "Quantitative" OR "Algorithm" OR "Telemetry" OR "Instrumented" OR "Virtual Reality") AND
Axial symptoms	("Gait" OR "Gait Disorders, Neurologic" OR "Posture" OR "Posture Balance" OR "Freezing of Gait" OR "Gait Disturbances" OR "Postural Instability" OR "Falls" OR "Fall") AND
Time range	("2005/01/01"[PDAT]: "2019/08/30"[PDAT])

For each selected study, we collected data about the technology, the algorithm used and its performance metrics like accuracy, sensitivity and specificity. In addition for studies, which declares only the regular accuracy ((true positives + true negatives)/(true positives + true negatives + false positives + false negatives)), the balanced accuracy ((sensitivity + specificity)/2) was calculated. This is because for unbalanced test sets, balanced accuracy is a better index for accuracy than regular accuracy [76].

3. Results

3.1. Discrimination of Parkinson's Disease from Healthy Subjects

According to the inclusion and exclusion criteria, after the literature search and studies screening, 10 studies were selected that focused on distinguishing Parkinson's disease patients from healthy subjects with gait analysis. To distinguish PD from healthy subjects, several technologies can be used (Tables 4 and 5). One study used the data collected from wireless inertial sensors (Micro-attitude and heading reference system (AHRS) model, MicroStrain, Inc, Williston, VT, USA) placed on the foot in PD patients and healthy subjects to detect peculiar gait features and distinguish PD patients from controls [77]. In particular, authors detected physical kinematic features of pitch, roll and yaw rotations of the foot during walking and used principal component analysis (PCA) to select the best features that were subsequently used for the SVM method to classify PD patients, with and without gait impairment, and healthy subjects. From 67 collected features, they selected 15 kinematic features divided in three categories: pitch, roll and yaw features. The proposed classification has very high sensitivity, specificity and positive predict values (93.3%, 95.8% and 97.7% respectively) to distinguish PD patients from healthy subjects [77].

Table 4. Parkinson's disease vs. healthy subjects discrimination.

Ref	Algorithm	N. Features	N. Patients/Healthy	Regular Accuracy	Balanced Accuracy	Sensitivity (%)	Specificity (%)
[77]	SVM	15		NA	94.6%	93.3%	95.8%
[78]	Decision tree	13	25/45	95%	92.3%	88.8%	95.8%
[78]	Neural Network	13	25/45	99%	100.0%	100.0%	100.0%
[79]	LDA	12	27/16	NA	87.0%	88.0%	86.0%
[80]	NA	3	10/17	96.3	97.1%	100.0%	94.1%
[81]	SVM	8	5/5	NA	90.0%	90.0%	90.0%
[82]	Random forest	23	10/10	NA	98.1%	98.5%	97.6%
[83]	Bayesian probability	2	18/33	92.2%	93.3%	94.4%	92.2%
[83]	Bayesian probability	2	18/33	94.1%	94.2%	94.4%	93.9%
[84]	SVM	19	40/40	85.0%	83.5%	85.0%	82.0%
[85]	SVM	13	29/18	95.7%	95.5%	94.4%	96.6%
[85]	Random forest	13	29/18	89.4%	89.3%	88.9%	89.7%
[85]	kNN	13	29/18	85.1%	84.8%	83.3%	86.2%
[85]	Decision tree	13	29/18	87.2%	87.6%	88.9%	86.2%
[86]	Tensor decomposition	16	93/72	100.0%	100.0%	100.0%	100.0%

Abbreviations: CoP: center of pressure; CV: coefficient of variation; kNN: k-nearest neighbor, LDA: linear discriminant analysis; NA: not available; SVM: support vector machine; VGRF: vertical ground reaction force.

Table 5. Parkinson's disease vs. healthy subjects discrimination features selected.

Ref	Algorithm	Features
[77]	SVM	**Pitch** - Pitch range of motion - Maximum angle of dorsiflexion - Maximum angle of plantar flexion - Plantar flexion SD - Single-step maximum of maximum angle of plantar flexion **Roll** - Roll range of motion - Maximum positive roll angle - Maximum negative roll angle **Yaw** - Yaw range of motion - Maximum positive yaw angle - Maximum negative yaw angle - Overall 3D SD - Maximum cadence **Additional** - Single-step maximum of maximum negative roll angle - Single-step minimum of maximum negative roll angle

Table 5. *Cont.*

Ref	Algorithm	Features
[78]	- Decision tree - Neural Network	- Absolute difference between i) average distance between right elbow and right hip and ii) average distance between right wrist and left hip. - Average angle of the right elbow. - Quotient between maximal angle of the left knee and maximal angle of the right knee. - Difference between maximal and minimal angle of the right knee. - Difference between maximal and minimal height of the left shoulder. - Difference between maximal and minimal height of the right shoulder. - Quotient between i) difference between maximal and minimal height of left ankle and ii) maximal and minimal height of right ankle. - Absolute difference between i) difference between maximal and minimal speed (magnitude of velocity) of the left ankle and ii) difference between maximal and minimal speed of the right ankle. - Absolute difference between i) average distance between right shoulder and right elbow and ii) average distance between left shoulder and right wrist. - Average speed (magnitude of velocity) of the right wrist. - Frequency of angle of the right elbow passing average angle of the right elbow - Average angle between (i) vector between right shoulder and right hip and (ii) vector between right shoulder and right wrist. - Difference between average height of the right shoulder and average height of the left shoulder.
[79]	LDA	**Step features** - Step duration - Rise gradient of swing phase - Fall gradient of swing phase - Standard deviation of minima - Maxima minima difference **Signal sequence** - Variance - Integral - Entropy **Frequency analysis** - Dominant frequency - Energy ratio - Energy in band 0.5–3 - Energy in band 3–8
[80]	NA	- High intensity, - Periodicity, - Biphasicity

Table 5. *Cont.*

Ref	Algorithm	Features
[81]	SVM	**EMG statistics** - Variance - Skewness - Kurtosis - RMS Energy **EMG frequency** - Dominant Frequency - Mean Frequency - Median Frequency - Total Power
[82]	Random forest	- Mean - Standard deviation - 25th percentile - 75th percentile - Inter-quartile range - Median - Mode - Data range (maximum – minimum) - Skewness - Kurtosis - Mean squared energy - Entropy - Cross-correlation between the acceleration in x and y axis - Mutual information between the acceleration in x and y axis - Cross-entropy between the acceleration in x and y axis - Extent of randomness in body motion - Instantaneous changes in energy due to body motion - Autoregression coefficient at time lag1 - Zero-crossing rate - Dominant frequency component - Radial distance - Polar angle - Azimuth angle
[83]	Bayesian probability	- Stride length, - Gait speed
[83]	Bayesian probability	- Stride length, - Age

Table 5. *Cont.*

Ref	Algorithm	Features
[84]	SVM	- Step time - Step time asymmetry - Stance % of cycle - Swing time - Swing time CV - Stride time - Stride time CV - Stride time asymmetry - Single support time CV - Heel off on time - Heel off on std- deviation - Double support time - Double support time CV - Double support load % of cycle - Step length asymmetry - Stride length - Stride length CV - Heel-to-heel support base - Heel-to-heel support base CV
[85]	- SVM - Random forest - kNN - Decision tree	- CV of swing time - CV of stride time - Mean CoP of x-coordinate - Standard deviation CoP of x-coordinate - Mean CoP of y-coordinate - Standard deviation CoP of y-coordinate - Mean peak force at heel strike - Mean peak force at toe strike - Standard deviation of peak forces at heel strike - Standard deviation of peak forces at toe strike - Mean kurtosis - Mean skewness - Mean Peak power of VGRF signal
[86]	Tensor decomposition	- VGRF measurements from 8 sensors for the foot

Another study compared two machine learning algorithms (decision-tree and neural networks) to differentiate healthy subjects gait patterns in different disease conditions in an elderly population (including patients affected by PD, hemiplegia, leg pain and back pain). Authors used movements data obtained from 12 retroreflective tags placed on the body captured by an infrared camera (Smart IR motion capture system) [78]. They studied 45 healthy controls and 25 PD patients. Predictors were based on velocity and calculated body distances (i.e., difference between average distance between right elbow and right hip and average distance between right wrist and left hip or the angle between two body segments). Global classification accuracy was high for both systems and reached over 95% for decision tree and more than 99% for neural network [78].

Moreover, Barth, et al. [79] demonstrated a good sensitivity and specificity (88% and 86% respectively) to differentiate healthy subjects and early PD patients using only a single mobile inertial sensor (gyroscope and accelerometer, integrated in the SHIMMER Company system) placed over the shoes. The patients performed standardized gait tests and from this data step, signal sequence and frequency features were extrapolated and used as predictors for the linear discriminant analysis (LDA). Moreover, they demonstrated that this system was able to distinguish between mild and severe gait pattern with high sensibility and specificity (100%).

Another approach was used by Yoneyama, et al. [80]: they used a single accelerometer placed on the waist and performed gait analysis to compare a continuous 24-h assessment of 10 PD patients and 17 healthy controls [80]. They used a gait detection algorithm based on the gait cycle (i.e., stride-to-stride time interval) and gait-induced acceleration relationship using 3 features of gait (high intensity, periodicity and biphasicity of gait) that introduced a set of indices in order to quantify subject's walking mode and to assess daily gait characteristics. All the calculated indices were smaller in the PD group, and the proposed method was able to distinguish the PD gait from the normal gait with 100% sensitivity, 94.1% specificity and 96.3% accuracy. These results suggest that the afore-mentioned systems could differentiate normal subjects from those with movement disorders [80].

To easily and objectively assess the difference between PD patients and healthy subjects, Kugler, et al. [81] proposed a classification algorithm based on data collected from surface wireless EMG (Delsys Trigno, Delsys Inc., Boston, MA, USA) positioned on two inferior limbs muscles during the performance of standardized gait tests in five PD patients and five healthy subjects [81]. Furthermore, data from accelerometers (Trigno sensor) placed on heels were collected and used for step segmentation. Statistical and frequency features from EMG signals were used to train an SVM algorithm for step detection. The proposed step detection method reached 98.9% sensitivity and 99.3% specificity and the classification accuracy to distinguish between PD and healthy subjects reached 90% sensitivity and 90% specificity (average value).

Arora, et al. [82] investigated the feasibility and the accuracy of smartphones' built-in tri-axial accelerometer (in LG Optimus S) developing an app to objectively assess PD patients and distinguish them from healthy subjects [82]. They studied 10 PD patients and 10 controls for 1 month and during execution of the gait tests to extract 23 features of frequency and time domain from accelerometer and subsequently used a random forest method to distinguish between PD and controls. This system reached a 98% balanced accuracy, 98.5% sensitivity and 97.6% specificity.

Another study proposed the use of a Bayesian gait recognition method based on data acquisition by video infrared camera system (Microsoft Kinect depth sensors) [83]. This system consisted of an infrared projector and two infrared cameras (on left and right) that follow the structured light principle. The collected data are converted in the depth frame matrix, depth frame contour, image frame matrix and skeleton numbering. Then the acquired data were further analyzed through MATLAB software. Eighteen PD patients, eighteen healthy subjects and fifteen healthy students were assessed and probabilistically classified according to their detected gait featured (stride length and gait speed) through skeletal tracking. This Bayesian system used the stride length, gait speed and age as features and was able to distinguish between PD patients and controls with 92.2% accuracy combining stride length and gait speed, and 94.1% accuracy combining stride length and patient/healthy subjects' age.

Djurić-Jovičić, et al. [84] used an electronic walkaway to distinguish between the PD patient and controls. Authors compared 40 de novo PD patients and 40 controls while walking selecting three different tasks: normal pace walking, dual motor task (as walking and carrying a glass of water) and walking during mental task execution. The most relevant predictor variables were selected (19 features) including the stride length, stride length coefficient of variation (CV), swing time, step time asymmetry and heel-to-heel base support CV. These features were selected with the random forests algorithm and the classification accuracy of these selected features was tested with the support vector machine. The overall accuracy combining the three conditions was 85% to identify de novo PD patients from healthy subjects, with a sensitivity of 85% and a specificity of 82%. Their study also found that step time asymmetry and the support base CV are the most relevant factors that contribute to global system accuracy.

Alam, et al. [85] proposed a novel mathematical method to assess the gait in PD patients and controls using data from eight sensors below each foot and extrapolated features from vertical ground reaction force (VGRF) data recorded during subjects walking [85]. Twenty-nine PD patients and eighteen healthy subjects were enrolled in this study. Three different algorithms (sequential forward selection, minimum redundancy maximum relevancy feature selection (MRMR) and the mutual information-based feature ranking method) were applied to select the best features extracted from VGRF data and 13 features (like CV swing time, CV stride time and centre of pression data) were chosen. Finally, four different machine learning classifiers (SVM, k-nearest neighbor-kNN, random forest and decision trees) were compared to distinguish the gait pattern between healthy subjects and PD patients. The accuracy of the machine learning methods ranged from 85.21% (kNN) to 95.7% (SVM cubic kernel). SVM with cubic kernel showed a sensitivity of 94.4% and a specificity of 96.6%.

Finally, Pham and Yan [86] used a tensor decomposition algorithm called canonical polyadic decomposition (CPD) also known as Parallel Factor Analysis (PARAFAC), a generalization of PCA, to differentiate the multisensors time series of the gait between PD and controls in a previous published dataset of 93 PD patients and 72 controls [86]. Data were collected from load sensors (Ultraflex Computer Dyno Graphy, Infotronic Inc., Vriezenveen, NL, USA) placed on each shoe that recorded force in a function of time. Tensor-decomposition factors of control and PD patients showed a distinct relationship. This system used the full length of the VGRF time without considering the minimum number of strides required for effective tensor decomposition analysis of the gait dynamics. This system can be applied for very short time duration signals and can resolve the problem of obtaining several trials for stable and trustable results. Authors showed 100% of accuracy, sensitivity and specificity to distinguish PD and controls.

3.2. Parkinson's Disease Motor Status Discrimination

According to the inclusion and exclusion criteria, after a literature search and studies screening, only three studies were selected and these papers aimed at identifying different motor statuses in Parkinson's disease (Tables 6 and 7). Sensors can be used to discriminate the different motor status in individual PD patients, in a single disease's stage, monitoring the motor fluctuations, or can be used in a longitudinal way to monitor the motor status changes during the disease evolution.

Table 6. Parkinson's disease motor status discrimination.

Ref	Algorithm	N. Features	N. Patients	Regular Accuracy	Balanced Accuracy	Sensitivity (%)	Specificity (%)
[79]	LDA	12	27	NA	100.00%	100.00%	100.00%
[87]	SVM	1	12	91.81%	90.80%	92.52%	89.07%
[88]	NA	NA	41	NA	92.50%	97.00%	88.00%

Abbreviations: LDA: linear discriminant analysis; NA: not available; SVM: support vector machine.

Table 7. Parkinson's disease motor status discrimination features selected.

Ref	Algorithm	Features
[79]	LDA	**Step features** - Step duration - Rise gradient of swing phase - Fall gradient of swing phase - Standard deviation of minima - Maxima minima difference **Signal sequence** - Variance - Integral - Entropy **Frequency analysis** - Dominant frequency - Energy ratio - Energy in band 0.5–3 - Energy in band 3–8
[87]	SVM	- Motion fluency value
[88]	NA	NA

Samà, et al. [87] focused their research on automatic detection of bradykinesia, the cardinal symptom of PD [87]. They proposed a mathematical algorithm to automatically identify bradykinesia in PD patients at home using an SVM classifier (that detects strides) based on data collected from a single accelerometer placed on the waist combined with a video-recording of the examination, they correlated the stride frequency with the UPDRS bradykinesia score. Their methods showed a high accuracy (>90%) to identify bradykinesia, a high sensitivity and specificity (92.52% and 89.07%, respectively) and a good correlation with UPDRS specific items.

To objectively detect the motor on–off fluctuations, an integrated system like REMPARK (personal health device for the remote and autonomous management of Parkinson's disease, FP7 project REMPARK ICT-287677) was used [88]. The REMPARK system consists of an algorithm added in an app inside a smartphone that used the data from an accelerometer placed on the iliac crest. This system was developed for longitudinal evaluation. In this study, 41 PD patients were enrolled for a 3-day monitoring. For the on/off state discrimination, authors developed an algorithm, which analyzed gait [89] and dyskinesias [90]. The algorithm responses were compared to a self-reported on–off diary. The REMPARK system showed 97% sensitivity in detecting off states and 88% specificity in detecting on phases compared to diaries.

Barth, et al. [79] demonstrated, in a cohort of 14 early stage PD patients and 13 intermediate PD stage patients, that a mobile and light inertial sensor (gyroscope and accelerometer, integrated in the SHIMMER system) placed over the shoes allows differentiation between the two groups with 100% sensitivity and specificity with using a linear discriminant analysis (LDA) classifier that combines step, signal sequence and frequency features as predictors [79].

4. Discussion

Several studies aimed at detecting specific patterns of gait alterations in Parkinson's disease by using a quantitative technology-based assessment [28,29,91,92]. Gait impairment is an evolving condition throughout the progression of the disease and different patterns of gait disturbances can be detected in early, mild to moderate and advanced stages [28].

Early specific alterations include reduced amplitude of arm swing, smoothness of locomotion and increased interlimb asymmetry [29]. Impaired muscle contraction, rigidity and postural instability contribute to reduced forward limb propulsion, which, in turn, can negatively affect spatiotemporal gait parameters, such as speed and step length [29]. In particular, reduced step length seems to be a specific feature of Parkinson's disease gait [92]. Sensor-based observations showed that the increased variability in gait reflects increased gait instability that can be detected early in the disease and can be a useful marker of disease progression [91,93]. To carry out two tasks at the same time, a paradigm known as dual-task interference, is particularly complex in PD patients because of two independent effects influencing gait: the first is an age-associated reduction in gait performance unrelated to pathology, and the second one is a PD-specific effect due to a dual-task coordination deficit interfering with postural control. The latter suggests reduced stability and ability to adapt to PD patients under dual-task conditions [93]. Arm swing outcomes provide a sensitive measure of decline in gait function in PD under dual-task conditions [91]. On the other hand, one of the most representative early feature of Parkinsonian gait, reduced speed, is not disease specific [28].

In the mild-to-moderate stage, symptoms spread bilaterally so that asymmetry might decrease [94]. Gait problems worsen and shuffling steps, increased double-limb support and increased cadence become common [28]. Motor automaticity becomes further impaired, resulting in fragmented motor function, such as defragmentation of turns (i.e., turning en block) and problems with gait initiation [30].

Further worsening in gait characterizes the advanced stage of the disease, with more frequent freezing of gait (FOG) and motor blocks, reduced balance and postural control, motor fluctuations and dyskinesia [28].

The use of several sensors technologies to objectively assess Parkinson's disease (PD) symptoms has exponentially increased in the last twenty years. Despite different features such as analysis of the face [95,96], speech [97,98], bradykinesia [15–17], rigidity [17–20] and tremor [21–23] being explored for objective PD evaluation, gait analysis has received widespread attention as part of an objective assessment for PD patient examination. Gait analysis is based on capturing movement with a motion capture device that can be wearable or non-wearable. The analysis of the acquired signals used different statistical or machine learning algorithms like the support vector machine (SVM), dynamic neural network (DNN), naïve Bayes, random forests or decision tree. For machine learning algorithms, the more complex the algorithm, the more likely it is able to determine an optimal decision boundary and hence improved accuracy. However, this improvement in accuracy comes at the cost of reduced interpretability. Among the algorithms included in this survey, decision trees are the most interpretable. All the algorithms used data derived from several features of the gait pattern, which can be grouped into three parameters groups: spatiotemporal, kinematic and kinetic [85]. Spatial parameters measure the physical distance between two steps (like strength length); temporal parameters evaluate the time spent to complete a gait cycle (like the cadence, duration of swing and stance phase) and kinematic parameters evaluate the movement of an object without consideration of its cause while kinetic parameters measure the force that cause the movement (like the ground reaction force during walking) [85].

The majority of these studies assessed if the motion capture device and associated algorithms can distinguish between PD patients and healthy subjects. Other studies investigated if these tools could help to classify different motor status of PD or identify various disease stages.

Among studies focused on discriminating PD vs. healthy subjects, various studies showed high accuracy (more than 90%; Table 4). In addition, regarding studies focused on motor status discrimination, Bayes, et al. [88] were able to discriminate the on/off state by merging an algorithm, which analyzed gait and dyskinesia; instead Barth, et al. [79] were able to differentiate the early vs. intermediate PD stage, with 100% sensitivity and specificity (Table 6). However, it should be considered that all these algorithms need to be validated on larger and representative populations in order to avoid overfitting the problem, which makes the algorithm valid only for the analyzed sample. In addition, the simplest algorithms that provide acceptable accuracy are preferable (i.e., logistic regression and decision trees) over more complex algorithms (e.g., SVM and neural networks) that may provide slightly higher accuracy but are less interpretable.

5. Conclusions

The present overview showed that among the high volume of literature, published on the topic of objective gait analysis in PD, only few studies showed accurate algorithms that can potentially be clinically useful for diagnosis and symptoms monitoring. However, none of those studies have been independently validated or tested on a large scale.

Author Contributions: Writing—original draft preparation, L.d.B.; A.D.S.; M.L.C.; A.D.L.; S.A.S. and L.R.; writing—review and editing: L.d.B., S.A.S. and V.D.L. All authors have read and agreed to the published version of the manuscript.

Funding: This research received no external funding.

Conflicts of Interest: The authors declare no conflict of interest.

References

1. Gibb, W.R.G.; Lees, A.J. The relevance of the Lewy body to the pathogenesis of idiopathic Parkinson's disease. *J. Neurol. Neurosurg. Psychiatry* **1988**, *51*, 745–752. [CrossRef] [PubMed]
2. Rizzo, G.; Copetti, M.; Arcuti, S.; Martino, D.; Fontana, A.; Logroscino, G. Accuracy of clinical diagnosis of Parkinson disease A systematic review and meta-analysis. *Neurology* **2016**, *86*, 566–576. [CrossRef] [PubMed]
3. Fahn, S. Parkinson disease, the effect of levodopa, and the ELLDOPA trial. Earlier vs Later L-DOPA. *Arch. Neurol.* **1999**, *56*, 529–535. [CrossRef]

4. Olanow, C.W.; Kieburtz, K.; Odin, P.; Espay, A.J.; Standaert, D.G.; Fernandez, H.H.; Vanagunas, A.; Othman, A.A.; Widnell, K.L.; Robieson, W.Z.; et al. Continuous intrajejunal infusion of levodopa-carbidopa intestinal gel for patients with advanced Parkinson's disease: A randomised, controlled, double-blind, double-dummy study. *Lancet Neurol.* **2014**, *13*, 141–149. [CrossRef]
5. Salomone, G.; Marano, M.; di Biase, L.; Melgari, J.M.; Di Lazzaro, V. Dopamine dysregulation syndrome and punding in levodopa-carbidopa intestinal gel (LCIG) infusion: A serious but preventable complication. *Parkinsonism Relat. Disord.* **2015**, *21*, 1124–1125. [CrossRef] [PubMed]
6. Melgari, J.M.; Salomone, G.; di Biase, L.; Marano, M.; Scrascia, F.; Di Lazzaro, V. Dyskinesias during levodopa-carbidopa intestinal gel (LCIG) infusion: Management inclinical practice. *Parkinsonism Relat. Disord.* **2015**, *21*, 327–328. [CrossRef] [PubMed]
7. Manson, A.J.; Turner, K.; Lees, A.J. Apomorphine monotherapy in the treatment of refractory motor complications of Parkinson's disease: Long-term follow-up study of 64 patients. *Mov. Disord. Off. J. Mov. Disord. Soc.* **2002**, *17*, 1235–1241. [CrossRef] [PubMed]
8. Rizzone, M.; Fasano, A.; Daniele, A.; Zibetti, M.; Merola, A.; Rizzi, L.; Piano, C.; Piccininni, C.; Romito, L.; Lopiano, L. Long-term outcome of subthalamic nucleus DBS in Parkinson's disease: From the advanced phase towards the late stage of the disease? *Parkinsonism Relat. Disord.* **2014**, *20*, 376–381. [CrossRef] [PubMed]
9. Di Biase, L.; Fasano, A. Low-frequency deep brain stimulation for Parkinson's disease: Great expectation or false hope? *Mov. Disord.* **2016**, *31*, 962–967. [CrossRef]
10. Kern, D.S.; Picillo, M.; Thompson, J.A.; Sammartino, F.; di Biase, L.; Munhoz, R.P.; Fasano, A. Interleaving Stimulation in Parkinson's Disease, Tremor, and Dystonia. *Stereotact. Funct. Neurosurg.* **2018**, *96*, 379–391. [CrossRef]
11. Sandoe, C.; Krishna, V.; Basha, D.; Sammartino, F.; Tatsch, J.; Picillo, M.; di Biase, L.; Poon, Y.Y.; Hamani, C.; Reddy, D.; et al. Predictors of deep brain stimulation outcome in tremor patients. *Brain Stimul.* **2018**, *11*, 592–599. [CrossRef]
12. Tinkhauser, G.; Pogosyan, A.; Debove, I.; Nowacki, A.; Shah, S.A.; Seidel, K.; Tan, H.; Brittain, J.S.; Petermann, K.; di Biase, L.; et al. Directional local field potentials: A tool to optimize deep brain stimulation. *Mov. Disord. Off. J. Mov. Disord. Soc.* **2018**, *33*, 159–164. [CrossRef]
13. Fahn, S.; Elton, R.L. Unified Parkinson's disease rating scale. *Recent Dev. Parkinson's Dis.* **1987**, *2*, 153–163.
14. Goetz, C.G.; Tilley, B.C.; Shaftman, S.R.; Stebbins, G.T.; Fahn, S.; Martinez-Martin, P.; Poewe, W.; Sampaio, C.; Stern, M.B.; Dodel, R. Movement Disorder Society-sponsored revision of the Unified Parkinson's Disease Rating Scale (MDS-UPDRS): Scale presentation and clinimetric testing results. *Mov. Disord.* **2008**, *23*, 2129–2170. [CrossRef] [PubMed]
15. Stamatakis, J.; Ambroise, J.; Cremers, J.; Sharei, H.; Delvaux, V.; Macq, B.; Garraux, G. Finger tapping clinimetric score prediction in Parkinson's disease using low-cost accelerometers. *Comput. Intell. Neurosci.* **2013**, *2013*, 717853. [CrossRef] [PubMed]
16. Summa, S.; Tosi, J.; Taffoni, F.; Di Biase, L.; Marano, M.; Rizzo, A.C.; Tombini, M.; Di Pino, G.; Formica, D. Assessing bradykinesia in Parkinson's disease using gyroscope signals. In Proceedings of the 2017 International Conference on Rehabilitation Robotics (ICORR), London, UK, 17–20 July 2017; pp. 1556–1561.
17. Di Biase, L.; Summa, S.; Tosi, J.; Taffoni, F.; Marano, M.; Cascio Rizzo, A.; Vecchio, F.; Formica, D.; Di Lazzaro, V.; Di Pino, G.; et al. Quantitative Analysis of Bradykinesia and Rigidity in Parkinson's Disease. *Front. Neurol.* **2018**, *9*, 121. [CrossRef]
18. Endo, T.; Okuno, R.; Yokoe, M.; Akazawa, K.; Sakoda, S. A novel method for systematic analysis of rigidity in Parkinson's disease. *Mov. Disord. Off. J. Mov. Disord. Soc.* **2009**, *24*, 2218–2224. [CrossRef]
19. Kwon, Y.; Park, S.H.; Kim, J.W.; Ho, Y.; Jeon, H.M.; Bang, M.J.; Koh, S.B.; Kim, J.H.; Eom, G.M. Quantitative evaluation of parkinsonian rigidity during intra-operative deep brain stimulation. *Bio-Med. Mater. Eng.* **2014**, *24*, 2273–2281. [CrossRef]
20. Raiano, L.; Di Pino, G.; Di Biase, L.; Tombini, M.; Tagliamonte, N.L.; Formica, D. PDMeter: A Wrist Wearable Device for an at-home Assessment of the Parkinson's Disease Rigidity. *IEEE Trans. Neural Syst. Rehabil. Eng.* **2020**. [CrossRef]
21. Deuschl, G.; Krack, P.; Lauk, M.; Timmer, J. Clinical neurophysiology of tremor. *J. Clin. Neurophysiol.* **1996**, *13*, 110–121. [CrossRef]

22. Di Pino, G.; Formica, D.; Melgari, J.-M.; Taffoni, F.; Salomone, G.; di Biase, L.; Caimo, E.; Vernieri, F.; Guglielmelli, E. Neurophysiological bases of tremors and accelerometric parameters analysis. In Proceedings of the 2012 4th IEEE RAS & EMBS International Conference on Biomedical Robotics and Biomechatronics (BioRob), Rome, Italy, 24–27 June 2012; pp. 1820–1825.
23. Di Biase, L.; Brittain, J.S.; Shah, S.A.; Pedrosa, D.J.; Cagnan, H.; Mathy, A.; Chen, C.C.; Martin-Rodriguez, J.F.; Mir, P.; Timmerman, L.; et al. Tremor stability index: A new tool for differential diagnosis in tremor syndromes. *Brain A J. Neurol.* **2017**, *140*, 1977–1986. [CrossRef] [PubMed]
24. Moore, S.T.; MacDougall, H.G.; Gracies, J.M.; Cohen, H.S.; Ondo, W.G. Long-term monitoring of gait in Parkinson's disease. *Gait Posture* **2007**, *26*, 200–207. [CrossRef] [PubMed]
25. Schlachetzki, J.C.M.; Barth, J.; Marxreiter, F.; Gossler, J.; Kohl, Z.; Reinfelder, S.; Gassner, H.; Aminian, K.; Eskofier, B.M.; Winkler, J.; et al. Wearable sensors objectively measure gait parameters in Parkinson's disease. *PLoS ONE* **2017**, *12*, e0183989. [CrossRef] [PubMed]
26. Tosi, J.; Summa, S.; Taffoni, F.; di Biase, L.; Marano, M.; Rizzo, A.C.; Tombini, M.; Schena, E.; Formica, D.; Di Pino, G. Feature Extraction in Sit-to-Stand Task Using M-IMU Sensors and Evaluatiton in Parkinson's Disease. In Proceedings of the 2018 IEEE International Symposium on Medical Measurements and Applications (MeMeA), Rome, Italy, 11–13 June 2018; pp. 1–6. [CrossRef]
27. Suppa, A.; Kita, A.; Leodori, G.; Zampogna, A.; Nicolini, E.; Lorenzi, P.; Rao, R.; Irrera, F. L-DOPA and freezing of gait in Parkinson's disease: Objective assessment through a wearable wireless system. *Front. Neurol.* **2017**, *8*, 406. [CrossRef] [PubMed]
28. Mirelman, A.; Bonato, P.; Camicioli, R.; Ellis, T.D.; Giladi, N.; Hamilton, J.L.; Hass, C.J.; Hausdorff, J.M.; Pelosin, E.; Almeida, Q.J. Gait impairments in Parkinson's disease. *Lancet Neurol.* **2019**, *18*, 697–708. [CrossRef]
29. Pistacchi, M.; Gioulis, M.; Sanson, F.; De Giovannini, E.; Filippi, G.; Rossetto, F.; Marsala, S.Z. Gait analysis and clinical correlations in early Parkinson's disease. *Funct. Neurol.* **2017**, *32*, 28–34. [CrossRef]
30. Son, M.; Youm, C.; Cheon, S.; Kim, J.; Lee, M.; Kim, Y.; Kim, J.; Sung, H. Evaluation of the turning characteristics according to the severity of Parkinson disease during the timed up and go test. *Aging Clin. Exp. Res.* **2017**, *29*, 1191–1199. [CrossRef]
31. Dicharry, J. Kinematics and kinetics of gait: From lab to clinic. *Clin. Sports Med.* **2010**, *29*, 347–364. [CrossRef]
32. Webster, J.B.; Darter, B.J. Principles of Normal and Pathologic Gait. In *Atlas of Orthoses and Assistive Devices*; Elsevier: London, UK, 2019; pp. 49–62.
33. Alvaro Muro-de-la-Herran, B.G.-Z.; Mendez-Zorrilla, A. Gait Analysis Methods: An Overview of Wearable and Non-Wearable Systems, Highlighting Clinical Applications. *Sensors* **2014**, *14*, 3362–3394. [CrossRef]
34. Tao, W.; Liu, T.; Zheng, R.; Feng, H. Gait analysis using wearable sensors. *Sensors* **2012**, *12*, 2255–2283. [CrossRef]
35. Dominguez, G.; Cardiel, E.; Arias, S.; Rogeli, P. A Digital Goniometer Based on Encoders for Measuring Knee-Joint Position in an Orthosis. In Proceedings of the 2013 World Congress on Nature and Biologically Inspired Computing (NaBIC), Fargo, ND, USA, 12–14 August 2013; pp. 1–4.
36. Bae, J.; Tomizuka, M. A tele-monitoring system for gait rehabilitation with an inertial measurement unit and a shoe-type ground reaction force sensor. *Mechatronics* **2013**, *23*, 646–651. [CrossRef]
37. Davis, R.B. Clinical gait analysis. *IEEE Eng. Med. Biol.* **1988**, *7*, 35–40. [CrossRef] [PubMed]
38. Rainoldi, A.; Melchiorri, G.; Caruso, I. A method for positioning electrodes during surface EMG recordings in lower limb muscles. *J. Neurosci. Methods* **2004**, *134*, 37–43. [CrossRef] [PubMed]
39. Qi, Y.; Soh, C.B.; Gunawan, E.; Low, K.S.; Maskooki, A. Using Wearable UWB Radios to Measure Foot Clearance During Walking. In Proceedings of the 2013 35th Annual International Conference of the IEEE Engineering in Medicine and Biology Society (EMBC), Osaka, Japan, 3–7 July 2013; pp. 5199–5202.
40. Wahab, Y.; Bakar, N.A. Gait Analysis Measurement for Sport Application Based on Ultrasonic System. In Proceedings of the 2011 IEEE 15th International Symposium on Consumer Electronics (ISCE), Singapore, 14–17 June 2011; pp. 20–24.
41. Leusmann, P.; Mollering, C.; Klack, L.; Kasugai, K.; Ziefle, M.; Rumpe, B. Your Floor Knows Where You Are: Sensing and Acquisition of Movement Data. *2011 IEEE 12th Int. Conf. Mob. Data Manag.* **2011**, *2*, 61–66.
42. Vera-Rodriguez, R.; Mason, J.S.; Fierrez, J.; Ortega-Garcia, J. Comparative analysis and fusion of spatiotemporal information for footstep recognition. *IEEE Trans. Pattern Anal. Mach. Intell.* **2013**, *35*, 823–834. [CrossRef] [PubMed]

43. Washita, Y.; Kurazume, R.; Ogawara, K. Expanding Gait Identification Methods from Straightto Curved Trajectories. In Proceedings of the 2013 IEEE Workshop on Applications of Computer Vision (WACV), Tampa, FL, USA, 15–17 January 2013; pp. 193–199.
44. Pratheepan, Y.; Condell, J.V.; Prasad, G. The Use of Dynamic and Static Characteristics of Gait for Individual Identification. In Proceedings of the 13th International Machine Vision and Image Processing Conference, Dublin, Ireland, 2–4 September 2009; pp. 111–116.
45. Colyer, S.L.; Evans, M.; Cosker, D.P.; Salo, A.I.T. A Review of the Evolution of Vision-Based Motion Analysis and the Integration of Advanced Computer Vision Methods Towards Developing a Markerless System. *Sports Med. -Open* **2018**, *4*, 24. [CrossRef]
46. Kolb, A.; Barth, E.; Koch, R.; Larsen, R. Time-of-Flight Cameras in Computer Graphics. *Comput. Graph. Forum* **2010**, *29*, 141–159. [CrossRef]
47. Samson, W.; van Hamme, A.; Sanchez, S.; Chèze, L.; van Sint Jan, S.; Feipel, V. Dynamic footprint analysis by time-of-flight camera. *Comput. Methods Biomech. Biomed. Engin.* **2012**, *15*, 180–182. [CrossRef]
48. Derawi, M.O.; Ali, H.; Cheikh, F.A. Gait Recognition Using Time-of-Flight Sensor. Available online: http://subs.emis.de/LNI/Proceedings/Proceedings191/187.pdf (accessed on 17 February 2014).
49. Narayan, A.; Gomatam, M.; Sasi, S. Multimodal gait recognition based on stereo vision and 3D template matching. In Proceedings of the International Conference on Imaging Science, Systems and Technology, CISST'04, Las Vegas, NV, USA, 21–24 June 2004; pp. 405–410.
50. Liu, H.; Cao, Y.; Wang, Z. Automatic Gait Recognition from a Distance. In Proceedings of the Control and Decision Conference (CCDC), Xuzhou, China, 26–28 May 2010; pp. 2777–2782.
51. Clark, R.A.; Pua, Y.H.; Bryant, A.L.; Hunt, M.A. Validity of the Microsoft Kinect for providing lateral trunk lean feedback during gait retraining. *Gait Posture* **2013**, *38*, 1064–1066. [CrossRef]
52. Xue, Z.; Ming, D.; Song, W.; Wan, B.; Jin, S. Infrared gait recognition based on wavelet transform and support vector machine. *Pattern Recognit.* **2010**, *43*, 2904–2910. [CrossRef]
53. Salarian, A.; Russmann, H.; Vingerhoets, F.J.G.; Dehollaini, C.; Blanc, Y.; Burkhard, P.R.; Aminian, K. Gait assessment in Parkinson's disease: Toward an ambulatory system for long-term monitoring. *IEEE Trans. Biomed. Eng.* **2004**, *51*, 1434–1443. [CrossRef] [PubMed]
54. Bonora, G.; Mancini, M.; Carpinella, I.; Chiari, L.; Horak, F.B.; Ferrarin, M. Gait initiation is impaired in subjects with Parkinson's disease in the OFF state: Evidence from the analysis of the anticipatory postural adjustments through wearable inertial sensors. *Gait Posture* **2017**, *51*, 218–221. [CrossRef]
55. Ghislieri, M.; Gastaldi, L.; Pastorelli, S.; Tadano, S.; Agostini, V. Wearable Inertial Sensors to Assess Standing Balance: A Systematic Review. *Sensors* **2019**, *19*, 4075. [CrossRef] [PubMed]
56. Memar, S.; Delrobaei, M.; Pieterman, M.; McIsaac, K.; Jog, M. Quantification of whole-body bradykinesia in Parkinson's disease participants using multiple inertial sensors. *J. Neurol. Sci.* **2018**, *387*, 157–165. [CrossRef]
57. Shah, J.; Pillai, L.; Williams, D.K.; Doerhoff, S.M.; Larson-Prior, L.; Garcia-Rill, E.; Virmani, T. Increased foot strike variability in Parkinson's disease patients with freezing of gait. *Parkinsonism Relat. Disord.* **2018**, *53*, 58–63. [CrossRef] [PubMed]
58. Roy, B.D. Clinical Gait Analysis. *Clin. Gait Anal.* **2006**, *111*, 35–40.
59. Geroin, C.; Squintani, G.; Morini, A.; Donato, F.; Smania, N.; Gandolfi, M.; Tamburin, S.; Fasano, A.; Tinazzi, M. Pisa syndrome in Parkinson's disease: Electromyographic quantification of paraspinal and non-paraspinal muscle activity. *Funct. Neurol.* **2017**, *32*, 143–151. [CrossRef]
60. Lee, Y.; Park, J.-Y.; Choi, Y.-W.; Park, H.-K.; Cho, S.-H.; Cho, S.H.; Lim, Y.-H. A novel non-contact heart rate monitor using impulse-radio ultra-wideband (IR-UWB) radar technology. *Sci. Rep.* **2018**, *8*, 1–10. [CrossRef]
61. Yim, D.; Lee, W.H.; Kim, J.I.; Kim, K.; Ahn, D.H.; Lim, Y.H.; Cho, S.H.; Park, H.K.; Cho, S.H. Quantified Activity Measurement for Medical Use in Movement Disorders through IR-UWB Radar Sensor. *Sensors* **2019**, *19*, 688. [CrossRef]
62. Chong, C.C.; Watanabe, F.; Win, M.Z. Effect of bandwidth on UWB ranging error. *IEEE Wirel. Commun. Netw. Conf. WCNC* **2007**, 1561–1566. [CrossRef]
63. Ashhar, K.; Cheong Boon, S.; Keng He, K. A wearable ultrasonic sensor network for analysis of bilateral gait symmetry. In Proceedings of the 2017 39th Annual International Conference of the IEEE Engineering in Medicine and Biology Society (EMBC), Jeju Island, Korea, 11–15 July 2017; pp. 4455–4458.
64. Iwashita, Y.; Kurazume, R.; Ogawara, K. Expanding gait identification methods from straight to curved trajectories. *2013 IEEE Workshop Appl. Comput. Vis.* **2013**, 193–199. [CrossRef]

65. Arias-Enriquez, O.; Chacon-Murguia, M.I.; Sandoval-Rodriguez, R. Kinematic Analysis of Gait Cycle Using a Fuzzy System for Medical Diagnosis. In Proceedings of the 2012 Annual Meeting of the North American Fuzzy Information Processing Society (NAFIPS), Berkeley, CA, USA, 6–8 August 2012; pp. 1–6.
66. Tucker, C.S.; Behoora, I.; Nembhard, H.B.; Lewis, M.; Sterling, N.W.; Huang, X. Machine learning classification of medication adherence in patients with movement disorders using non-wearable sensors. *Comput. Boil. Med.* **2015**, *66*, 120–134. [CrossRef] [PubMed]
67. Beam, A.L.; Kohane, I.S. Big data and machine learning in health care. *Jama* **2018**, *319*, 1317–1318. [CrossRef] [PubMed]
68. Valliani, A.A.-A.; Ranti, D.; Oermann, E.K. Deep Learning and Neurology: A Systematic Review. *Neurol. Ther.* **2019**, *8*, 351–365. [CrossRef]
69. Esteva, A.; Robicquet, A.; Ramsundar, B.; Kuleshov, V.; DePristo, M.; Chou, K.; Cui, C.; Corrado, G.; Thrun, S.; Dean, J. A guide to deep learning in healthcare. *Nat. Med.* **2019**, *25*, 24–29. [CrossRef]
70. Pathak, A.N.; Sehgal, M.; Christopher, D. A study on selective data mining algorithms. *Int. J. Comput. Sci. Issues (IJCSI)* **2011**, *8*, 479.
71. Cortes, C.; Vapnik, V. Support-vector networks. *Mach. Learn.* **1995**, *20*, 273–297. [CrossRef]
72. Hyde, K.K.; Novack, M.N.; LaHaye, N.; Parlett-Pelleriti, C.; Anden, R.; Dixon, D.R.; Linstead, E. Applications of supervised machine learning in autism spectrum disorder research: A review. *Rev. J. Autism Dev. Disord.* **2019**, *6*, 128–146. [CrossRef]
73. Bishop, C.M. *Pattern Recognition and Machine Learning*; Springer-Verlag New York Inc.: Secaucus, NJ, USA, 2006.
74. Wan, E.A.; Van der Merwe, R. Kalman filtering and neural networks. *Unscented Kalman Filter* **2001**, *7*, 221–277.
75. Sánchez-Ferro, Á.; Elshehabi, M.; Godinho, C.; Salkovic, D.; Hobert, M.A.; Domingos, J.; Uem, J.M.; Ferreira, J.J.; Maetzler, W. New methods for the assessment of Parkinson's disease (2005 to 2015): A systematic review. *Mov. Disord.* **2016**, *31*, 1283–1292. [CrossRef]
76. Brodersen, K.H.; Ong, C.S.; Stephan, K.E.; Buhmann, J.M. The Balanced Accuracy and Its Posterior Distribution. In Proceedings of the 2010 20th International Conference on Pattern Recognition, Istanbul, Turkey, 23–26 August 2010; pp. 3121–3124.
77. Tien, I.; Glaser, S.D.; Aminoff, M.J. Characterization of gait abnormalities in Parkinson's disease using a wireless inertial sensor system. In Proceedings of the 2010 Annual International Conference of the IEEE Engineering in Medicine and Biology, Buenos Aire, Argentina, 30 August–4 September 2010; pp. 3353–3356.
78. Pogorelc, B.; Gams, M. Diagnosing health problems from gait patterns of elderly. In Proceedings of the 2010 Annual International Conference of the IEEE Engineering in Medicine and Biology, Buenos Aire, Argentina, 30 August–4 September 2010; pp. 2238–2241.
79. Barth, J.; Klucken, J.; Kugler, P.; Kammerer, T.; Steidl, R.; Winkler, J.; Hornegger, J.; Eskofier, B. Biometric and mobile gait analysis for early diagnosis and therapy monitoring in Parkinson's disease. In Proceedings of the 2011 Annual International Conference of the IEEE Engineering in Medicine and Biology Society, Boston, MA, USA, 30 August–3 September 2011; pp. 868–871.
80. Yoneyama, M.; Kurihara, Y.; Watanabe, K.; Mitoma, H. Accelerometry-based gait analysis and its application to parkinson's disease assessment-Part 2: A new measure for quantifying walking behavior. *IEEE Trans. Neural Syst. Rehabil. Eng.* **2013**, *21*, 999–1005. [CrossRef] [PubMed]
81. Kugler, P.; Jaremenko, C.; Schlachetzki, J.; Winkler, J.; Klucken, J.; Eskofier, B. Automatic recognition of Parkinson's disease using surface electromyography during standardized gait tests. In Proceedings of the 2013 35th Annual International Conference of the IEEE Engineering in Medicine and Biology Society (EMBC), Osaka, Japan, 3–7 July 2013; pp. 5781–5784.
82. Arora, S.; Venkataraman, V.; Donohue, S.; Biglan, K.M.; Dorsey, E.R.; Little, M.A. High accuracy discrimination of Parkinson's disease participants from healthy controls using smartphones. In Proceedings of the 2014 IEEE International Conference on Acoustics, Speech and Signal Processing (ICASSP), Florence, Italy, 4–9 May 2014; pp. 3641–3644.
83. Procházka, A.; Vyšata, O.; Vališ, M.; upa, O.; Schätz, M.; Mařík, V. Bayesian classification and analysis of gait disorders using image and depth sensors of Microsoft Kinect. *Digit. Signal Process. A Rev. J.* **2015**, *47*, 169–177. [CrossRef]

84. Djurić-Jovičić, M.; Belić, M.; Stanković, I.; Radovanović, S.; Kostić, V.S. Selection of gait parameters for differential diagnostics of patients with de novo Parkinson's disease. *Neurol. Res.* **2017**, *39*, 853–861. [CrossRef] [PubMed]
85. Alam, M.N.; Garg, A.; Munia, T.T.K.; Fazel-Rezai, R.; Tavakolian, K. Vertical ground reaction force marker for Parkinson's disease. *PLoS ONE* **2017**, *12*, e0175951. [CrossRef]
86. Pham, T.D.; Yan, H. Tensor decomposition of gait dynamics in Parkinson's disease. *IEEE Trans. Biomed. Eng.* **2018**, *65*, 1820–1827.
87. Samà, A.; Pérez-López, C.; Rodríguez-Martín, D.; Català, A.; Moreno-Aróstegui, J.M.; Cabestany, J.; de Mingo, E.; Rodríguez-Molinero, A. Estimating bradykinesia severity in Parkinson's disease by analysing gait through a waist-worn sensor. *Comput. Biol. Med.* **2017**, *84*, 114–123. [CrossRef]
88. Bayés, À.; Samá, A.; Prats, A.; Pérez-López, C.; Crespo-Maraver, M.; Moreno, J.M.; Alcaine, S.; Rodriguez-Molinero, A.; Mestre, B.; Quispe, P.; et al. A "HOLTER" for Parkinson's disease: Validation of the ability to detect on-off states using the REMPARK system. *Gait Posture* **2018**, *59*, 1–6. [CrossRef]
89. Rodriguez-Molinero, A.; Sama, A.; Perez-Martinez, D.A.; Perez Lopez, C.; Romagosa, J.; Bayes, A.; Sanz, P.; Calopa, M.; Galvez-Barron, C.; de Mingo, E.; et al. Validation of a portable device for mapping motor and gait disturbances in Parkinson's disease. *JMIR mHealth uHealth* **2015**, *3*, e9. [CrossRef] [PubMed]
90. Perez-Lopez, C.; Sama, A.; Rodriguez-Martin, D.; Moreno-Arostegui, J.M.; Cabestany, J.; Bayes, A.; Mestre, B.; Alcaine, S.; Quispe, P.; Laighin, G.O.; et al. Dopaminergic-induced dyskinesia assessment based on a single belt-worn accelerometer. *Artif. Intell. Med.* **2016**, *67*, 47–56. [CrossRef]
91. Baron, E.I.; Koop, M.M.; Streicher, M.C.; Rosenfeldt, A.B.; Alberts, J.L. Altered kinematics of arm swing in Parkinson's disease patients indicates declines in gait under dual-task conditions. *Park. Relat. Disord.* **2018**, *48*, 61–67. [CrossRef]
92. Yang, Y.-R.; Lee, Y.-Y.; Cheng, S.-J.; Lin, P.-Y.; Wang, R.-Y. Relationships between gait and dynamic balance in early Parkinson's disease. *Gait Posture* **2008**, *27*, 611–615. [CrossRef]
93. Rochester, L.; Galna, B.; Lord, S.; Burn, D. The nature of dual-task interference during gait in incident Parkinson's disease. *Neuroscience* **2014**, *265*, 83–94. [CrossRef]
94. Galna, B.; Lord, S.; Burn, D.J.; Rochester, L. Progression of gait dysfunction in incident Parkinson's disease: Impact of medication and phenotype. *Mov. Disord. Off. J. Mov. Disord. Soc.* **2015**, *30*, 359–367. [CrossRef]
95. Langevin, R.; Ali, M.R.; Sen, T.; Snyder, C.; Myers, T.; Dorsey, E.R.; Hoque, M.E. The PARK Framework for Automated Analysis of Parkinson's Disease Characteristics. *Proc. ACM Interact. Mob. Wearable Ubiquitous Technol.* **2019**, *3*, 1–22. [CrossRef]
96. Wu, P.; Gonzalez, I.; Patsis, G.; Jiang, D.; Sahli, H.; Kerckhofs, E.; Vandekerckhove, M. Objectifying facial expressivity assessment of Parkinson's patients: Preliminary study. *Comput. Math. Methods Med.* **2014**, *2014*, 427826. [CrossRef] [PubMed]
97. Tsanas, A.; Little, M.A.; McSharry, P.E.; Ramig, L.O. Nonlinear speech analysis algorithms mapped to a standard metric achieve clinically useful quantification of average Parkinson's disease symptom severity. *J. R. Soc. Interface* **2011**, *8*, 842–855. [CrossRef] [PubMed]
98. Tsanas, A.; Little, M.A.; McSharry, P.E.; Ramig, L.O. Accurate telemonitoring of Parkinson's disease progression by noninvasive speech tests. *IEEE Trans. Biomed. Eng.* **2010**, *57*, 884–893. [CrossRef] [PubMed]

Review

Fifteen Years of Wireless Sensors for Balance Assessment in Neurological Disorders

Alessandro Zampogna [1,†], Ilaria Mileti [2,†], Eduardo Palermo [2], Claudia Celletti [3], Marco Paoloni [3], Alessandro Manoni [4], Ivan Mazzetta [4], Gloria Dalla Costa [5], Carlos Pérez-López [6,7], Filippo Camerota [3], Letizia Leocani [5], Joan Cabestany [6,7], Fernanda Irrera [4] and Antonio Suppa [1,8,*]

1 Department of Human Neurosciences, Sapienza University of Rome, 00185 Rome, Italy; alessandro.zampogna@uniroma1.it
2 Department of Mechanical and Aerospace Engineering, Sapienza University of Rome, 00184 Rome, Italy; ilaria.mileti@uniroma1.it (I.M.); eduardo.palermo@uniroma1.it (E.P.)
3 Department of Physical Medicine and Rehabilitation, Sapienza University of Rome, 00161 Rome, Italy; clacelletti@gmail.com (C.C.); marco.paoloni@uniroma1.it (M.P.); filippo.camerota@uniroma1.it (F.C.)
4 Department of Information Engineering, Electronics and Telecommunications, Sapienza University of Rome, 00184 Rome, Italy; alessandro.manoni@uniroma1.it (A.M.); ivan.mazzetta@uniroma1.it (I.M.); fernanda.irrera@uniroma1.it (F.I.)
5 Department of Neurorehabilitation and Experimental Neurophysiology Unit, INSPE-Institute of Experimental Neurology, University Vita-Salute and Hospital San Raffaele, 20132 Milan, Italy; dallacosta.gloria@hsr.it (G.D.C.); leocani.letizia@hsr.it (L.L.)
6 Technical Research Centre for Dependency Care and Autonomous Living (CETpD), Universitat Politècnica de Catalunya, Vilanova I la Geltrú, 08800 Barcelona, Spain; carlos.perez-lopez@upc.edu (C.P.-L.); joan.cabestany@upc.edu (J.C.)
7 Sense4Care, 08940 Cornellà de Llobregat, Spain
8 IRCCS Neuromed, 86077 Pozzilli (IS), Italy
* Correspondence: antonio.suppa@uniroma1.it; Tel.: +39-06-4991-4544
† The authors equally contributed to this work.

Received: 3 April 2020; Accepted: 3 June 2020; Published: 7 June 2020

Abstract: Balance impairment is a major mechanism behind falling along with environmental hazards. Under physiological conditions, ageing leads to a progressive decline in balance control per se. Moreover, various neurological disorders further increase the risk of falls by deteriorating specific nervous system functions contributing to balance. Over the last 15 years, significant advancements in technology have provided wearable solutions for balance evaluation and the management of postural instability in patients with neurological disorders. This narrative review aims to address the topic of balance and wireless sensors in several neurological disorders, including Alzheimer's disease, Parkinson's disease, multiple sclerosis, stroke, and other neurodegenerative and acute clinical syndromes. The review discusses the physiological and pathophysiological bases of balance in neurological disorders as well as the traditional and innovative instruments currently available for balance assessment. The technical and clinical perspectives of wearable technologies, as well as current challenges in the field of teleneurology, are also examined.

Keywords: wireless sensors; wearables; balance; posturography; Alzheimer's disease; Parkinson's disease; multiple sclerosis; cerebellar ataxia; stroke; vestibular syndrome

1. Introduction

Countries are globally experiencing a demographic shift in the distribution of the population towards older ages [1] and every year up to 35% of people aged 65 and over fall, often requiring hospital admission after mild to severe injuries [2]. Falls account for 40% of all injury-related deaths [2], and even when non-fatal, commonly cause a "post-fall syndrome", a psychomotor regression condition responsible for psychological, postural and gait dysfunction in elderly [3]. In terms of the economic burden of falls, in 2015 the estimated medical costs attributable to fatal and non-fatal falls increased to 50 billion dollars in the United States [4]. Falls represent a major public health concern and have an enormous economic impact on society, thus requiring the development of effective strategies to prevent underlying causes. Among these, balance impairment is one of the leading determinants of falls along with ecological factors, such as environmental hazards [5]. Ageing significantly impacts on postural ability due to age-related changes in the sensorimotor and cognitive function [6]. Moreover, balance impairment frequently affects patients with neurological disorders who are twice as likely to fall compared to an age-matched healthy population [7].

To date, a history of falls is the strongest predictor of future falls [8,9], thus underscoring the need for predictive measures to determine early preventive interventions. However, clinical assessment is subjective and is not sensitive enough to identify early balance control dysfunction [10]. Conversely, traditional laboratory evaluation, including posturography through force platforms and optoelectronic systems, is objective and sensitive enough to identify subtle abnormalities but does not always reflect real-life situations. Over the last 15 years, advancements in healthcare technology have allowed analysing physiological measures of motor and non-motor behaviour objectively and unobtrusively [11]. Indeed, the availability of wearable devices has opened to the instrumental evaluation of clinical phenomena in free-living conditions. Accordingly, several authors have made a great effort to use wireless sensors in the study of balance impairment in patients with neurological disorders, thus offering new solutions for diagnosis and rehabilitation [12].

Despite several previous reviews discussing specific technical or clinical aspects of balance assessment through wearables, this narrative review aims to discuss the whole topic of balance evaluation, through wireless sensors, in patients with neurological disorders. Accordingly, in this review, we first introduce the physiology and pathophysiology of balance, including the main mechanisms underlying postural dysfunction in several neurological disorders, and report clinical tools commonly used for balance assessment. We then summarise the instrumental assessment of balance, including static and dynamic posturography. Moreover, we analyse wearable technologies available for balance assessment in neurological disorders. Finally, we speculate about prospects and challenges of wireless sensors for balance assessment in teleneurology and telerehabilitation.

2. Physiology and Pathophysiology of Balance

Balance is the ability to maintain body orientation in space under static and dynamic conditions [13], respectively intended as postural stability at rest and in response to active movement or external perturbations. Over the course of evolution, the complexity of this function greatly increased with the acquisition of vertical posture and bipedalism in humans, representing the main transformation in primates [14]. A composite sensorimotor-control system based on a closed-loop circuit dynamically coordinates body segments according to environmental hazards through feedback and feed-forward strategies [15].

The central nervous system oversees balance maintenance by integrating sensory inputs from the peripheral nervous system (e.g., receptors and nerves) and motor outputs to the musculoskeletal system [15,16] (Figure 1). Brainstem nuclei, along with basal ganglia, the cerebellum, and other subcortical structures (e.g., thalamus) play crucial roles in the integration of sensory cues from the somatosensory, vestibular, and visual systems, which continuously provide an overall representation of body movement, acceleration, and position in space [15,17] (Figure 1A). By encoding an internal postural model based on reciprocal connections with the parietal cortex, the cerebellum contributes to dynamic balance control through postural responses that serve as an error-correction mechanism [16] (Figure 1B). Finally, the cerebral cortex oversees attentional and visuospatial balance requirements and manages anticipatory postural adjustments (APAs) before and during voluntary movements [18]. Cognitive-motor processes are responsible for postural optimisation based on prior experience, current context, and learning through long-latency components of postural responses [19].

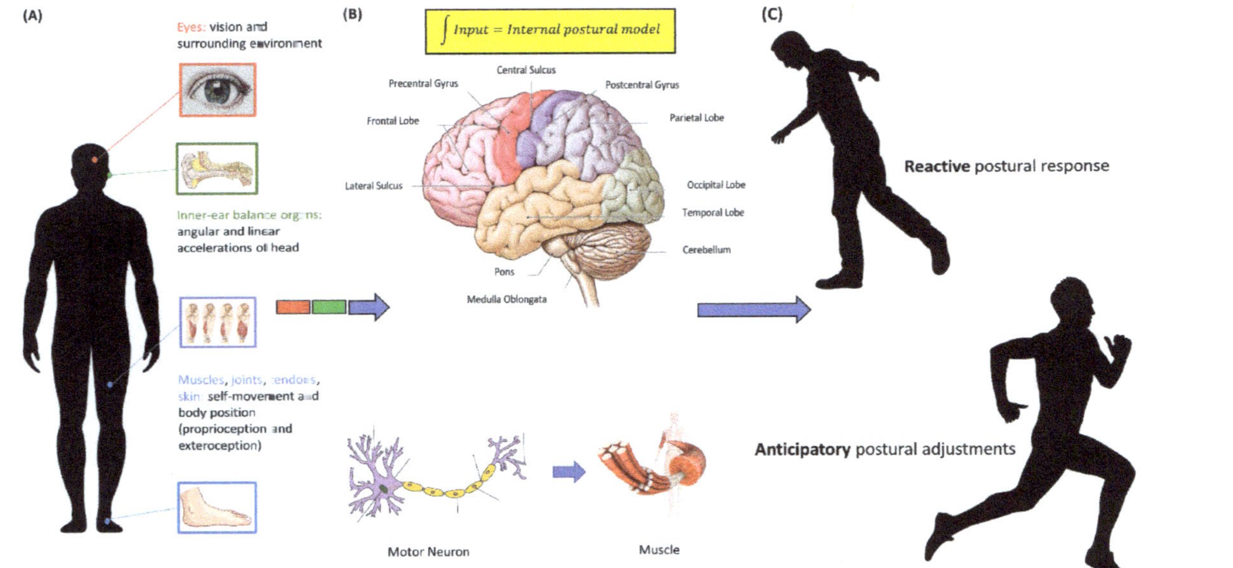

Figure 1. Physiology of balance. (**A**) The visual system provides information on the surrounding environment; the vestibular system, consisting of the two inner-ear balance organs and several nervous structures (nerves and central nuclei), encodes angular and linear accelerations of the head to support the clear vision and balance control via rapid eye movements (vestibulo-ocular reflexes) and postural reflexes (vestibulo-spinal reflexes); the somatosensory system senses self-movement and body position through specialised sensory receptors located in the muscles (muscle spindles), joints (Ruffini endings, Pacinian corpuscles, and Golgi-like receptors), tendons (Golgi tendon organs), and skin (Merkel cells, Ruffini endings, Meissner corpuscles, and Pacinian corpuscles) [20,21]. (**B**) Multisensory signals from visual, vestibular and somatosensory receptors are integrated in the central nervous system to provide an internal postural model and in turn, descending motor commands to muscles. (**C**) Reactive postural strategies and anticipatory postural adjustments allow balance control under environmental circumstances (e.g., external postural perturbations) and motor initiative (e.g., voluntary movement), respectively.

The main goal of physiological mechanisms underlying balance control is the maintenance of postural stability by managing the spatio-temporal relationship between the body's centre of mass (COM) and base of support (BOS) [22]. While reactive postural responses compensate for unexpected external perturbations, proactive postural responses allow balance control under expected external perturbations or self-produced balance disturbances through a motor prediction strategy [22]. When an external balance perturbation occurs, different postural strategies are adopted to maintain the COM projection within the BOS. Indeed, minor postural perturbations are usually counteracted by corrective strategies involving body rotations around the ankle (ankle strategy) or hip (hip strategy) that move the COM projection. Conversely, major postural disturbances require a broadening or displacement of the BOS in order to maintain the COM projection within the BOS (protective strategy) [22] (Figure 1C).

Three main pathophysiological mechanisms are responsible for balance dysfunction: (i) abnormal acquisition, transmission, or perception of sensory signals (Figure 1A); (ii) abnormal sensorimotor integration and motor planning (Figure 1B); (iii) impaired transmission of motor output or musculoskeletal system damage [23] (Figure 1B,C). In patients with impaired afferent sensory information (e.g., somatosensory, vestibular or visual inputs), balance control requires compensatory strategies including attentional resources [24] and sensory reweighting [25].

Ageing is commonly associated with a progressive loss of sensorimotor function, including structural and functional changes in the somatosensory, visual, and vestibular systems, along with a decline in central neural processing and muscle strength [6]. Accordingly, ageing leads to slower reaction times and reduced limits of stability, thus worsening balance control mainly under cognitive loads and unexpected postural perturbations [6,26].

Patients with neurological disorders may manifest balance dysfunction as a result of impairment of at least one physiological component responsible for balance control significantly increasing the risk of falls compared to age-matched healthy subjects [7]. Pathophysiological mechanisms leading to balance impairment in various neurological disorders are summarized in Table 1 along with the main nervous system structures underpinning postural dysfunction. Understanding the physiological mechanisms underlying balance control in humans is the necessary background to measure balance objectively, through conventional as well as wearable technologies, in patients with neurological disorders.

Table 1. Balance impairment in neurological disorders.

	Disease Definition	Nervous Structures Involved	Pathophysiological Mechanisms	Main Clinical Consequence
Alzheimer's disease	Neurodegenerative dementia associated with progressive cognitive and functional dysfunction [27]	Cerebral cortex and subcortical structures, prominently involving nucleus accumbens and putamen [28]	Cognitive impairment, abnormal sensorimotor function and vision, peripheral sensory loss, muscle weakness [29–32]	Hallucinations, inattention, abnormal sensory reweighting
Parkinson's disease	Neurodegenerative movement disorder associated with progressive motor and cognitive dysfunction [33]	Basal ganglia, locus coeruleus and pedunculopontine nucleus [34]	Impaired scaling of postural responses [35], abnormal central proprioceptive-motor integration [36], reduced kinaesthesia [37], axial rigidity [38], cognitive dysfunction [39]	Postural instability, disrupted trunk-legs coordination, freezing of gait
Multiple sclerosis	Acquired demyelinating disease of the central nervous system [40]	Cortico-spinal tract, cerebellum, proprioceptive pathways, vestibular system, brainstem structures for eye movement control [41]	Abnormal sensorimotor, visual, cerebellar, vestibular and cognitive functions [41], muscle weakness and spasticity [42]	Abnormal coordination and sensory reweighting, reduced attentional resources, strength impairment
Huntington's disease	Neurodegenerative disease with autosomal dominant pattern of inheritance [43], associated with cognitive and motor impairment, psychiatric disorders and involuntary movements (chorea) [44]	Basal ganglia, prominently interesting caudate and putamen [45]	Involuntary movements, trunk muscles weakness, hip flexor tightness, impairment in visual and vestibular integration, ocular pursuit movements and proprioception [46]	Chorea, abnormal sensory reweighting, increased stride variability
Cerebellar ataxia	Acquired or hereditary, as well as acute or progressive, disorder associated with dysfunction of cerebellum and/or its connections [47]	Cerebellum (primarily vermis and anterior lobe) and/or its connections, including spinocerebellar tracts [47]	Impaired coordination of movements	Axial motor impairment and asynergic movement

Table 1. *Cont.*

	Disease Definition	Nervous Structures Involved	Pathophysiological Mechanisms	Main Clinical Consequence
Stroke	Acute neurologic syndrome due to the interruption of blood supply to a part of the central nervous system by an ischemic or haemorrhagic vascular injury [48]	Cortico-spinal tract, cerebellum, proprioceptive pathways, vestibular system and brainstem structures [49]	Somatosensory and motor dysfunction [50,51], spasticity [52], visual and perceptual disorders [53,54], including impaired perception of upright body position, cognitive impairment [55]	Hemispatial neglect, strength impairment, abnormal coordination, sensory reweighting
Traumatic brain injury	Acute blunt head traumas or acceleration forces to the head [56]	Vestibular nuclei, cerebellar peduncles, medial lemniscus, dentato-rubro-thalamic and cortico-reticular pathways [57]	Impairment in cognitive and motor functionality [58]	Dizziness, visual-spatial deficits and inattention
Neuropathies	Acute or progressive disorders of the peripheral nervous system, associate with the disruption of nerve action potentials transmission [59]	Peripheral nervous system (nerves)	Sensory and/or motor impairment [59], retinopathy, vestibular and muscle impairment [60], sensory ataxia	Proprioception and strength impairment
Vestibular syndromes	Acute or chronic disorders of the inner-ear balance organs and/or their nervous structures [61] (e.g., Meniere's disease, benign positional vertigo, bilateral vestibular loss, vestibular neuritis, posterior circulation strokes)	Vestibular system (i.e., inner-ear balance organs, vestibular nerve and central nuclei)	Abnormal spatial orientation and motion perception [62], ataxia, eye movement abnormalities [61]	Dizziness and vertigo

3. Clinical Assessment of Balance

The clinical assessment aims at recognizing balance impairment and identifying possible underlying causes [63]. Neurological examination routinely involves several clinical manoeuvres, including the Romberg's test [64], the pull test [65], and the tandem gait test [66], designed to examine individual balance performance qualitatively (Table 2). In addition to these clinical manoeuvres, several standardized scales and tests provide a semiquantitative evaluation of balance (Table 2). A secondary task during motor performance (i.e., dual task) is commonly used to assess the involvement of cognitive function in balance control.

Table 2. Standardised clinical tests and scales for balance assessment.

Clinical Test or Scale	Aim of the Test/Scale	Procedures	Outcome Measures
Romberg test [64]	Postural ability and pathophysiological mechanisms	The subject stands with feet close together, arms by the side, and with eyes open, and then closes eyes while maintaining the same position (removal of vision possibly compensatory proprioceptive deficits)	Unbalance and fall
Pull test [65]	Postural ability	The subject undergoes a sudden body displacement by a quick and forceful pull on the shoulders during upright stance	Number of backward steps or falling (qualitative)
Tandem gait test [66]	Postural ability	The subject walks a straight line while touching the heel of one foot to the toe of the other (narrowed base of support)	Unbalance, falls or need to enlarge the base of support
One-leg stance test [67]	Postural ability	The subject stands unassisted on one leg with opened eyes and arms on the hips as long as possible	Time of performance in seconds
Timed up and go test [68]	Gait and postural ability	The subject sits on a chair, stands up, walks 3 m, turns around, walks back and sits down	Time of performance in seconds
Tinetti balance and mobility scale - Performance-oriented mobility assessment [69]	Gait and postural ability	The subject performs postural and walking motor tasks reflecting common daily activities, such as rising from a chair, maintaining upright stance after a nudge, walking and turning (total 24 items consisting of 14 balance items and 10 gait items)	Total score (sum of gait and balance scores) by using a 2/3-point ordinal scale for each item
Functional reach test [70]	Postural ability	The subject reaches as far forward as he can with arms at 90° flexion, keeping feet on the floor	Maximum distance (cm) that the subject can reach forward beyond arm's length
Berg balance scale [71]	Postural ability	The subject performs functional activities reflecting different components of postural control, such as reaching, bending, transferring and standing (total 14 items)	Total score by using a 5-point ordinal scale for each item
Activities of balance confidence scale [72]	Postural ability	The subject performs a self-report questionnaire on subjective impact of balance dysfunction on 16 daily activities, such as walking in different environmental and postural conditions (total 16 items)	Average score in percentage (each item rated from 0% to 100% of balance confidence)
Physiological profile assessment [73]	Pathophysiological mechanisms	The subject performs different sensorimotor tasks to assess vision (e.g., dual contrast visual acuity chart), lower limb sensation (e.g., tests of proprioception), legs strength, step reaction times, vestibular function (e.g., visual field dependence) and postural sway	Falls risk assessment based on the scores of sensorimotor tasks
Balance evaluation systems test [74]	Pathophysiological mechanisms	The subject performs several motor tasks reflecting different systems underlying balance control (e.g., stance on a firm or foam surface, stepping over obstacles, alternate stair touching); (total 36 items categorised into 6 underlying systems: "Biomechanical Constraints," "Stability Limits/Verticality," "Anticipatory Postural Adjustments," "Postural Responses," "Sensory Orientation" and "Stability in Gait")	Total score in percentage referring to the partial score of systems that involve a 4-point ordinal scale for each item

When considering the clinical assessment of balance, several issues should be taken into account. First, the clinical assessment unlikely detects early postural abnormalities since it identifies balance impairment when significant pathological changes in the nervous system have already occurred. Second, the clinical assessment provides qualitative rather than quantitative evaluations of postural ability, thus representing a subjective tool. Third, standardised clinical scales or indices, such as the Berg balance scale [75] or the dynamic gait index [76], are semiquantitative evaluations of balance, but are time-consuming and suffer from floor and ceiling effects. Lastly, the clinical setting usually involves rather predictable environments with poor ecological value. As a result, evaluation through instrumental tools, such as wearable sensors, would contribute to providing more sensitive, objective, multidimensional, long-term and ecological measures.

4. Static and Dynamic Posturography

Posturography refers to the instrumental assessment of balance [77–79] under static or dynamic conditions [80,81]. Static posturography examines body postural sway while subjects maintain a static stance on a non-movable surface [79,81]. During the upright stance, the human body can be considered an unstable system in which force gravity and body inertia generate torques to be balanced [82]. Indeed, the vertical projection of the whole body mass constantly varies over time, deviating from the ankle joint centre of rotation [83]. Human standing balance can be represented by a reduced number of joints resembling an unstable single-link inverted pendulum [84].

Unlike static evaluation, dynamic posturography includes several postural tests and ad-hoc instruments designed to assess balance under experimentally-induced external perturbations [85]. External disturbances are often designed to simulate environmental hazards occurring in daily activities including a set of visual and motor challenges [85,86]. Postural responses to external perturbations can be assessed by a non-motorised movable platform, such as the Biomechanical Ankle Platform System [87], or more complex commercial robotic systems, such as the Equitest system (Neurocom International, Clackmas, OR, USA) [85], the Balance Master (Micromedical Technologies, Chatham, IL, USA) [88], or Caren (Motek, Amsterdam, the Netherlands) [89].

Several non-commercial robotic platforms have been recently designed to provide various patterns of mechanical perturbation [90–94]. Common approaches include unidirectional [95] or multidirectional [85,96,97] disturbances, such as rotational [93,98–101] and translational perturbations [96,102–104], or forces applied to specific body segments [105,106]. Abrupt perturbations allow the examination of reactive postural responses, whereas continuous and oscillatory perturbations are used for the assessment of anticipatory postural strategies [101,102,104,107,108]. Postural perturbations can be also defined as predictable or unpredictable according to the subject's awareness. The predictability/unpredictability of a specific perturbation allows the experimental investigation of reactive or anticipatory postural strategies [105,106,109]. Mechanical perturbations are often merged with visual, vestibular, and proprioceptive disturbances such as visual scene movements, imposed head accelerations, galvanic vestibular stimulation, and tendon vibration [81,85,89,110,111]. The most common tests used are the Sensory Organization Test (SOT) [112], the Motor Control Test (MCT) [113], and the Adaptation Test (AT) [81]. In the SOT, subjects are elicited through visual, vestibular, and proprioceptive modifications of the support surface and visual surroundings to create sensory conflict conditions. The MCT consists of antero-posterior perturbations at different intensity levels, while in the AT subjects experience toes-up and toes-down rotations.

Several biomechanical parameters quantify balance dysfunction [114,115] by referring to two main variables: the centre of pressure (COP) and COM [116]. The COP is the application point of the total ground reaction force vector, whereas the COM refers to the average position in 3D space of all body segment positions according to their specific masses [116]. COM can be considered representative of the movements of the entire human body [116]. Several indices considering acceleration, velocity, displacement of single or multiple body segments, joint angles, and muscle activity can be measured using both traditional and wearable instrumentation (Table 3).

Table 3. Main biomechanical parameters for balance assessment through traditional and wearable instrumentation.

Name	Meaning	Static	Dynamic
RANGE	Range of acceleration/displacement in the AP, ML, and V direction. Impaired motor strategies report high values of Range Index	[114,117,118]	[111,119]
STD	Standard deviation of reference body landmarks. It is an index of average amplitude of body displacements.		[102,104,120,121]
DIST	Mean distance from the centre of acceleration/displacement trajectory. It is an index of desertion. In static evaluation, high values indicate poor motor control.	[114,117,122,123]	
RMS	Root mean square of the acceleration/displacement in AP, ML, and V direction. High values represent larger dispersion and poor motor control.	[114,117,122–127]	[128]
MEAN	Average acceleration/velocity/displacement in the AP, ML, V direction. High values represent unstable postural adjustments and poor motor control.	[118,122,127]	
PATH	Total length of the acceleration/displacement in static condition larger values represent poor motor control.	[114,117]	[26,102,128]
MV	Mean velocity. It is the first derivative of the acceleration signal in the AP, ML and V direction. Impaired motor strategies report High values of Mean Velocity Index.	[114,117]	
AREA	Total area that encapsulates the total sway path in AP and ML directions. In a static condition, higher values represent poor motor control.	[114,117,118,123,127]	
EA95	95% ellipse sway area It is the ellipse area that encapsulates the 95% of the sway path in the AP and ML direction. High values represent poor motor control.	[114,117,126,127]	
JERK	Time derivative of the acceleration signal. It represents the range of changes in the acceleration signal. High values represent accelerating and decelerating pattern attesting more unstable condition and poor motor control.	[114,117,118,122,125]	
Cross-correlation	Cross-correlation between displacements of two body points. It is an index of coupling between the motion behaviour of two body segments or between the movable platform and the human body		[102,104,120,129]
PWR	Total power of the power spectrum of the acceleration signal.	[114,123]	[102]
F95 or F50	Frequency below which is present the 95% or 50% of the total power. High values indicate a larger amount of postural adjustments and poor motor control.	[114,118]	

Table 3. *Cont.*

Name	Meaning	Static	Dynamic
CF	Centroidal frequency of the signal in the AP, ML and V direction. It is the frequency at which the power is balanced, i.e., the total power above this frequency is equal to the one below. Poor motor control is identified by low values of CF.	[114,117,122]	
FD	Frequency dispersion. It is a measure of the variability of the frequencies of the power spectral density. Values close to zero indicate pure sinusoidal patterns of the signal and a more stable motor control.	[114,117,118,122]	
Entropy	It is the power spectrum entropy of the signal. It is an index of movement smoothness and the inability to regulate postural fluctuations.	[127,130]	
Magnitude	It the area below the EMG curve over a specific range of time, starting from the onset of the perturbation. Mostly this index of muscular intensity is computed during the early response (0–200 ms), the intermediate response (201–400 ms) and the late response (401–600 ms). Impaired postural strategies report lower values of muscle activation.		[111,119]
Onset latency	Time delay between onset of perturbation and muscle activation. It represents how fast a muscle reacts after a perturbation. Impaired balancing strategies report high values of onset latency.		[86,90,111,119]
Time to peak	Time between the onset of perturbation and the maximum activation of the muscle or the maximum peak of joint angle. It indicates how quickly a muscle/joint reaches its maximal value. In dynamic evaluation, lower values indicate high capability in counteracting perturbation.		[86,90,111,119,129,131]
Coactivation	It is the ratio between the magnitude of the agonist and antagonist muscles activity. Impaired postural strategies present an increased coactivation of agonist-antagonist muscles.		[86,90]
Peak angle	Peak of the angular displacement of two adjacent body segment.		[86,129,131]
APAs–CPAs	Anticipatory and compensatory postural adjustments. EMG activity and principal component analysis are estimated over four-time windows in relation to perturbation onset, i.e., APA1 (from −250 ms to −100 ms); APA2 (from −100 ms to +50 ms); CPA1 (from +50 ms to +200 ms); CPA2 (from 200 ms to +350 ms). Impaired motor control reports smaller and delayed APAs during unexpected perturbation.		[95,105,106,109]

AP: antero-posterior; APA: anticipatory postural adjustment; CPA: compensatory postural adjustment; EMG: electromyography; ML: medio-lateral; V: vertical.

Overall, classical laboratory posturography through force plates and optoelectronic systems provides reliable, accurate, and comprehensive measurements for balance assessment. However, these techniques are generally expensive, encumbering, and also require supervised settings as well as technical expertise, thus precluding their use for long-term monitoring in daily life situations. Accordingly, current research on posturography has recently moved on wearable technologies [132–137] possibly providing objective, long-term and free-living monitoring of postural ability at a negligible cost.

5. Wearable Technologies

Recent advances in microelectronics have led to the production of small flexible sensors, even integrated into clothing ("e-textile") [138], thus making wearable devices suitable for free-living applications [139]. To date, the main wearable technologies available for balance assessment include mechanical devices, such as inertial and pressure sensors, and physiological devices, such as surface electromyography sensors (sEMG) (Figure 2). Wireless inertial sensors are the most used solution in wearable systems and have been widely adopted for balance and gait assessment [115,140–142]. Half of the previous studies used commercial inertial measurement unit (IMU) sensors including triaxial accelerometers and gyroscopes, and half adopted stand-alone accelerometers [143] or gyroscopes [144]. The combination of triaxial accelerometers, triaxial gyroscopes and magnetometers compose magnetic and inertial measurement units. Sensor placement depends on the specific postural task under investigation [115]. For instance, wearable sensors can be placed over the waist or trunk in order to measure postural sway and trunk acceleration. Other possible body locations include the lower limbs, sternum, upper limbs and forehead. Triaxial sensors can capture spatio-temporal and 3D kinematic data including joint and segment angles [145–147]. Overall, the combination of accelerometers, gyroscopes, and magnetometers provides accurate information on body spatial orientation and motion (Figure 2A). Besides inertial devices, wearable sEMG sensors evaluate specific patterns of muscle activation during static and dynamic postural perturbations. sEMG, therefore, allows a better understanding of physiological mechanisms responsible for balance control [148,149] (Figure 2B). Lastly, wearable pressure sensors are instrumented insoles placed or integrated into the shoe to measure pressure changes between the foot and ground [150]. The accuracy of this discrete sensor system is comparable to non-wearable technologies such as the laboratory force platform (Figure 2C). In addition to mechanical and physiological devices, there are wearable sensors able to continuously monitor the concentration of specific biochemical markers in biofluids, through miniaturized and flexible devices [151]. These innovative sensors would open to interesting prospects also referring to the assessment of balance. For instance, monitoring L-Dopa or dopamine concentration by microneedle patches would be a helpful tool to correlate postural ability with dopaminergic treatments in patients with Parkinson's disease [152,153]. Currently, several wearable sensors, mostly including inertial devices, are available on the market for approved clinical use in balance assessment [154], also including self-adhesive biosensors (for further details see www.clinicaltrials.gov).

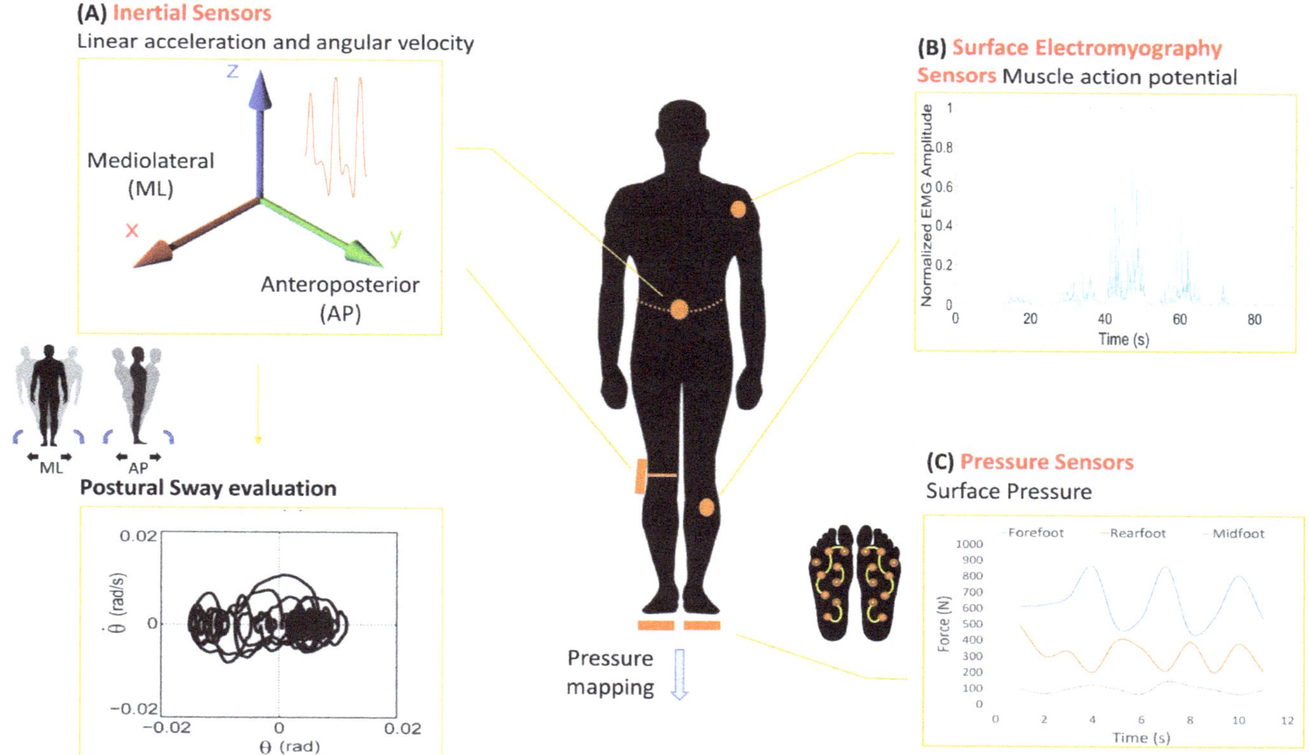

Figure 2. Wearable technologies. Three main types of wireless sensors are available for motion analysis and balance assessment, including mechanical (inertial and pressure sensors) (**A**,**C**) and physiological (surface electromyography sensors) (**B**) wearable devices. AP: antero-posterior; ML: medio-lateral.

The large volume of data produced by wearable sensors requires the development of specialised algorithms and machine learning algorithms to select clinically-valuable measures [138]. Owing to the considerable processing capacity of wearable devices, embedded algorithmic sets can be used for the online and remote execution, but at the expense of the battery charge duration. To optimise the performance of these algorithms in recognising clinical phenomena, a common approach leverages the so-called "sensor fusion", which consists of the combination of sensory data and signals derived from distinct sources so that the resulting information is more accurate (e.g., integration of inertial and electromyography signals) [155]. Accordingly, the emerging trends in wearables are moving towards the design of integrated sensors, including devices composed of IMUs and sEMG [148], to be user-friendly, waterproof and unobtrusive. Table 4 summarises the strengths, limitations and challenges of each type of wireless sensors currently used for balance assessment. Moreover, Table S1 reports all the previously published reviews on balance assessment through wearable devices in healthy subjects and patients affected by various medical conditions.

Table 4. Strengths, limitations and challenges of wireless sensors currently available for balance assessment.

Wireless Sensor	Strengths	Limitations	Challenges
IMU	Low cost and high accuracy	Possible magnetic interferences, errors of misalignment, orthogonality and offset and energy consumption	New algorithms for position and orientation correction
sEMG	Noninvasive analysis and unobtrusiveness	Crosstalk due to adjacent muscles, skin-electrode interface noise and electrode positioning	New implantable EMG sensors and dry electrodes composed of conductive fabric
Pressure	Outdoor measurements and easy integrability	Low comfortability during gait, limited sensitive area and high cost	New capacitive sensors composed of fabric

IMU: Inertial Measurement Unit; **sEMG**: surface electromyography

6. Literature Research Strategy and Criteria

Literature research of studies investigating balance impairment through the use of wireless sensors in neurological disorders was performed using the following databases: MEDLINE, Scopus, PubMed, Web of Science, EMBASE and the Cochrane Library. Literature criteria included the following terms: "wireless sensors" OR "wearables" OR "inertial measurement unit" OR "surface electromyography" OR "pressure sensors" AND "neurological disorders" OR "Alzheimer's disease" OR "stroke" OR "Parkinson's disease" OR "multiple sclerosis" OR "vestibular disorders" OR "cerebellar ataxia" OR "traumatic brain injury" OR "Huntington's disease" OR "neuropathy" AND "balance" OR "posturography" OR "postural control." Eligible studies were experimental studies published from January 2005 to March 2020, examining balance through wireless sensors in patients suffering from the above reported neurological disorders. The reference lists of retrieved articles were also manually searched for additional studies. Reviews, reports, conference proceedings, and articles in languages other than English were not considered in the evaluation of eligible studies.

7. Wearable Technologies in Neurological Disorders

Previous studies using wearable sensors have investigated balance impairment in Parkinson's disease [114,122,124,125,156–168], multiple sclerosis [118,146,169–177], stroke [52,178–184], traumatic brain injuries [123,126,185–189], cerebellar ataxia [130,190–195], vestibular syndromes [196–199], neuropathies [199–201], Alzheimer's disease [32,202,203], and Huntington's disease [46,204]. Most of these studies have compared patients affected by neurological disorders with healthy subjects. However, a minority of authors [52,167,176,178,180,187] have analysed postural ability only in a group of patients with neurological disorders without including a control group.

Concerning the type of sensors used for balance assessment, most of the existing studies have applied inertial devices, primarily accelerometers and gyroscopes. Several authors [46,162,163,183,184] have even used inertial sensors installed in common tablet computers and smartphones. Conversely, no authors have used pressure sensors, while only a few have adopted wireless sEMG sensors [166–168] to analyse balance impairment in patients with Parkinson's disease. Strengths and limitations of each type of sensor are shown in Table 4. Each type of sensor technology would be implemented by addressing some challenges, including the elaboration of new algorithms, the development of implantable EMG tools and, finally, the use of unobtrusive "e-textile" devices (see Table 4). Also, future studies would benefit from the integration of various sensor technologies (i.e., sensor fusion) to optimize the measure of balance dysfunction in patients with neurological disorders.

Regarding the number and body location of sensors, authors have used 1 to 8 inertial devices and multiple body segments, including the upper (10 studies) and lower limbs (21 studies), head (1 study), trunk (18 studies), and waist (48 studies), depending on the static or dynamic postural task chosen for balance assessment. Indeed, some authors who investigated postural evaluation during gait (e.g., [122,146,161,169,172,175]) and instrumented versions of clinical tests, such as the push and release test [171] and the Fukuda Stepping Test [182], have usually applied more sensors than those evaluating static balance during upright stance (e.g., [52,114,125,157,158,163,177,183–185,188,191–194,199,203,204]. However, despite one study [204], all authors have included the lumbo-sacral region as the main location of inertial sensors for the analysis of postural sway, according to the COM position. Conversely, multiple sEMG sensors have been placed mainly on lower limbs to monitor muscle activity during postural perturbations [166–168]. The number of sensors and their placement on the body is a relevant issue for balance assessment, also requiring to consider a proper cost and energy-benefit analysis, as well as the efforts for patients and caregivers. The number of sensors to be used depends on the specific clinical phenomenon under investigation (e.g., postural sway for balance control) and the need for maintaining high-quality measurements, through appropriate sampling rate and estimated energy consumption. Indeed, though more informative, a high number of devices would be computationally demanding and expensive, as well as uncomfortable to be applied in a domestic environment.

Considering the accuracy of sensors in balance assessment, some authors [52,114,156,157,159,162,164,171,173,174,186,191,193,194,200] have compared wearable device measurements with those of standardised laboratory measurement systems, such as force plates and 3D motion-capture systems. These authors have agreed on the moderate or strong correlation between specific inertial indices (e.g., root mean square of acceleration time series [114], acceleration peaks of anticipatory postural adjustments [156,159], time to reach stability [171]) and COP or optical measures, thus suggesting an accurate performance of inertial wearable devices compared to standardised instrumentations in the laboratory. However, validation studies in unsupervised settings are warranted to further support the reliability of wireless sensors for balance assessment in domestic environments.

Most authors [32,52,114,118,123–126,130,146,157,158,163,165,169,170,173,174,176–178,183,185–195,197–204] have performed a static balance evaluation by analysing maintenance of the upright stance with different amplitudes of the BOS (e.g., side-by-side, tandem, single-leg stance). These protocols have also included the assessment of sensory and cognitive contribution to balance control by removing visual and/or proprioceptive cues (e.g., closed eyes, foam surface) and by increasing cognitive load (e.g., dual-task). Moreover, a large number of authors [46,118,122,146,156,159–162,164,166–169,171–173,175,179–182,190,196] have investigated dynamic postural control, mostly through the use of walking tasks, instrumented versions of clinical tests (e.g., Timed-Up and Go, stand and walk, and push and release tests), and external or self-triggered postural perturbations. Although several authors [46,123,125,157,161,166–168,171–173,178–181,190,196,202] have assessed balance during tasks possibly reflecting daily postural challenges, all research protocols have been conducted in a laboratory setting. However, since supervised laboratory settings only partially reflect challenging "real-life" situations, these studies do not provide firm conclusions about the application of wireless sensors in a domestic environment.

Concerning biomechanical measures, previous studies have used filtered acceleration signals by inertial sensors to measure body sway in all the neurological disorders here considered, but have evaluated APAs during gait initiation only in patients with Parkinson's disease. Overall, these measures have shown increased postural sway in patients with neurological disorders and decreased APAs during gait initiation in patients with Parkinson's disease, as compared to age-matched healthy subjects. These parameters have also identified subclinical postural abnormalities (e.g., in vestibular syndromes) correlating with the amount of clinical disability [114,118,124,146,163,165,170,171,175,176,184,190,195]. A few authors [166–168] have measured muscle postural synergies with sEMG sensors in patients with Parkinson's disease. Given that no studies have directly compared biomechanical indices in patients with different neurological disorders, it is unclear whether any of the measures may discriminate the various conditions. These findings overall have shown that wireless sensors can accurately quantify several kinematic measures, including the time and frequency COM dynamics [114,174,200], the 3-D trajectory of body sway angles [191], the joint range of motion [205], the stepping latency [171], and the APAs [159]. Conversely, the evaluation of kinetic measures, including the analysis of internal forces and moments acting on human joints, by wearable systems remains quite challenging [206]. Although the novel approach by wearables would help to partially overcome this issue with inertial and pressure sensors, inverse dynamics techniques, through motion capture systems and force platforms, are currently more suitable to achieve these measures. Moreover, to date, other dynamic variables, including the joint power and the energy cost of a movement, have not yet been evaluated by wearable sensors. Specifically concerning APAs, in addition to inertial measurements, wearable technologies would also allow long-term APAs recordings, through wearable sEMG, in more ecological environments. However, APAs recordings through wearable sEMG would require advanced algorithms for pattern recognition to achieve consistent observation. A further consideration concerns the generalizability to more ecological environments of behavioural measures observed in the laboratory setting. Unlike motor performance under "real-world" postural perturbations, experimental measures under a supervised laboratory setting would improve per se patients' motor behaviour owing to unspecific and disease-unrelated factors, such as attentional and emotional aspects. The appropriate selection of a standardised measure for balance assessment would promote more consistent evaluation among the various neurological disorders. Table 5 provides an overall overview of the methodological approaches and findings from studies here examined. Also, a more detailed description of these studies is shown in Table S2. Finally, Figure 3 shows the positive trend of published studies on wireless sensors for balance assessment in the various neurological disorders.

Table 5. Sensor-based balance evaluation in neurological disorders.

Disease and Number of Studies	Studies with a Control Group	Type and Main Locations of Sensors	Other Measurements	Main Experimental Setups	Main Postural Measures	Main Findings	Clinical-Behavioural Correlations
Alzheimer's disease N = 3 [32,202,203]	N = 3 [32,202,203]	1 to 5 IMUs on trunk, waist, legs and thighs	Not performed	Upright stance with open or closed eyes, different BOS amplitudes and surfaces (e.g., firm and foam), as well as during virtual perturbations	Pitch and roll angles; COM displacement; sway velocity, area and path; RMS acceleration	Lower minimal roll angle, larger COM displacement, higher sway area and RMS acceleration in AD than HS	Not significant or not performed
Parkinson's disease N = 17 [114,122,124,125, 156–168]	N = 16 [114,122,124,125, 156–166,168]	1 to 8 IMUs on trunk, waist, wrists, thighs, shanks and feet; 10 to 22 sEMG on lower limb muscles, lumbar erector spinae, thoracic erector spinae and rectus abdominis	Force plate (COP measures) and infrared optical system	Gait initiation; upright stance with open or closed eyes, different BOS amplitudes and surfaces (e.g., firm and foam), under and not under cognitive load; SOT; ISAW; self-triggered and external postural perturbations; OLS	IMUs: APAs; mean velocity; RMS acceleration; jerkiness; peak-to-peak sway; 95% ellipse area; strategy index. sEMG: amount of variance accounted for; synergy index; ASAs; modulation index	Correlation between inertial, COP and optical measures; hypometric APAs, higher mean velocity, acceleration size and jerkiness, larger peak-to peak sway and 95% ellipse area, predominant ankle strategy; lower VAF and synergy index, reduced ASAs and muscle modulation in PD than HS	Acceleration changes correlated with PIGD and UPDRS-III scores, strategy index with ABC scores, muscle modulation with postural ability and disease severity in PD
Multiple sclerosis N = 11 [118,146,169–177]	N = 10 [118,146,169–175, 177]	1 to 6 IMUs on trunk, waist, wrists, thighs, shanks and feet	Force plate (COP measures) and infrared optical system	Upright stance with open or closed eyes and different surfaces (e.g., firm and foam); walking tasks (e.g., TUG, timed 25-foot walk, 6-minute walk test); external perturbations (e.g., push and release test, backward perturbation)	RMS acceleration; mean velocity; sway jerk, path length, area; F95%; time to reach stability; coherence of acceleration between trunk and legs	Correlation between inertial and COP measures; larger sway acceleration amplitude, angular trunk range of motion in roll and yaw axes, sway path length and area, reduced ML sway jerk, higher F95%, longer time to reach stability and lower acceleration coherence between trunk and legs in MS than HS	Sway acceleration correlated with ABC and MSWS12 scores; RMS acceleration, displacement, mean frequency and time to reach stability correlated with EDSS scores

Table 5. *Cont.*

Disease and Number of Studies	Studies with a Control Group	Type and Main Locations of Sensors	Other Measurements	Main Experimental Setups	Main Postural Measures	Main Findings	Clinical-Behavioural Correlations
Huntington's disease N = 2 [46,204]	N = 2 [46,204]	1 to 2 IMUs on trunk and waist	Not performed	Upright stance with open or closed eyes and different BOS amplitudes; sitting, standing and walking	RMS acceleration; total, peak and mean angular excursion	Higher RMS acceleration; larger peak and total excursions in HD than HS	Not significant or not performed
Cerebellar ataxia N = 7 [130,190–195]	N = 7 [130,190–195]	1 to 6 IMUs on trunk, waist wrists, ankles and feet	Force plate (COP measures)	Upright stance with open or closed eyes and different surfaces (e.g., firm and foam); walking tasks and external perturbations (e.g., retropulsion test)	Trunk angular displacement and velocity, sway path length, area of the convex hull, convex polyhedron volume, entropy, 95% of the ellipse sway area	Correlation between inertial and COP measures; larger trunk angular displacement and velocity, sway path length, area of the convex hull, convex polyhedron volume, entropy and 95% of the ellipse sway area in CA than HS	Inertial measures (e.g., trunk angular displacement and velocity) correlated with ICARS scores, Tinetti's Mobility Index and SARA scores
Stroke N = 8 [52,178–184]	N = 5 [179,181–184]	1 to 5 IMUs on head, trunk, waist and shins	Force plate (COP measures)	Upright stance with open or closed eyes and different BOS amplitudes; walking tasks; functional reach test; Fukuda stepping test; OLS	Body displacement (time, velocity, acceleration); RMS acceleration	Higher maximum and minimum acceleration, LL trunk acceleration, angular velocity in ST than HS	Gyroscope data negatively correlated with Berg balance scale scores
Traumatic brain injury N = 7 [123,126,185–189]	N = 6 [123,126,185,186, 188,189]	1 IMU on waist	Force plate (COP measures)	Upright stance with open or closed eyes, different BOS amplitudes and surfaces (e.g., firm and foam); standard and modified balance error scoring system	RMS acceleration; sway amplitude, velocity, variability and frequency; ellipse and total sway area; 95% ellipsoid sway volume	Higher RMS, total power, mean distance, acceleration range, path length, ellipse and total sway area, 95% ellipsoid sway volume and area in TBI than HS	Self-reported symptoms (e.g., dizziness, headache) correlated with sway path length and postural sway area

Table 5. *Cont.*

Disease and Number of Studies	Studies with a Control Group	Type and Main Locations of Sensors	Other Measurements	Main Experimental Setups	Main Postural Measures	Main Findings	Clinical-Behavioural Correlations
Neuropathies N = 3 [199–201]	N = 3 [199–201]	1 to 2 IMUs on waist and shin	Force plate (COP measures)	Upright stance with open or closed eyes, different BOS amplitudes and surfaces (e.g., firm and foam)	RMS acceleration; range of acceleration; peak velocity; body sway area	Correlation between inertial and COP measures; higher RMS acceleration, acceleration range, and peak velocity; larger body sway area in NP than HS	Vibration perception threshold negatively correlated with postural control
Vestibular syndromes N = 4 [196–199]	N = 4 [196–199]	1 to 4 IMUs on head, trunk, waist and legs	Not performed	Upright stance with open or closed eyes, different BOS amplitudes and surfaces (e.g., firm and foam); walking tasks; shortened functional mobility test	Range of acceleration; peak velocity; RMS acceleration; mean power frequency; quotient of Romberg for inertial measures	Higher range of acceleration, peak velocity, RMS acceleration and quotient of Romberg for some inertial measures; smaller mean power frequency in VS than HS	Not significant or not performed

ABC: Activities-Specific Balance Confidence Scale; AD: patients with Alzheimer's disease; ASA: Anticipatory Synergy Adjustment; APA: anticipatory postural adjustment; BOS: base of support; CA: patients with cerebellar ataxia; COM: centre of mass; COP: centre of pressure; EDSS: Expanded Disability Status Scale; F95%: frequency comprising 95% of the signal; HD: patients with Huntington's disease; HS: healthy subjects; ICARS: International Cooperative Ataxia Rating Scale; IMU: Inertial Measurement Unit; ISAW: Instrumented Stand and Walk Test; LL: latero-lateral; ML: medio-lateral; MS: patients with multiple sclerosis; MSWS12: 12-Item Multiple Sclerosis Walking Scale; N: number; NP: patients with neuropathies; OLS: one-leg stance; PIGD: Postural Instability/Gait Difficulty score; PD: patients with Parkinson's disease; RMS: root mean square; SARA: Scale for the Assessment and Rating of Ataxia; sEMG: surface electromyographic sensors; SOT: Sensory Organisation Test; ST: patients with previous stroke; TBI: patients with previous traumatic brain injury; TUG: Timed-Up and Go test; UPDRS-III: Unified Parkinson's Disease Rating Scale—part III; VAF: variance accounted for; VS: patients with vestibular syndrome.

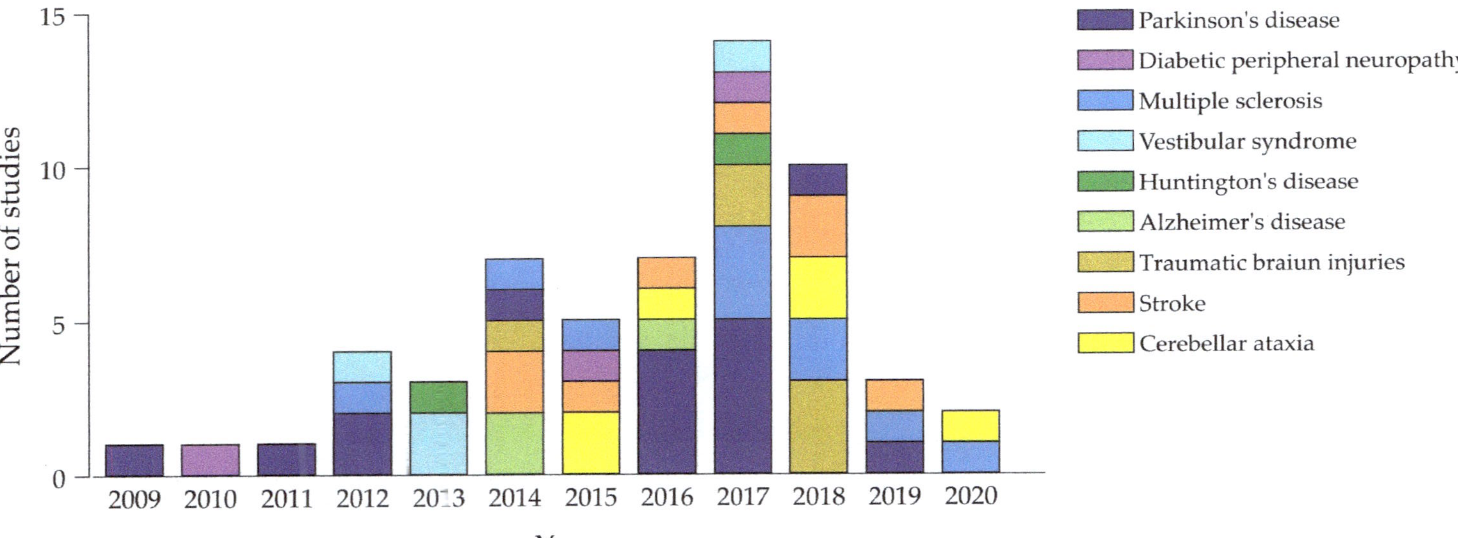

Figure 3. Trend of published studies on wireless sensors for balance assessment in neurological disorders. The figure includes studies reported in Table 5 and Table S2, based on the literature research through MEDLINE, Scopus, PubMed, Web of Science, EMBASE and the Cochrane Library (accessed on 31 March 2020). Note that the figure does not include a study published in 2005 [190] for graphical reasons.

8. Teleneurology and Telerehabilitation for Balance: Prospects and Challenges

Along with the ageing of the population, the prevalence of neurological disorders will also significantly increase in the next decades [207]. Accordingly, public health challenges will burden society and healthcare systems, which will face a heavy demand for the neurologic care of acute and chronic conditions. By allowing long-term monitoring for preventive and recovery strategies, wireless sensors will promote teleneurology and telerehabilitation and take some of the burden off of healthcare facilities.

Concerning the role of teleneurology for balance assessment through wireless sensors, so far, a few studies have addressed this topic in patients with neurological disorders. Nevertheless, several advantageous clinical prospects related to this issue should be considered. First, access to care for patients with balance impairment is quite challenging due to transportation difficulties and dependence on caregivers. Wireless sensors would be a sensitive and objective tool for the domestic measurement of balance control during the performance of validated instrumented tasks, such as maintenance of an upright stance. Moreover, other symptoms commonly associated with postural dysfunction, such as gait disorders [208], would also be measured, thus providing more detailed clinical information. Current evidence suggests that teleneurology promotes a reduction of patient and caregiver burden [209]. Second, medical visits in a hospital setting do not always reflect real-life situations, which commonly present insidious postural challenges. Therefore, the long-term monitoring of postural ability during common daily activities could provide ecological data on patient balance control in free-living conditions. This approach would help to identify early subclinical changes of balance, allow the objective assessment of fall risk and design individualised strategies for fall prevention (e.g., use of mobility aids and changes of environmental hazards). Third, the real-time identification of situations at high risk of falling would also allow patients to benefit from temporary preventive or rescue interventions. For instance, the detection of near-falls could be used for the automatic activation of protective tools, such as inflatable hip pads aimed to prevent fall-related injuries [210]. A further strategy would include the improvement of balance control by wearable-based sensory biofeedback, able to enhance patients' awareness and in turn, prevent falls [211,212].

Along with fall prevention strategies, rehabilitation is the main therapeutic approach for improving balance in patients with neurological disorders. The main goal of rehabilitation is to enhance individual postural skills, supporting patient independence in ecological settings. To this aim, by using information and communication technologies, telerehabilitation would provide rehabilitative services directly at home [213] with similar effectiveness to conventional therapy [214]. Wireless sensors would allow monitoring of individual postural ability in a domestic environment, increasing adherence to the rehabilitative programme, and thus promoting tailored therapeutic approaches [215]. Moreover, wireless sensors would also support home-based interactive rehabilitation programmes by providing real-time feedback during unsupervised training. Nowadays, the increasing use of mobile phones and other technological tools in multiple aspects of daily life is promoting a widespread technological education in the general population, including the elderly. Accordingly, in the next decades, user-friendly wearables will be increasingly used to increase adherence to telerehabilitation strategies. Owing to remote and continuous evaluation by physicians and physical therapists, telerehabilitation would reduce the number of periodic hospital admissions. However, some initial education to patients and caregivers concerning wearables applications for therapeutic purposes is likely required. So far, several clinical trials have already adopted sensor-based measurements to objectively evaluate balance and its response to pharmacological as well as non-pharmacological interventions [216] (for further details see www.clinicaltrials.gov). However, only a few authors [216–220] have examined the effectiveness of sensor-based balance training in patients with neurological disorders. Furthermore, most of these studies involved a laboratory or clinical setting supervised by experienced staff [216]. Hence, to reach some firm conclusion, new randomised controlled trials should assess large samples of patients in ecological settings, including the domestic environment [216].

The main current challenge is the technological migration of wireless sensors from the laboratory setting to a domestic and unsupervised environment. The technological feasibility of sensor systems primarily depends on the variables to be measured as well as on the computing-capacity integrated into the wearables. Unlike conventional laboratory systems, the domestic use of wearable sensors would imply some limitations such as autonomy and interface capabilities (e.g., interaction with the user, communication with external devices and servers for information sharing). Concerning IMUs, challenges include the calculation capacity, which mainly depends on the running algorithms thus influencing the selection of a specific device, processing characteristics, memory capacity and communication protocol. Overall, the technological migration of wireless sensors from the laboratory setting to a domestic environment would benefit from the identification of standardized and accurate measures. To this aim, understanding the physiological and pathophysiological mechanisms underlying balance is the background for selecting, measuring and interpreting the specific postural variables to be assessed. Also, the improvement of communication between wearable sensors and external devices, as well as the implementation of standardized and low energy-consuming algorithms are additional limitations to overcome. To support this migration process, current commercialization efforts are reducing sensor dimensions to ensure the unobtrusiveness of the devices, though maintaining safety and accuracy standards. "Real-world" evidence aimed at monitoring balance disorders through wireless sensors in ecological settings (e.g., patients' home or nursing home) will further clarify strengths and limitations in the telemedicine and telerehabilitation approaches.

Several open questions remain when considering teleneurology and telerehabilitation approaches. To date, only a few randomised controlled trials have addressed this topic in patients with neurological disorders, thus pointing to the weak internal validity of the current clinical evidence. Future studies should propose easier solutions to be applied in unsupervised settings without requiring technical expertise (e.g., issues related to data storage, access platforms and software/app usage). As a possible solution, machine-learning algorithms, including those using artificial neural networks (deep learning algorithms) [221], would be suitable tools for the automatic storage, interpretation and management of healthcare data [222–224]. Indeed, by learning from massive amounts of longitudinal data, machine-learning systems could lighten the burden of technical expertise and improve clinical decision making through a tailored approach. Another relevant point concerns some ethical issues, such as the security of the overwhelming amount of healthcare sensitive data derived from the use of wireless sensors, possibly leading to the generation of discriminatory profiles, manipulative marketing or data breaches [225]. Accordingly, limiting the wireless transmission to a small number of selected data (e.g., fall episodes) would help to preserve the confidentiality of a large amount of recorded information in case of privacy violation. Using proper encryption technology and increasing the users' awareness of privacy rights would help to address ethical issues. Nonetheless, strict regulations for data management should also be adopted to guarantee users' confidentiality and integrity [226]. The use of inertial sensors included in smartphones would address the issue of the cost and availability of wearable sensors [227].

9. Conclusions

Over the last 15 years, wearable devices have been largely used for the assessment of balance in patients affected by neurological disorders, providing valuable data compared with standard laboratory instrumentation. Indeed, a great experience in the use of wireless sensors for balance evaluation has been achieved in the laboratory setting. Conversely, much still needs to be done for the technological migration of wearable devices from the laboratory to the domestic unsupervised environment. This migration would open several valuable prospects, including teleneurology and telerehabilitation approaches.

Supplementary Materials: The following are available online at http://www.mdpi.com/1424-8220/20/11/3247/s1, Table S1: previous reviews on balance and fall risk assessment through wireless sensors and Table S2: sensor-based balance evaluation in neurological disorders. References [32,46,52,114,115,118,122–126,130,146,147,156–204,211,216] are cited in the supplementary materials.

Author Contributions: Conceptualization, A.S., A.Z. and I.M. (Ilaria Mileti); Methodology, A.S., F.I., J.C., A.Z. and I.M. (Ilaria Mileti); Electronic Database Search, A.Z., I.M. (Ilaria Mileti), E.P., C.C., M.P., A.M., I.M. (Ivan Mazzetta), G.D.C., L.L., C.P.-L. and J.C.; Quality Assessment, A.S., F.I., J.C., E.P., F.C. and L.L.; Writing – Original Draft Preparation, A.Z., I.M. (Ilaria Mileti), C.C., M.P., A.M., I.M. (Ivan Mazzetta), G.D.C., L.L., J.C. and C.P.-L.; Writing – Review & Editing, A.S., F.I., J.C., A.Z., I.M. (Ilaria Mileti) and A.M.; Supervision, A.S., F.I., J.C., E.P., F.C. All authors have read and agreed to the published version of the manuscript.

Funding: This research received no external funding.

Conflicts of Interest: The authors declare no conflicts of interest.

Abbreviations

APA	Anticipatory postural adjustment
AT	Adaptation Test
BOS	Base of support
IMU	Inertial measurements unit
COM	Centre of mass
COP	Centre of pressure
MCT	Motor Control Test
sEMG	Surface electromyography
SOT	Sensory Organisation Test

References

1. Ageing and Health. Available online: https://www.who.int/news-room/fact-sheets/detail/ageing-and-health (accessed on 16 March 2020).
2. World Health Organization (Ed.) *WHO Global Report on Falls Prevention in Older Age*; World Health Organization: Geneva, Switzerland, 2008; ISBN 978-92-4-156353-6.
3. Matheron, E.; Dubost, V.; Mourey, F.; Pfitzenmeyer, P.; Manckoundia, P. Analysis of postural control in elderly subjects suffering from Psychomotor Disadaptation Syndrome (PDS). *Arch. Gerontol. Geriatr.* **2010**, *51*, e19–e23. [CrossRef] [PubMed]
4. Florence, C.S.; Bergen, G.; Atherly, A.; Burns, E.; Stevens, J.; Drake, C. Medical Costs of Fatal and Nonfatal Falls in Older Adults. *J. Am. Geriatr. Soc.* **2018**, *66*, 693–698. [CrossRef] [PubMed]
5. Cuevas-Trisan, R. Balance Problems and Fall Risks in the Elderly. *Clin. Geriatr. Med.* **2019**, *35*, 173–183. [CrossRef] [PubMed]
6. Sturnieks, D.L.; St George, R.; Lord, S.R. Balance disorders in the elderly. *Neurophysiol. Clin.* **2008**, *38*, 467–478. [CrossRef] [PubMed]
7. Stolze, H.; Klebe, S.; Zechlin, C.; Baecker, C.; Friege, L.; Deuschl, G. Falls in frequent neurological diseases-prevalence, risk factors and aetiology. *J. Neurol.* **2004**, *251*, 79–84. [CrossRef] [PubMed]
8. Sai, A.J.; Gallagher, J.C.; Smith, L.M.; Logsdon, S. Fall predictors in the community dwelling elderly: A cross sectional and prospective cohort study. *J. Musculoskelet. Neuronal Interact.* **2010**, *10*, 142–150. [PubMed]
9. Pickering, R.M.; Grimbergen, Y.A.M.; Rigney, U.; Ashburn, A.; Mazibrada, G.; Wood, B.; Gray, P.; Kerr, G.; Bloem, B.R. A meta-analysis of six prospective studies of falling in Parkinson's disease. *Mov. Disord.* **2007**, *22*, 1892–1900. [CrossRef]
10. Bloem, B.R.; Steijns, J.A.G.; Smits-Engelsman, B.C. An update on falls. *Curr. Opin. Neurol.* **2003**, *16*, 15–26. [CrossRef]
11. Dorsey, E.R.; Glidden, A.M.; Holloway, M.R.; Birbeck, G.L.; Schwamm, L.H. Teleneurology and mobile technologies: The future of neurological care. *Nat. Rev. Neurol.* **2018**, *14*, 285–297. [CrossRef]
12. Bernhard, F.P.; Sartor, J.; Bettecken, K.; Hobert, M.A.; Arnold, C.; Weber, Y.G.; Poli, S.; Margraf, N.G.; Schlenstedt, C.; Hansen, C.; et al. Wearables for gait and balance assessment in the neurological ward-study design and first results of a prospective cross-sectional feasibility study with 384 inpatients. *BMC Neurol.* **2018**, *18*, 114. [CrossRef]

13. Pollock, A.S.; Durward, B.R.; Rowe, P.J.; Paul, J.P. What is balance? *Clin. Rehabil.* **2000**, *14*, 402–406. [CrossRef] [PubMed]
14. Le Huec, J.C.; Saddiki, R.; Franke, J.; Rigal, J.; Aunoble, S. Equilibrium of the human body and the gravity line: The basics. *Eur. Spine J.* **2011**, *20*, 558–563. [CrossRef] [PubMed]
15. Vassar, R.L.; Rose, J. Motor systems and postural instability. *Handb. Clin. Neurol.* **2014**, *125*, 237–251. [CrossRef] [PubMed]
16. Takakusaki, K. Functional Neuroanatomy for Posture and Gait Control. *J. Mov. Disord.* **2017**, *10*, 1–17. [CrossRef] [PubMed]
17. MacKinnon, C.D. Chapter 1—Sensorimotor anatomy of gait, balance, and falls. In *Handbook of Clinical Neurology*; Balance, G., Day, B.L., Lord, S.R., Eds.; Elsevier: Amsterdam, The Netherlands, 2018; Volume 159, pp. 3–26.
18. Varghese, J.P.; Merino, D.M.; Beyer, K.B.; McIlroy, W.E. Cortical control of anticipatory postural adjustments prior to stepping. *Neuroscience* **2016**, *313*, 99–109. [CrossRef]
19. Jacobs, J.V.; Horak, F.B. Cortical control of postural responses. *J. Neural. Transm.* **2007**, *114*, 1339–1348. [CrossRef]
20. Forbes, P.A.; Chen, A.; Blouin, J.-S. Sensorimotor control of standing balance. *Handb. Clin. Neurol.* **2018**, *159*, 61–83. [CrossRef]
21. Proske, U.; Gandevia, S.C. The proprioceptive senses: Their roles in signaling body shape, body position and movement, and muscle force. *Physiol. Rev.* **2012**, *92*, 1651–1697. [CrossRef]
22. Rogers, M.W.; Mille, M.-L. Balance perturbations. *Handb. Clin. Neurol.* **2018**, *159*, 85–105. [CrossRef]
23. Bronte-Stewart, H.M.; Minn, A.Y.; Rodrigues, K.; Buckley, E.L.; Nashner, L.M. Postural instability in idiopathic Parkinson's disease: The role of medication and unilateral pallidotomy. *Brain* **2002**, *125*, 2100–2114. [CrossRef]
24. Woollacott, M.; Shumway-Cook, A. Attention and the control of posture and gait: A review of an emerging area of research. *Gait Posture* **2002**, *16*, 1–14. [CrossRef]
25. Peterka, R.J. Sensory integration for human balance control. *Handb. Clin. Neurol.* **2018**, *159*, 27–42. [CrossRef]
26. Mileti, I.; Taborri, J.; Rossi, S.; Del Prete, Z.; Paoloni, M.; Suppa, A.; Palermo, E. Reactive Postural Responses to Continuous Yaw Perturbations in Healthy Humans: The Effect of Aging. *Sensors* **2019**, *20*, 63. [CrossRef]
27. Apostolova, L.G. Alzheimer Disease. *Continuum* **2016**, *22*, 419–434. [CrossRef] [PubMed]
28. Lee, Y.W.; Lee, H.; Chung, I.S.; Yi, H.A. Relationship between postural instability and subcortical volume loss in Alzheimer's disease. *Medicine* **2017**, *96*, e7286. [CrossRef] [PubMed]
29. Gordon, B.; Carson, K. The basis for choice reaction time slowing in Alzheimer's disease. *Brain Cognit.* **1990**, *13*, 148–166. [CrossRef]
30. Uhlmann, R.F.; Larson, E.B.; Koepsell, T.D.; Rees, T.S.; Duckert, L.G. Visual impairment and cognitive dysfunction in Alzheimer's disease. *J. Gen. Intern. Med.* **1991**, *6*, 126–132. [CrossRef] [PubMed]
31. Moreland, J.D.; Richardson, J.A.; Goldsmith, C.H.; Clase, C.M. Muscle weakness and falls in older adults: A systematic review and meta-analysis. *J. Am. Geriatr. Soc.* **2004**, *52*, 1121–1129. [CrossRef]
32. Gago, M.F.; Fernandes, V.; Ferreira, J.; Silva, H.; Rocha, L.; Bicho, E.; Sousa, N. Postural Stability Analysis with Inertial Measurement Units in Alzheimer's Disease. *Dement. Geriatr. Cognit. Dis. Extra* **2014**, *4*, 22–30. [CrossRef]
33. Kim, S.D.; Allen, N.E.; Canning, C.G.; Fung, V.S.C. Parkinson disease. *Handb. Clin. Neurol.* **2018**, *159*, 173–193. [CrossRef]
34. Crouse, J.J.; Phillips, J.R.; Jahanshahi, M.; Moustafa, A.A. Postural instability and falls in Parkinson's disease. *Rev. Neurosci.* **2016**, *27*, 549–555. [CrossRef] [PubMed]
35. Di Giulio, I.; St George, R.J.; Kalliolia, E.; Peters, A.L.; Limousin, P.; Day, B.L. Maintaining balance against force perturbations: Impaired mechanisms unresponsive to levodopa in Parkinson's disease. *J. Neurophysiol.* **2016**, *116*, 493–502. [CrossRef] [PubMed]
36. Jacobs, J.V.; Horak, F.B. Abnormal proprioceptive-motor integration contributes to hypometric postural responses of subjects with Parkinson's disease. *Neuroscience* **2006**, *141*, 999–1009. [CrossRef] [PubMed]
37. Maschke, M.; Gomez, C.M.; Tuite, P.J.; Konczak, J. Dysfunction of the basal ganglia, but not the cerebellum, impairs kinaesthesia. *Brain* **2003**, *126*, 2312–2322. [CrossRef]
38. Wright, W.G.; Gurfinkel, V.S.; Nutt, J.; Horak, F.B.; Cordo, P.J. Axial hypertonicity in Parkinson's disease: Direct measurements of trunk and hip torque. *Exp. Neurol.* **2007**, *208*, 38–46. [CrossRef]

39. Allcock, L.M.; Rowan, E.N.; Steen, I.N.; Wesnes, K.; Kenny, R.A.; Burn, D.J. Impaired attention predicts falling in Parkinson's disease. *Parkinsonism Relat. Disord.* **2009**, *15*, 110–115. [CrossRef]
40. What Is MS? Available online: http://www.nationalmssociety.org/What-is-MS (accessed on 15 March 2020).
41. Fjeldstad, C.; Pardo, G.; Bemben, D.; Bemben, M. Decreased postural balance in multiple sclerosis patients with low disability. *Int. J. Rehabil. Res.* **2011**, *34*, 53–58. [CrossRef]
42. Gunn, H.J.; Newell, P.; Haas, B.; Marsden, J.F.; Freeman, J.A. Identification of risk factors for falls in multiple sclerosis: A systematic review and meta-analysis. *Phys. Ther.* **2013**, *93*, 504–513. [CrossRef]
43. Vuong, K.; Canning, C.G.; Menant, J.C.; Loy, C.T. Gait, balance, and falls in Huntington disease. *Handb. Clin. Neurol.* **2018**, *159*, 251–260. [CrossRef]
44. Ross, C.A.; Aylward, E.H.; Wild, E.J.; Langbehn, D.R.; Long, J.D.; Warner, J.H.; Scahill, R.I.; Leavitt, B.R.; Stout, J.C.; Paulsen, J.S.; et al. Huntington disease: Natural history, biomarkers and prospects for therapeutics. *Nat. Rev. Neurol.* **2014**, *10*, 204–216. [CrossRef]
45. Medina, L.D.; Pirogovsky, E.; Salomonczyk, D.; Goldstein, J.; Panzera, R.; Gluhm, S.; Simmons, R.; Corey-Bloom, J.; Gilbert, P.E. Postural limits of stability in premanifest and manifest Huntington's disease. *J. Huntingt. Dis.* **2013**, *2*, 177–184. [CrossRef] [PubMed]
46. Kegelmeyer, D.A.; Kostyk, S.K.; Fritz, N.E.; Fiumedora, M.M.; Chaudhari, A.; Palettas, M.; Young, G.; Kloos, A.D. Quantitative biomechanical assessment of trunk control in Huntington's disease reveals more impairment in static than dynamic tasks. *J. Neurol. Sci.* **2017**, *376*, 29–34. [CrossRef] [PubMed]
47. Ashizawa, T.; Xia, G. Ataxia. *Continuum* **2016**, *22*, 1208–1226. [CrossRef] [PubMed]
48. Rodgers, H. Stroke. *Handb. Clin. Neurol.* **2013**, *110*, 427–433. [CrossRef] [PubMed]
49. Handelzalts, S.; Melzer, I.; Soroker, N. Analysis of Brain Lesion Impact on Balance and Gait Following Stroke. *Front. Hum. Neurosci.* **2019**, *13*, 149. [CrossRef] [PubMed]
50. Sullivan, J.E.; Hedman, L.D. Sensory dysfunction following stroke: Incidence, significance, examination, and intervention. *Top. Stroke Rehabil.* **2008**, *15*, 200–217. [CrossRef] [PubMed]
51. Tasseel-Ponche, S.; Yelnik, A.P.; Bonan, I.V. Motor strategies of postural control after hemispheric stroke. *Neurophysiol. Clin.* **2015**, *45*, 327–333. [CrossRef]
52. Rahimzadeh Khiabani, R.; Mochizuki, G.; Ismail, F.; Boulias, C.; Phadke, C.P.; Gage, W.H. Impact of Spasticity on Balance Control during Quiet Standing in Persons after Stroke. *Stroke Res. Treat.* **2017**, *2017*, 6153714. [CrossRef]
53. Jehkonen, M.; Ahonen, J.P.; Dastidar, P.; Koivisto, A.M.; Laippala, P.; Vilkki, J. How to detect visual neglect in acute stroke. *Lancet* **1998**, *351*, 727–728. [CrossRef]
54. Sand, K.M.; Midelfart, A.; Thomassen, L.; Melms, A.; Wilhelm, H.; Hoff, J.M. Visual impairment in stroke patients—A review. *Acta Neurol. Scand.* **2013**, *127*, 52–56. [CrossRef]
55. Delavaran, H.; Jönsson, A.-C.; Lövkvist, H.; Iwarsson, S.; Elmståhl, S.; Norrving, B.; Lindgren, A. Cognitive function in stroke survivors: A 10-year follow-up study. *Acta Neurol. Scand.* **2017**, *136*, 187–194. [CrossRef] [PubMed]
56. Hyder, A.A.; Wunderlich, C.A.; Puvanachandra, P.; Gururaj, G.; Kobusingye, O.C. The impact of traumatic brain injuries: A global perspective. *NeuroRehabilitation* **2007**, *22*, 341–353. [CrossRef] [PubMed]
57. Jang, S.H.; Yi, J.H.; Kwon, H.G. Injury of the inferior cerebellar peduncle in patients with mild traumatic brain injury: A diffusion tensor tractography study. *Brain Inj.* **2016**, *30*, 1271–1275. [CrossRef] [PubMed]
58. Lin, L.F.; Liou, T.H.; Hu, C.J.; Ma, H.P.; Ou, J.C.; Chiang, Y.H.; Chiu, W.T.; Tsai, S.H.; Chu, W.C. Balance function and sensory integration after mild traumatic brain injury. *Brain Inj.* **2015**, *29*, 41–46. [CrossRef]
59. Ramdharry, G. Peripheral nerve disease. *Handb. Clin. Neurol.* **2018**, *159*, 403–415. [CrossRef]
60. Nicholson, M.; King, J.; Smith, P.F.; Darlington, C.L. Vestibulo-ocular, optokinetic and postural function in diabetes mellitus. *Neuroreport* **2002**, *13*, 153–157. [CrossRef]
61. Young, A.S.; Rosengren, S.M.; Welgampola, M.S. Disorders of the inner-ear balance organs and their pathways. *Handb. Clin. Neurol.* **2018**, *159*, 385–401. [CrossRef]
62. Fernández, L.; Breinbauer, H.A.; Delano, P.H. Vertigo and Dizziness in the Elderly. *Front. Neurol.* **2015**, *6*, 144. [CrossRef]
63. Mancini, M.; Horak, F.B. The relevance of clinical balance assessment tools to differentiate balance deficits. *Eur. J. Phys. Rehabil. Med.* **2010**, *46*, 239–248.
64. Khasnis, A.; Gokula, R.M. Romberg's test. *J. Postgrad. Med.* **2003**, *49*, 169.
65. Hunt, A.L.; Sethi, K.D. The pull test: A history. *Mov. Disord.* **2006**, *21*, 894–899. [CrossRef] [PubMed]

66. Margolesky, J.; Singer, C. How tandem gait stumbled into the neurological exam: A review. *Neurol. Sci.* **2018**, *39*, 23–29. [CrossRef] [PubMed]
67. Fregly, A.R.; Graybiel, A. An ataxia test battery not requiring rails. *Aerosp. Med.* **1968**, *39*, 277–282. [PubMed]
68. Mathias, S.; Nayak, U.S.; Isaacs, B. Balance in elderly patients: The "get-up and go" test. *Arch. Phys. Med. Rehabil.* **1986**, *67*, 387–389. [PubMed]
69. Tinetti, M.E. Performance-oriented assessment of mobility problems in elderly patients. *J. Am. Geriatr. Soc.* **1986**, *34*, 119–126. [CrossRef] [PubMed]
70. Duncan, P.W.; Weiner, D.K.; Chandler, J.; Studenski, S. Functional reach: A new clinical measure of balance. *J. Gerontol.* **1990**, *45*, M192–M197. [CrossRef]
71. Berg, K.O.; Wood-Dauphinee, S.L.; Williams, J.I.; Maki, B. Measuring balance in the elderly: Validation of an instrument. *Can. J. Public Health* **1992**, *83*, S7–S11.
72. Powell, L.E.; Myers, A.M. The Activities-specific Balance Confidence (ABC) Scale. *J. Gerontol. A Biol. Sci. Med. Sci.* **1995**, *50A*, M28–M34. [CrossRef]
73. Lord, S.R.; Menz, H.B.; Tiedemann, A. A physiological profile approach to falls risk assessment and prevention. *Phys. Ther.* **2003**, *83*, 237–252. [CrossRef]
74. Horak, F.B.; Wrisley, D.M.; Frank, J. The Balance Evaluation Systems Test (BESTest) to differentiate balance deficits. *Phys. Ther.* **2009**, *89*, 484–498. [CrossRef]
75. Blum, L.; Korner-Bitensky, N. Usefulness of the Berg Balance Scale in stroke rehabilitation: A systematic review. *Phys. Ther.* **2008**, *88*, 559–566. [CrossRef] [PubMed]
76. Dye, D.C.; Eakman, A.M.; Bolton, K.M. Assessing the validity of the dynamic gait index in a balance disorders clinic: An application of Rasch analysis. *Phys. Ther.* **2013**, *93*, 809–818. [CrossRef] [PubMed]
77. Horak, F.B.; Shupert, C.L.; Mirka, A. Components of postural dyscontrol in the elderly: A review. *Neurobiol. Aging* **1989**, *10*, 727–738. [CrossRef]
78. Piirtola, M.; Era, P. Force platform measurements as predictors of falls among older people-a review. *Gerontology* **2006**, *52*, 1–16. [CrossRef] [PubMed]
79. Bronstein, A.M.; Pavlou, M. Chapter 16—Balance. In *Handbook of Clinical Neurology*; Neurological Rehabilitation; Barnes, M.P., Good, D.C., Eds.; Elsevier: Amsterdam, The Netherlands, 2013; Volume 110, pp. 189–208.
80. Visser, J.E.; Carpenter, M.G.; van der Kooij, H.; Bloem, B.R. The clinical utility of posturography. *Clin. Neurophysiol.* **2008**, *119*, 2424–2436. [CrossRef] [PubMed]
81. Gandolfi, M.; Geroin, C.; Picelli, A.; Smania, N.; Bartolo, M. Assessment of Balance Disorders. In *Advanced Technologies for the Rehabilitation of Gait and Balance Disorders*; Sandrini, G., Homberg, V., Saltuari, L., Smania, N., Pedrocchi, A., Eds.; Springer International Publishing: Cham, Switzerland, 2018; Volume 19, pp. 47–67.
82. Peterka, R.J. Sensorimotor integration in human postural control. *J. Neurophysiol.* **2002**, *88*, 1097–1118. [CrossRef] [PubMed]
83. Cenciarini, M.; Loughlin, P.J.; Sparto, P.J.; Redfern, M.S. Stiffness and damping in postural control increase with age. *IEEE Trans. Biomed. Eng.* **2010**, *57*, 267–275. [CrossRef]
84. Winter, D.A.; Patla, A.E.; Prince, F.; Ishac, M.; Gielo-Perczak, K. Stiffness control of balance in quiet standing. *J. Neurophysiol.* **1998**, *80*, 1211–1221. [CrossRef]
85. Chaudhry, H.; Bukiet, B.; Ji, Z.; Findley, T. Measurement of balance in computer posturography: Comparison of methods—A brief review. *J. Bodyw. Mov. Ther.* **2011**, *15*, 82–91. [CrossRef]
86. Parijat, P.; Lockhart, T.E. Effects of moveable platform training in preventing slip-induced falls in older adults. *Ann. Biomed. Eng.* **2012**, *40*, 1111–1121. [CrossRef]
87. Lee, A.J.Y.; Lin, W.-H. Twelve-week biomechanical ankle platform system training on postural stability and ankle proprioception in subjects with unilateral functional ankle instability. *Clin. Biomech.* **2008**, *23*, 1065–1072. [CrossRef] [PubMed]
88. Newstead, A.H.; Hinman, M.R.; Tomberlin, J.A. Reliability of the Berg Balance Scale and balance master limits of stability tests for individuals with brain injury. *J. Neurol. Phys. Ther.* **2005**, *29*, 18–23. [CrossRef] [PubMed]
89. Kalron, A.; Fonkatz, I.; Frid, L.; Baransi, H.; Achiron, A. The effect of balance training on postural control in people with multiple sclerosis using the CAREN virtual reality system: A pilot randomized controlled trial. *J. Neuroeng. Rehabil.* **2016**, *13*, 13. [CrossRef] [PubMed]

90. De Freitas, P.B.; Knight, C.A.; Barela, J.A. Postural reactions following forward platform perturbation in young, middle-age, and old adults. *J. Electromyogr. Kinesiol.* **2010**, *20*, 693–700. [CrossRef] [PubMed]
91. Halická, Z.; Lobotková, J.; Bzdúšková, D.; Hlavačka, F. Age-related changes in postural responses to backward platform translation. *Physiol. Res.* **2012**, *61*, 331–335. [CrossRef] [PubMed]
92. Kharboutly, H.; Ma, J.; Benali, A.; Thoumie, P.; Pasqui, V.; Bouzit, M. Design of multiple axis robotic platform for postural stability analysis. *IEEE Trans. Neural. Syst. Rehabil. Eng.* **2015**, *23*, 93–103. [CrossRef]
93. Taborri, J.; Mileti, I.; Del Prete, Z.; Rossi, S.; Palermo, E. Yaw Postural Perturbation Through Robotic Platform: Aging Effects on Muscle Synergies. In Proceedings of the 2018 7th IEEE International Conference on Biomedical Robotics and Biomechatronics (Biorob), Enschede, The Netherlands, 26–29 August 2018; pp. 916–921.
94. Dimitrova, D.; Horak, F.B.; Nutt, J.G. Postural muscle responses to multidirectional translations in patients with Parkinson's disease. *J. Neurophysiol.* **2004**, *91*, 489–501. [CrossRef]
95. Kanekar, N.; Aruin, A.S. Aging and balance control in response to external perturbations: Role of anticipatory and compensatory postural mechanisms. *Age* **2014**, *36*, 9621. [CrossRef]
96. Horak, F.B.; Nashner, L.M. Central programming of postural movements: Adaptation to altered support-surface configurations. *J. Neurophysiol.* **1986**, *55*, 1369–1381. [CrossRef]
97. Grin, L.; Frank, J.; Allum, J.H.J. The effect of voluntary arm abduction on balance recovery following multidirectional stance perturbations. *Exp. Brain Res.* **2007**, *178*, 62–78. [CrossRef]
98. Chen, C.L.; Lee, J.Y.; Horng, R.F.; Lou, S.Z.; Su, F.C. Development of a three-degrees-of-freedom moveable platform for providing postural perturbations. *Proc. Inst. Mech. Eng. H* **2009**, *223*, 87–97. [CrossRef] [PubMed]
99. Cappa, P.; Patanè, F.; Rossi, S.; Petrarca, M.; Castelli, E.; Berthoz, A. Effect of changing visual condition and frequency of horizontal oscillations on postural balance of standing healthy subjects. *Gait Posture* **2008**, *28*, 615–626. [CrossRef] [PubMed]
100. Rossi, S.; Gazzellini, S.; Petrarca, M.; Patanè, F.; Salfa, I.; Castelli, E.; Cappa, P. Compensation to whole body active rotation perturbation. *Gait Posture* **2014**, *39*, 621–624. [CrossRef] [PubMed]
101. Amori, V.; Petrarca, M.; Patané, F.; Castelli, E.; Cappa, P. Upper body balance control strategy during continuous 3D postural perturbation in young adults. *Gait Posture* **2015**, *41*, 19–25. [CrossRef] [PubMed]
102. Corna, S.; Tarantola, J.; Nardone, A.; Giordano, A.; Schieppati, M. Standing on a continuously moving platform: Is body inertia counteracted or exploited? *Exp. Brain Res.* **1999**, *124*, 331–341. [CrossRef] [PubMed]
103. Brown, L.A.; Jensen, J.L.; Korff, T.; Woollacott, M.H. The translating platform paradigm: Perturbation displacement waveform alters the postural response. *Gait Posture* **2001**, *14*, 256–263. [CrossRef]
104. De Nunzio, A.M.; Nardone, A.; Schieppati, M. Head stabilization on a continuously oscillating platform: The effect of a proprioceptive disturbance on the balancing strategy. *Exp. Brain Res.* **2005**, *165*, 261–272. [CrossRef]
105. Santos, M.J.; Kanekar, N.; Aruin, A.S. The role of anticipatory postural adjustments in compensatory control of posture: 1. Electromyographic analysis. *J. Electromyogr. Kinesiol.* **2010**, *20*, 388–397. [CrossRef]
106. Santos, M.J.; Kanekar, N.; Aruin, A.S. The role of anticipatory postural adjustments in compensatory control of posture: 2. Biomechanical analysis. *J. Electromyogr. Kinesiol.* **2010**, *20*, 398–405. [CrossRef]
107. Schmid, M.; Bottaro, A.; Sozzi, S.; Schieppati, M. Adaptation to continuous perturbation of balance: Progressive reduction of postural muscle activity with invariant or increasing oscillations of the center of mass depending on perturbation frequency and vision conditions. *Hum. Mov. Sci.* **2011**, *30*, 262–278. [CrossRef]
108. Mileti, I.; Taborri, J.; Rossi, S.; Prete, Z.D.; Paoloni, M.; Suppa, A.; Palermo, E. Measuring age-related differences in kinematic postural strategies under yaw perturbation. In Proceedings of the 2018 IEEE International Symposium on Medical Measurements and Applications, MeMeA, Rome, Italy, 11–13 June 2018; p. 8438804.
109. Claudino, R.; dos Santos, E.C.C.; Santos, M.J. Compensatory but not anticipatory adjustments are altered in older adults during lateral postural perturbations. *Clin. Neurophysiol.* **2013**, *124*, 1628–1637. [CrossRef] [PubMed]
110. Inglis, J.T.; Shupert, C.L.; Hlavacka, F.; Horak, F.B. Effect of galvanic vestibular stimulation on human postural responses during support surface translations. *J. Neurophysiol.* **1995**, *73*, 896–901. [CrossRef] [PubMed]

111. Paquette, C.; Fung, J. Temporal facilitation of gaze in the presence of postural reactions triggered by sudden surface perturbations. *Neuroscience* **2007**, *145*, 505–519. [CrossRef] [PubMed]
112. Ford-Smith, C.D.; Wyman, J.F.; Elswick, R.K.; Fernandez, T.; Newton, R.A. Test-retest reliability of the sensory organization test in noninstitutionalized older adults. *Arch. Phys. Med. Rehabil.* **1995**, *76*, 77–81. [CrossRef]
113. Hale, L.; Miller, R.; Barach, A.; Skinner, M.; Gray, A. Motor Control Test responses to balance perturbations in adults with an intellectual disability. *J. Intellect. Dev. Disabil.* **2009**, *34*, 81–86. [CrossRef] [PubMed]
114. Mancini, M.; Salarian, A.; Carlson-Kuhta, P.; Zampieri, C.; King, L.; Chiari, L.; Horak, F.B. ISway: A sensitive, valid and reliable measure of postural control. *J. Neuroeng. Rehabil.* **2012**, *9*, 59. [CrossRef]
115. Ghislieri, M.; Gastaldi, L.; Pastorelli, S.; Tadano, S.; Agostini, V. Wearable Inertial Sensors to Assess Standing Balance: A Systematic Review. *Sensors* **2019**, *19*, 4075. [CrossRef]
116. Winter, D.A. *Biomechanics and Motor Control of Human Movement*, 4th ed.; Wiley: Hoboken, NJ, USA, 2009; ISBN 978-0-470-39818-0.
117. Park, J.-H.; Mancini, M.; Carlson-Kuhta, P.; Nutt, J.G.; Horak, F.B. Quantifying effects of age on balance and gait with inertial sensors in community-dwelling healthy adults. *Exp. Gerontol.* **2016**, *85*, 48–58. [CrossRef]
118. Craig, J.J.; Bruetsch, A.P.; Lynch, S.G.; Horak, F.B.; Huisinga, J.M. Instrumented balance and walking assessments in persons with multiple sclerosis show strong test-retest reliability. *J. Neuroeng. Rehabil.* **2017**, *14*, 43. [CrossRef]
119. Chen, C.L.; Lou, S.Z.; Wu, H.W.; Wu, S.K.; Yeung, K.T.; Su, F.C. Postural responses to yaw rotation of support surface. *Gait Posture* **2013**, *37*, 296–299. [CrossRef]
120. Nardone, A.; Grasso, M.; Tarantola, J.; Corna, S.; Schieppati, M. Postural coordination in elderly subjects standing on a periodically moving platform. *Arch. Phys. Med. Rehabil.* **2000**, *81*, 1217–1223. [CrossRef] [PubMed]
121. De Nunzio, A.M.; Nardone, A.; Schieppati, M. The control of equilibrium in Parkinson's disease patients: Delayed adaptation of balancing strategy to shifts in sensory set during a dynamic task. *Brain Res. Bull.* **2007**, *74*, 258–270. [CrossRef] [PubMed]
122. Curtze, C.; Nutt, J.G.; Carlson-Kuhta, P.; Mancini, M.; Horak, F.B. Objective Gait and Balance Impairments Relate to Balance Confidence and Perceived Mobility in People with Parkinson Disease. *Phys. Ther.* **2016**, *96*, 1734–1743. [CrossRef] [PubMed]
123. King, L.A.; Mancini, M.; Fino, P.C.; Chesnutt, J.; Swanson, C.W.; Markwardt, S.; Chapman, J.C. Sensor-Based Balance Measures Outperform Modified Balance Error Scoring System in Identifying Acute Concussion. *Ann. Biomed. Eng.* **2018**, *45*, 2135–2145. [CrossRef]
124. Baston, C.; Mancini, M.; Rocchi, L.; Horak, F. Effects of Levodopa on Postural Strategies in Parkinson's disease. *Gait Posture* **2016**, *46*, 26–29. [CrossRef]
125. Chen, T.; Fan, Y.; Zhuang, X.; Feng, D.; Chen, Y.; Chan, P.; Du, Y. Postural sway in patients with early Parkinson's disease performing cognitive tasks while standing. *Neurol. Res.* **2018**, *40*, 491–498. [CrossRef]
126. Baracks, J.; Casa, D.J.; Covassin, T.; Sacko, R.; Scarneo, S.E.; Schnyer, D.; Yeargin, S.W.; Neville, C. Acute Sport-Related Concussion Screening for Collegiate Athletes Using an Instrumented Balance Assessment. *J. Athl. Train.* **2018**, *53*, 597–605. [CrossRef]
127. Greene, B.R.; McGrath, D.; Walsh, L.; Doheny, E.P.; McKeown, D.; Garattini, C.; Cunningham, C.; Crosby, L.; Caulfield, B.; Kenny, R.A. Quantitative falls risk estimation through multi-sensor assessment of standing balance. *Physiol. Meas.* **2012**, *33*, 2049–2063. [CrossRef]
128. Whitney, S.L.; Roche, J.L.; Marchetti, G.F.; Steed, D.P.; Furman, G.R.; Redfern, M.S.; Consulting, C. A comparison of accelerometry and center of pressure measures during computerized dynamic posturography: A measure of balance. *Gait Posture* **2016**, *33*, 594–599. [CrossRef]
129. Sozzi, S.; Nardone, A.; Schieppati, M. Vision does not necessarily stabilize the head in space during continuous postural perturbations. *Front. Neurol.* **2019**, *10*, 1–13. [CrossRef]
130. Nguyen, N.; Phan, D.; Pathirana, P.N.; Horne, M.; Power, L.; Szmulewicz, D. Quantification of Axial Abnormality Due to Cerebellar Ataxia with Inertial Measurements. *Sensors* **2018**, *18*, 2791. [CrossRef] [PubMed]
131. Szturm, T.; Fallang, B. Effects of Varying Acceleration of Platform Translation and Toes-Up Rotations on the Pattern and Magnitude of Balance Reactions in Humans. *J. Vestib. Res.* **1998**, *8*, 381–397. [CrossRef] [PubMed]

132. Kim, J.H.; Sienko, K.H. The Design of a Cell-Phone Based Balance-Training Device. *J. Med. Devices* **2009**, *3*. [CrossRef]
133. Weiss, A.; Herman, T.; Plotnik, M.; Brozgol, M.; Maidan, I.; Giladi, N.; Gurevich, T.; Hausdorff, J.M. Can an accelerometer enhance the utility of the Timed Up & Go Test when evaluating patients with Parkinson's disease? *Med. Eng. Phys.* **2010**, *32*, 119–125. [CrossRef] [PubMed]
134. Giggins, O.M.; Sweeney, K.T.; Caulfield, B. Rehabilitation exercise assessment using inertial sensors: A cross-sectional analytical study. *J. Neuroeng. Rehabil.* **2014**, *11*, 158. [CrossRef]
135. Leardini, A.; Lullini, G.; Giannini, S.; Berti, L.; Ortolani, M.; Caravaggi, P. Validation of the angular measurements of a new inertial-measurement-unit based rehabilitation system: Comparison with state-of-the-art gait analysis. *J. Neuroeng. Rehabil.* **2014**, *11*, 136. [CrossRef]
136. Grimm, B.; Bolink, S. Evaluating physical function and activity in the elderly patient using wearable motion sensors. *EFORT Open Rev.* **2016**, *1*, 112–120. [CrossRef]
137. Horak, F.; King, L.; Mancini, M. Role of body-worn movement monitor technology for balance and gait rehabilitation. *Phys. Ther.* **2015**, *95*, 461–470. [CrossRef]
138. Tokuçoğlu, F. Monitoring Physical Activity with Wearable Technologies. *Noro Psikiyatr. Ars.* **2018**, *55*, S63–S65. [CrossRef]
139. Dobkin, B.H.; Martinez, C. Wearable Sensors to Monitor, Enable Feedback, and Measure Outcomes of Activity and Practice. *Curr. Neurol. Neurosci. Rep.* **2018**, *18*, 87. [CrossRef]
140. Tien, I.; Glaser, S.D.; Aminoff, M.J. Characterization of gait abnormalities in Parkinson's disease using a wireless inertial sensor system. *Conf. Proc. IEEE Eng. Med. Biol. Soc.* **2010**, *2010*, 3353–3356. [CrossRef] [PubMed]
141. Buganè, F.; Benedetti, M.G.; D'Angeli, V.; Leardini, A. Estimation of pelvis kinematics in level walking based on a single inertial sensor positioned close to the sacrum: Validation on healthy subjects with stereophotogrammetric system. *Biomed. Eng. Online* **2014**, *13*, 146. [CrossRef] [PubMed]
142. Mason, B.S.; Rhodes, J.M.; Goosey-Tolfrey, V.L. Validity and reliability of an inertial sensor for wheelchair court sports performance. *J. Appl. Biomech.* **2014**, *30*, 326–331. [CrossRef] [PubMed]
143. Shahzad, A.; Ko, S.; Lee, S.; Lee, J.A.; Kim, K. Quantitative Assessment of Balance Impairment for Fall-Risk Estimation Using Wearable Triaxial Accelerometer. *IEEE Sens. J.* **2017**, *17*, 6743–6751. [CrossRef]
144. Gouwanda, D.; Gopalai, A.A.; Khoo, B.H. A Low Cost Alternative to Monitor Human Gait Temporal Parameters–Wearable Wireless Gyroscope. *IEEE Sens. J.* **2016**, *16*, 9029–9035. [CrossRef]
145. Adkin, A.L.; Bloem, B.R.; Allum, J.H.J. Trunk sway measurements during stance and gait tasks in Parkinson's disease. *Gait Posture* **2005**, *22*, 240–249. [CrossRef]
146. Spain, R.I.; St George, R.J.; Salarian, A.; Mancini, M.; Wagner, J.M.; Horak, F.B.; Bourdette, D. Body-worn motion sensors detect balance and gait deficits in people with multiple sclerosis who have normal walking speed. *Gait Posture* **2012**, *35*, 573–578. [CrossRef]
147. Hubble, R.P.; Naughton, G.A.; Silburn, P.A.; Cole, M.H. Wearable sensor use for assessing standing balance and walking stability in people with Parkinson's disease: A systematic review. *PLoS ONE* **2015**, *10*, e0123705. [CrossRef]
148. Mazzetta, I.; Gentile, P.; Pessione, M.; Suppa, A.; Zampogna, A.; Bianchini, E.; Irrera, F. Stand-Alone Wearable System for Ubiquitous Real-Time Monitoring of Muscle Activation Potentials. *Sensors* **2018**, *18*, 1748. [CrossRef]
149. Mileti, I.; Zampogna, A.; Taborri, J.; Martelli, F.; Rossi, S.; Del Prete, Z.; Paoloni, M.; Suppa, A.; Palermo, E. Parkinson's disease and Levodopa effects on muscle synergies in postural perturbation. In Proceedings of the 2019 IEEE International Symposium on Medical Measurements and Applications (MeMeA), Istanbul, Turkey, 26–28 June 2019; pp. 1–6.
150. Jiang, S.; Pang, Y.; Wang, D.; Yang, Y.; Yang, Z.; Yang, Y.; Ren, T.L. Gait Recognition Based on Graphene Porous Network Structure Pressure Sensors for Rehabilitation Therapy. In Proceedings of the 2018 IEEE International Conference on Electron Devices and Solid State Circuits (EDSSC), Shenzhen, China, 6–8 June 2018; pp. 1–2.
151. Kim, J.; Campbell, A.S.; de Ávila, B.E.-F.; Wang, J. Wearable biosensors for healthcare monitoring. *Nat. Biotechnol.* **2019**, *37*, 389–406. [CrossRef]
152. Li, G.; Wen, D. Wearable biochemical sensors for human health monitoring: Sensing materials and manufacturing technologies. *J. Mater. Chem. B* **2020**, *8*, 3423–3436. [CrossRef] [PubMed]

153. Goud, K.Y.; Moonla, C.; Mishra, R.K.; Yu, C.; Narayan, R.; Litvan, I.; Wang, J. Wearable Electrochemical Microneedle Sensor for Continuous Monitoring of Levodopa: Toward Parkinson Management. *ACS Sens.* **2019**, *4*, 2196–2204. [CrossRef] [PubMed]
154. Mancini, M.; King, L.; Salarian, A.; Holmstrom, L.; McNames, J.; Horak, F.B. Mobility Lab to Assess Balance and Gait with Synchronized Body-worn Sensors. *J. Bioeng. Biomed. Sci.* **2011**, *12*, 007. [CrossRef]
155. Mazzetta, I.; Zampogna, A.; Suppa, A.; Gumiero, A.; Pessione, M.; Irrera, F. Wearable Sensors System for an Improved Analysis of Freezing of Gait in Parkinson's Disease Using Electromyography and Inertial Signals. *Sensors* **2019**, *19*, 948. [CrossRef]
156. Mancini, M.; Zampieri, C.; Carlson-Kuhta, P.; Chiari, L.; Horak, F.B. Anticipatory postural adjustments prior to step initiation are hypometric in untreated Parkinson's disease: An accelerometer-based approach. *Eur. J. Neurol.* **2009**, *16*, 1028–1034. [CrossRef]
157. Mancini, M.; Horak, F.B.; Zampieri, C.; Carlson-Kuhta, P.; Nutt, J.G.; Chiari, L. Trunk accelerometry reveals postural instability in untreated Parkinson's disease. *Parkinsonism Relat. Disord.* **2011**, *17*, 557–562. [CrossRef]
158. Maetzler, W.; Mancini, M.; Liepelt-Scarfone, I.; Müller, K.; Becker, C.; van Lummel, R.C.; Ainsworth, E.; Hobert, M.; Streffer, J.; Berg, D.; et al. Impaired trunk stability in individuals at high risk for Parkinson's disease. *PLoS ONE* **2012**, *7*, e32240. [CrossRef]
159. Mancini, M.; Chiari, L.; Holmstrom, L.; Salarian, A.; Horak, F.B. Validity and reliability of an IMU-based method to detect APAs prior to gait initiation. *Gait Posture* **2016**, *43*, 125–131. [CrossRef]
160. Baston, C.; Mancini, M.; Schoneburg, B.; Horak, F.; Rocchi, L. Postural strategies assessed with inertial sensors in healthy and parkinsonian subjects. *Gait Posture* **2014**, *40*, 70–75. [CrossRef]
161. De Souza Fortaleza, A.C.; Mancini, M.; Carlson-Kuhta, P.; King, L.A.; Nutt, J.G.; Chagas, E.F.; Freitas, I.F.; Horak, F.B. Dual task interference on postural sway, postural transitions and gait in people with Parkinson's disease and freezing of gait. *Gait Posture* **2017**, *56*, 76–81. [CrossRef]
162. Ozinga, S.J.; Linder, S.M.; Alberts, J.L. Use of Mobile Device Accelerometry to Enhance Evaluation of Postural Instability in Parkinson Disease. *Arch. Phys. Med. Rehabil.* **2017**, *98*, 649–658. [CrossRef] [PubMed]
163. Ozinga, S.J.; Koop, M.M.; Linder, S.M.; Machado, A.G.; Dey, T.; Alberts, J.L. Three-dimensional evaluation of postural stability in Parkinson's disease with mobile technology. *NeuroRehabilitation* **2017**, *41*, 211–218. [CrossRef] [PubMed]
164. Bonora, G.; Mancini, M.; Carpinella, I.; Chiari, L.; Horak, F.B.; Ferrarin, M. Gait initiation is impaired in subjects with Parkinson's disease in the OFF state: Evidence from the analysis of the anticipatory postural adjustments through wearable inertial sensors. *Gait Posture* **2017**, *51*, 218–221. [CrossRef] [PubMed]
165. Bonora, G.; Mancini, M.; Carpinella, I.; Chiari, L.; Ferrarin, M.; Nutt, J.G.; Horak, F.B. Investigation of Anticipatory Postural Adjustments during One-Leg Stance Using Inertial Sensors: Evidence from Subjects with Parkinsonism. *Front. Neurol.* **2017**, *8*, 361. [CrossRef]
166. Falaki, A.; Huang, X.; Lewis, M.M.; Latash, M.L. Impaired Synergic Control of Posture in Parkinson's Patients without Postural Instability. *Gait Posture* **2016**, *44*, 209–215. [CrossRef]
167. Falaki, A.; Huang, X.; Lewis, M.M.; Latash, M.L. Dopaminergic modulation of multi-muscle synergies in postural tasks performed by patients with Parkinson's disease. *J. Electromyogr. Kinesiol.* **2017**, *33*, 20–26. [CrossRef]
168. Lang, K.C.; Hackney, M.E.; Ting, L.H.; McKay, J.L. Antagonist muscle activity during reactive balance responses is elevated in Parkinson's disease and in balance impairment. *PLoS ONE* **2019**, *14*, e0211137. [CrossRef]
169. Spain, R.I.; Mancini, M.; Horak, F.B.; Bourdette, D. Body-worn sensors capture variability, but not decline, of gait and balance measures in multiple sclerosis over 18 months. *Gait Posture* **2014**, *39*, 958–964. [CrossRef]
170. Solomon, A.J.; Jacobs, J.V.; Lomond, K.V.; Henry, S.M. Detection of postural sway abnormalities by wireless inertial sensors in minimally disabled patients with multiple sclerosis: A case-control study. *J. Neuroeng. Rehabil.* **2015**, *12*, 74. [CrossRef]
171. El-Gohary, M.; Peterson, D.; Gera, G.; Horak, F.B.; Huisinga, J.M. Validity of the Instrumented Push and Release Test to Quantify Postural Responses in Persons With Multiple Sclerosis. *Arch. Phys. Med. Rehabil.* **2017**, *98*, 1325–1331. [CrossRef]

172. Witchel, H.J.; Oberndorfer, C.; Needham, R.; Healy, A.; Westling, C.E.I.; Guppy, J.H.; Bush, J.; Barth, J.; Herberz, C.; Roggen, D.; et al. Thigh-Derived Inertial Sensor Metrics to Assess the Sit-to-Stand and Stand-to-Sit Transitions in the Timed Up and Go (TUG) Task for Quantifying Mobility Impairment in Multiple Sclerosis. *Front. Neurol.* **2018**, *9*, 684. [CrossRef] [PubMed]
173. Huisinga, J.; Mancini, M.; Veys, C.; Spain, R.; Horak, F. Coherence analysis of trunk and leg acceleration reveals altered postural sway strategy during standing in persons with multiple sclerosis. *Hum. Mov. Sci.* **2018**, *58*, 330–336. [CrossRef] [PubMed]
174. Sun, R.; Moon, Y.; McGinnis, R.S.; Seagers, K.; Motl, R.W.; Sheth, N.; Wright, J.A.; Ghaffari, R.; Patel, S.; Sosnoff, J.J. Assessment of Postural Sway in Individuals with Multiple Sclerosis Using a Novel Wearable Inertial Sensor. *Digit. Biomark.* **2018**, *2*, 1–10. [CrossRef] [PubMed]
175. Arpan, I.; Fino, P.C.; Fling, B.W.; Horak, F. Local dynamic stability during long-fatiguing walks in people with multiple sclerosis. *Gait Posture* **2020**, *76*, 122–127. [CrossRef] [PubMed]
176. Chitnis, T.; Glanz, B.I.; Gonzalez, C.; Healy, B.C.; Saraceno, T.J.; Sattarnezhad, N.; Diaz-Cruz, C.; Polgar-Turcsanyi, M.; Tummala, S.; Bakshi, R.; et al. Quantifying neurologic disease using biosensor measurements in-clinic and in free-living settings in multiple sclerosis. *NPJ Digit. Med.* **2019**, *2*, 123. [CrossRef] [PubMed]
177. Gera, G.; Fling, B.W.; Horak, F.B. Cerebellar White Matter Damage Is Associated with Postural Sway Deficits in People with Multiple Sclerosis. *Arch. Phys. Med. Rehabil.* **2020**, *101*, 258–264. [CrossRef]
178. Perez-Cruzado, D.; González-Sánchez, M.; Cuesta-Vargas, A.I. Parameterization and reliability of single-leg balance test assessed with inertial sensors in stroke survivors: A cross-sectional study. *Biomed. Eng. Online* **2014**, *13*, 127. [CrossRef]
179. Merchán-Baeza, J.A.; González-Sánchez, M.; Cuesta-Vargas, A.I. Comparison of kinematic variables obtained by inertial sensors among stroke survivors and healthy older adults in the Functional Reach Test: Cross-sectional study. *Biomed. Eng. Online* **2015**, *14*, 49. [CrossRef]
180. Merchán-Baeza, J.A.; González-Sánchez, M.; Cuesta-Vargas, A.I. Reliability in the parameterization of the functional reach test in elderly stroke patients: A pilot study. *Biomed. Res. Int.* **2014**, *2014*, 637671. [CrossRef]
181. Iosa, M.; Bini, F.; Marinozzi, F.; Fusco, A.; Morone, G.; Koch, G.; Martino Cinnera, A.; Bonnì, S.; Paolucci, S. Stability and Harmony of Gait in Patients with Subacute Stroke. *J. Med. Biol. Eng.* **2016**, *36*, 635–643. [CrossRef]
182. Belluscio, V.; Bergamini, E.; Iosa, M.; Tramontano, M.; Morone, G.; Vannozzi, G. The iFST: An instrumented version of the Fukuda Stepping Test for balance assessment. *Gait Posture* **2018**, *60*, 203–208. [CrossRef] [PubMed]
183. Hou, Y.R.; Chiu, Y.L.; Chiang, S.L.; Chen, H.Y.; Sung, W.H. Feasibility of a smartphone-based balance assessment system for subjects with chronic stroke. *Comput. Methods Programs Biomed.* **2018**, *161*, 191–195. [CrossRef] [PubMed]
184. Hou, Y.R.; Chiu, Y.L.; Chiang, S.-L.; Chen, H.Y.; Sung, W.H. Development of a Smartphone-Based Balance Assessment System for Subjects with Stroke. *Sensors* **2019**, *20*, 88. [CrossRef]
185. Furman, G.R.; Lin, C.C.; Bellanca, J.L.; Marchetti, G.F.; Collins, M.W.; Whitney, S.L. Comparison of the balance accelerometer measure and balance error scoring system in adolescent concussions in sports. *Am. J. Sports Med.* **2013**, *41*, 1404–1410. [CrossRef] [PubMed]
186. Doherty, C.; Zhao, L.; Ryan, J.; Komaba, Y.; Inomata, A.; Caulfield, B. Quantification of postural control deficits in patients with recent concussion: An inertial-sensor based approach. *Clin. Biomech.* **2017**, *42*, 79–84. [CrossRef] [PubMed]
187. Alkathiry, A.A.; Sparto, P.J.; Freund, B.; Whitney, S.L.; Mucha, A.; Furman, J.M.; Collins, M.W.; Kontos, A.P. Using Accelerometers to Record Postural Sway in Adolescents With Concussion: A Cross-Sectional Study. *J. Athl. Train.* **2018**, *53*, 1166–1172. [CrossRef] [PubMed]
188. Gera, G.; Chesnutt, J.; Mancini, M.; Horak, F.B.; King, L.A. Inertial Sensor-Based Assessment of Central Sensory Integration for Balance After Mild Traumatic Brain Injury. *Mil. Med.* **2018**, *183*, 327–332. [CrossRef]
189. King, L.A.; Horak, F.B.; Mancini, M.; Pierce, D.; Priest, K.C.; Chesnutt, J.; Sullivan, P.; Chapman, J.C. Instrumenting the Balance Error Scoring System for use with patients reporting persistent balance problems after mild traumatic brain injury. *Arch. Phys. Med. Rehabil.* **2014**, *95*, 353–359. [CrossRef]
190. Van de Warrenburg, B.P.C.; Bakker, M.; Kremer, B.P.H.; Bloem, B.R.; Allum, J.H.J. Trunk sway in patients with spinocerebellar ataxia. *Mov. Disord.* **2005**, *20*, 1006–1013. [CrossRef]

191. Hejda, J.; Cakrt, O.; Socha, V.; Schlenker, J.; Kutilek, P. 3-D trajectory of body sway angles: A technique for quantifying postural stability. *Biocybern. Biomed. Eng.* **2015**, *35*, 185–191. [CrossRef]
192. Kutílek, P.; Socha, V.; Čakrt, O.; Svoboda, Z. Assessment of postural stability in patients with cerebellar disease using gyroscope data. *J. Bodyw. Mov. Ther.* **2015**, *19*, 421–428. [CrossRef] [PubMed]
193. Melecky, R.; Socha, V.; Kutilek, P.; Hanakova, L.; Takac, P.; Schlenker, J.; Svoboda, Z. Quantification of Trunk Postural Stability Using Convex Polyhedron of the Time-Series Accelerometer Data. *J. Healthc. Eng.* **2016**, *2016*. [CrossRef] [PubMed]
194. Adamová, B.; Kutilek, P.; Cakrt, O.; Svoboda, Z.; Viteckova, S.; Smrcka, P. Quantifying postural stability of patients with cerebellar disorder during quiet stance using three-axis accelerometer. *Biomed. Signal Process. Control* **2018**, *40*, 378–384. [CrossRef]
195. Widener, G.L.; Conley, N.; Whiteford, S.; Gee, J.; Harrell, A.; Gibson-Horn, C.; Block, V.; Allen, D.D. Changes in standing stability with balance-based torso-weighting with cerebellar ataxia: A pilot study. *Physiother. Res. Int.* **2020**, *25*, e1814. [CrossRef]
196. Cohen, H.S.; Mulavara, A.P.; Peters, B.T.; Sangi-Haghpeykar, H.; Bloomberg, J.J. Tests of walking balance for screening vestibular disorders. *J. Vestib. Res.* **2012**, *22*, 95–104. [CrossRef]
197. Kapoula, Z.; Gaertner, C.; Yang, Q.; Denise, P.; Toupet, M. Vergence and Standing Balance in Subjects with Idiopathic Bilateral Loss of Vestibular Function. *PLoS ONE* **2013**, *8*, e66652. [CrossRef]
198. Kim, S.C.; Kim, M.J.; Kim, N.; Hwang, J.H.; Han, G.C. Ambulatory balance monitoring using a wireless attachable three-axis accelerometer. *J. Vestib. Res.* **2013**, *23*, 217–225. [CrossRef]
199. D'Silva, L.J.; Kluding, P.M.; Whitney, S.L.; Dai, H.; Santos, M. Postural sway in individuals with type 2 diabetes and concurrent benign paroxysmal positional vertigo. *Int. J. Neurosci.* **2017**, *127*, 1065–1073. [CrossRef]
200. Najafi, B.; Horn, D.; Marclay, S.; Crews, R.T.; Wu, S.; Wrobel, J.S. Assessing postural control and postural control strategy in diabetes patients using innovative and wearable technology. *J. Diabetes Sci. Technol.* **2010**, *4*, 780–791. [CrossRef]
201. Toosizadeh, N.; Mohler, J.; Armstrong, D.G.; Talal, T.K.; Najafi, B. The influence of diabetic peripheral neuropathy on local postural muscle and central sensory feedback balance control. *PLoS ONE* **2015**, *10*, e0135255. [CrossRef]
202. Gago, M.F.; Yelshyna, D.; Bicho, E.; Silva, H.D.; Rocha, L.; Lurdes Rodrigues, M.; Sousa, N. Compensatory Postural Adjustments in an Oculus Virtual Reality Environment and the Risk of Falling in Alzheimer's Disease. *Dement. Geriatr. Cogniy. Dis. Extra* **2016**, *6*, 252–267. [CrossRef] [PubMed]
203. Hsu, Y.L.; Chung, P.C.J.; Wang, W.H.; Pai, M.C.; Wang, C.Y.; Lin, C.W.; Wu, H.L.; Wang, J.S. Gait and balance analysis for patients with Alzheimer's disease using an inertial-sensor-based wearable instrument. *IEEE J. Biomed. Health Inform.* **2014**, *18*, 1822–1830. [CrossRef] [PubMed]
204. Dalton, A.; Khalil, H.; Busse, M.; Rosser, A.; van Deursen, R.; Ólaighin, G. Analysis of gait and balance through a single triaxial accelerometer in presymptomatic and symptomatic Huntington's disease. *Gait Posture* **2013**, *37*, 49–54. [CrossRef] [PubMed]
205. Palermo, E.; Rossi, S.; Marini, F.; Patanè, F.; Cappa, P. Experimental evaluation of accuracy and repeatability of a novel body-to-sensor calibration procedure for inertial sensor-based gait analysis. *Measurement* **2014**, *52*, 145–155. [CrossRef]
206. Ancillao, A.; Tedesco, S.; Barton, J.; O'flynn, B. Indirect measurement of ground reaction forces and moments by means of wearable inertial sensors: A systematic review. *Sensors* **2018**, *18*, 2564. [CrossRef] [PubMed]
207. Carroll, W.M. The global burden of neurological disorders. *Lancet Neurol.* **2019**, *18*, 418–419. [CrossRef]
208. Suppa, A.; Kita, A.; Leodori, G.; Zampogna, A.; Nicolini, E.; Lorenzi, P.; Rao, R.; Irrera, F. l-DOPA and Freezing of Gait in Parkinson's Disease: Objective Assessment through a Wearable Wireless System. *Front. Neurol.* **2017**, *8*, 406. [CrossRef]
209. Hatcher-Martin, J.M.; Adams, J.L.; Anderson, E.R.; Bove, R.; Burrus, T.M.; Chehrenama, M.; Dolan O'Brien, M.; Eliashiv, D.S.; Erten-Lyons, D.; Giesser, B.S.; et al. Telemedicine in neurology: Telemedicine Work Group of the American Academy of Neurology update. *Neurology* **2020**, *94*, 30–38. [CrossRef]
210. Wu, G.; Xue, S. Portable preimpact fall detector with inertial sensors. *IEEE Trans. Neural. Syst. Rehabil. Eng.* **2008**, *16*, 178–183. [CrossRef]
211. Ma, C.Z.H.; Wong, D.W.C.; Lam, W.K.; Wan, A.H.P.; Lee, W.C.C. Balance improvement effects of biofeedback systems with state-of-the-art wearable sensors: A systematic review. *Sensors* **2016**, *16*, 434. [CrossRef]

212. Erra, C.; Mileti, I.; Germanotta, M.; Petracca, M.; Imbimbo, I.; De Biase, A.; Rossi, S.; Ricciardi, D.; Pacilli, A.; Di Sipio, E.; et al. Immediate effects of rhythmic auditory stimulation on gait kinematics in Parkinson's disease ON/OFF medication. *Clin. Neurophysiol.* **2019**, *130*, 1789–1797. [CrossRef] [PubMed]
213. Altilio, R.; Liparulo, L.; Panella, M.; Proietti, A.; Paoloni, M. Multimedia and Gaming Technologies for Telerehabilitation of Motor Disabilities [Leading Edge]. *IEEE Technol. Soc. Mag.* **2015**, *34*, 23–30. [CrossRef]
214. Porciuncula, F.; Roto, A.V.; Kumar, D.; Davis, I.; Roy, S.; Walsh, C.J.; Awad, L.N. Wearable Movement Sensors for Rehabilitation: A Focused Review of Technological and Clinical Advances. *PM R* **2018**, *10*, S220–S232. [CrossRef] [PubMed]
215. Allen, J.L.; McKay, J.L.; Sawers, A.; Hackney, M.E.; Ting, L.H. Increased neuromuscular consistency in gait and balance after partnered, dance-based rehabilitation in Parkinson's disease. *J. Neurophysiol.* **2017**, *118*, 363–373. [CrossRef] [PubMed]
216. Gordt, K.; Gerhardy, T.; Najafi, B.; Schwenk, M. Effects of Wearable Sensor-Based Balance and Gait Training on Balance, Gait, and Functional Performance in Healthy and Patient Populations: A Systematic Review and Meta-Analysis of Randomized Controlled Trials. *Gerontology* **2017**, *64*, 74–89. [CrossRef] [PubMed]
217. Grewal, G.S.; Schwenk, M.; Lee-Eng, J.; Parvaneh, S.; Bharara, M.; Menzies, R.A.; Talal, T.K.; Armstrong, D.G.; Najafi, B. Sensor-Based Interactive Balance Training with Visual Joint Movement Feedback for Improving Postural Stability in Diabetics with Peripheral Neuropathy: A Randomized Controlled Trial. *Gerontology* **2015**, *61*, 567–574. [CrossRef]
218. Schwenk, M.; Grewal, G.S.; Holloway, D.; Moucha, A.; Garland, L.; Najafi, B. Interactive sensor-based balance training in older cancer patients with chemotherapy-induced peripheral neuropathy: A randomized controlled trial. *Gerontology* **2016**, *62*, 553–563. [CrossRef] [PubMed]
219. Schwenk, M.; Sabbagh, M.; Lin, I.; Morgan, P.; Grewal, G.S.; Mohler, J.; Coon, D.W.; Najafi, B. Sensor-based balance training with motion feedback in people with mild cognitive impairment. *J. Rehabil. Res. Dev.* **2016**, *53*, 945–958. [CrossRef]
220. Carpinella, I.; Cattaneo, D.; Bonora, G.; Bowman, T.; Martina, L.; Montesano, A.; Ferrarin, M. Wearable Sensor-Based Biofeedback Training for Balance and Gait in Parkinson Disease: A Pilot Randomized Controlled Trial. *Arch. Phys. Med. Rehabil.* **2017**, *98*, 622–630.e3. [CrossRef]
221. Esteva, A.; Robicquet, A.; Ramsundar, B.; Kuleshov, V.; DePristo, M.; Chou, K.; Cui, C.; Corrado, G.; Thrun, S.; Dean, J. A guide to deep learning in healthcare. *Nat. Med.* **2019**, *25*, 24–29. [CrossRef]
222. Rajkomar, A.; Dean, J.; Kohane, I. Machine Learning in Medicine. *N. Engl. J. Med.* **2019**, *380*, 1347–1358. [CrossRef] [PubMed]
223. Kubota, K.J.; Chen, J.A.; Little, M.A. Machine learning for large-scale wearable sensor data in Parkinson's disease: Concepts, promises, pitfalls, and futures. *Mov. Disord.* **2016**, *31*, 1314–1326. [CrossRef] [PubMed]
224. Handelman, G.S.; Kok, H.K.; Chandra, R.V.; Razavi, A.H.; Lee, M.J.; Asadi, H. eDoctor: Machine learning and the future of medicine. *J. Intern. Med.* **2018**, *284*, 603–619. [CrossRef]
225. Cilliers, L. Wearable devices in healthcare: Privacy and information security issues. *Health Inf. Manag.* **2019**, 1833358319851684. [CrossRef] [PubMed]
226. Guideline on Data Integrity. Available online: https://www.who.int/medicines/areas/quality_safety/quality_assurance/QAS19_819_data_integrity.pdf?ua=1 (accessed on 23 May 2020).
227. Mourcou, Q.; Fleury, A.; Diot, B.; Vuillerme, N. iProprio: A smartphone-based system to measure and improve proprioceptive function. *Conf. Proc. IEEE Eng. Med. Biol. Soc.* **2016**, *2016*, 2622–2625. [CrossRef] [PubMed]

Article

Reactive Postural Responses to Continuous Yaw Perturbations in Healthy Humans: The Effect of Aging

Ilaria Mileti [1,*], Juri Taborri [2], Stefano Rossi [2], Zaccaria Del Prete [1], Marco Paoloni [3], Antonio Suppa [4,5] and Eduardo Palermo [1]

1 Department of Mechanical and Aerospace Engineering, Sapienza University of Rome, 00184 Rome, Italy; zaccaria.delprete@uniroma1.it (Z.D.P.); eduardo.palermo@uniroma1.it (E.P.)
2 Department of Economics, Engineering, Society and Business Organization (DEIM), University of Tuscia, 01100 Viterbo, Italy; juri.taborri@unitus.it (J.T.); stefano.rossi@unitus.it (S.R.)
3 Department of Physical Medicine and Rehabilitation, Sapienza University of Rome, 00185 Rome, Italy; marco.paoloni@uniroma1.it
4 Department of Human Neurosciences, Sapienza University of Rome, 00185 Rome, Italy; antonio.suppa@uniroma1.it
5 IRCCS Neuromed Institute, 86077 Pozzilli (IS), Italy
* Correspondence: ilaria.mileti@uniroma1.it; Tel.: +39-06-44585-585

Received: 31 October 2019; Accepted: 18 December 2019; Published: 20 December 2019 

Abstract: Maintaining balance stability while turning in a quasi-static stance and/or in dynamic motion requires proper recovery mechanisms to manage sudden center-of-mass displacement. Furthermore, falls during turning are among the main concerns of community-dwelling elderly population. This study investigates the effect of aging on reactive postural responses to continuous yaw perturbations on a cohort of 10 young adults (mean age 28 ± 3 years old) and 10 older adults (mean age 61 ± 4 years old). Subjects underwent external continuous yaw perturbations provided by the RotoBit^{1D} platform. Different conditions of visual feedback (eyes opened and eyes closed) and perturbation intensity, i.e., sinusoidal rotations on the horizontal plane at different frequencies (0.2 Hz and 0.3 Hz), were applied. Kinematics of axial body segments was gathered using three inertial measurement units. In order to measure reactive postural responses, we measured body-absolute and joint absolute rotations, center-of-mass displacement, body sway, and inter-joint coordination. Older adults showed significant reduction in horizontal rotations of body segments and joints, as well as in center-of-mass displacement. Furthermore, older adults manifested a greater variability in reactive postural responses than younger adults. The abnormal reactive postural responses observed in older adults might contribute to the well-known age-related difficulty in dealing with balance control during turning.

Keywords: aging; reactive postural responses; yaw perturbation; kinematics; postural stability; dynamic posturography

1. Introduction

Falls are among the most common leading causes of accidental death, hospitalization, or injuries, such as broken bones, head, and spinal cord injuries [1]. Since falling still represents a challenging social, medical, and economical matter [2], several research efforts have been spent in understanding the mechanisms leading to falls in the elderly [3]. A possible theory for explaining the increased risk of falling in the elderly population is that age compromises balance capability. However, understanding the effects of aging on balance is still an unanswered question.

Exploring dynamic balance control of upright stance through imposed external perturbations has the main objective to identify deficit in reactive postural responses. According to the amplitude of COM dislocation and the intensity of perturbation, different reactive strategies, such as hip and ankle strategies and/or stepping responses, prevent imbalance [4,5].

Various imposed posture manipulations, such as mechanical, visual, and vestibular perturbations, have been designed to reproduce free-living imbalance conditions and to investigate risk of fall [6–9], especially in elderly [10–14]. In a mechanical perspective, popular approaches focused on rotational [8,13,15–17] and/or translational support surface movements [9,18,19] or on body release paradigms [20,21] to simulate common postural disturbances (e.g., standing on a bus, slipping on a slippery surface, and falling due to a sudden boost). In this context, the majority of the literature studies focused on the postural responses under forward/backward perturbations to assess risk of fall in the elderly. Across this range of perturbation paradigms, older adults exhibited greater difficulty than young adults in recovering loss of balance through protective postural strategies, such as stepping, especially in case of a forward fall. Accordingly, a lower maximum body lean angle [20], smaller peak knee extensor torques, and larger peak extensors torques at hip and ankle joints [21] were reported in older adults during step response in forward perturbations. Conversely, when considering backward perturbations, older adults showed shorter reaction times than during forward perturbations to avoid fall [22].

Age-related postural deficits were also investigated considering different perturbation paradigms such as sinusoidal translating perturbation in the anterior-posterior direction [14], random rotations around sagittal axis [17], slippery floor surface [23], and mixed forward/backward and left/right platform translations [24]. More specifically, Nardone et al. found a greater head stabilization strategy and a looser coupling between head and hip motion in the eyes opened condition in older adults [14]. While, in the study of Cenciarini et al., older adults presented significantly higher active stiffness as compared to young adults to maintain body balance and to counteract destabilizations effects [17]. Moreover, impaired motor patterns in older adults have been also linked to an abnormal postural sway under forward/backward and left/right perturbation directions as attested in the study of Liaw et al. [24].

Although much has been learned about postural responses of older adults exposed to the before mentioned perturbation paradigms, the motor strategies involved in maintaining balance under perturbations around the vertical axis are still unclear. In healthy subjects, reactive postural responses during rotational perturbations around the vertical axis imposed by a robotic platform have been investigated only by few studies [10,11,15]. Earlier postural responses of the distal compared to the proximal body segment were observed both in kinematics [10] and in muscle activity [11]. The evaluation of reactive postural responses in term of COM displacement control, coordination of body segments and body sway response during this specific rotation would offer an ecological analysis of fall-inducing factors in everyday life. In fact, in the elderly, falls frequently occur within a narrowing familiar environment, such as the kitchen and the bathroom, during activities required for basic mobility, such as turning in place [25].

Over the last decade, measuring motor impairments through portable and wearable devices have demonstrated to be of great impact in managing neuromotor and aging deficits [26–28]. In fact, the several advantages of wearable sensors, such as the low cost, the high portability, the limited size and the ease-of-use, encourage the use of these technologies in clinical setting as a useful tool for monitoring assessment [29,30]. In this context, the aim of this study is to investigate the effects of aging on reactive postural responses to rotational perturbations around the vertical axis, by using a rotating platform and wearable inertial sensors in different visual conditions. Clarifying changes of dynamic postural control in older adults during rotational perturbations around the vertical axis would help to understand the mechanisms leading to falls when turning. This in turn, would be useful for the design of effective strategies for falls prevention.

2. Materials and Methods

2.1. Subjects

A cohort of ten healthy older adults (six females and four males, mean age 61 ± 4 years old, mean body mass 68 ± 9 kg, and mean height 170 ± 6 cm) and a cohort of ten young adults (five females and five males, mean age 28 ± 3 years old, mean body mass 63 ± 14 kg, and mean height 169 ± 11 cm) were enrolled. All participants were community-dwelling, medically stable, and able to walk and stand independently without aids. Subjects with intellectual, vestibular and/or visual deficits, neuromuscular diseases, orthopedic and/or neurological surgery interventions in the last three years were excluded from this study. All participants gave written consent before being included in the experimental session. The protocol was designed and conducted in accordance with the Ethical Standard of the 1964 Declaration of Helsinki.

2.2. Experimental Setup

The RotoBit^{1D} was used in this study to provide sinusoidal perturbations around the vertical axis [15]. It is a rigid, round, flat robotic platform with a diameter of 0.5 m that allows a comfortable upright bipedal stance without narrowing feet. The mechanical design consists of a servo motor (SANYO DENKI) with maximum torque of 1.96 Nm, an incremental encoder, a toothed belt (PowerGrip HDT), speed reducer, and a polyethylene rotating disk. The robotic platform was computer-controlled by an ad-hoc LabVIEW software program (v.2014, National Instruments, Austin, TX, USA). Two sinusoidal perturbations around vertical axis were designed with fixed peak amplitude of ± 55° C and frequencies of (i) 0.2 Hz and (ii) 0.3 Hz, namely lower (L) and higher (H) frequency, respectively. The peak angular velocity was 80 °C/s and 100 °C/s for the lower and higher frequency, respectively. To avoid sudden variation in starting/stopping velocity, a sigmoidal wave was added at the start/end of the sinusoidal trajectory. In pilot trials, we chose rotation parameters as those able to provide a high perturbation intensity without requiring stepping response of subjects. Inertial Measurement Units (IMUs) (MTw, Xsens Technologies—NL) including a 3-axes accelerometer (± 160 m/s^2 FS), a 3-axes gyroscope (± 1200 °C/s FS), and a 3-axes magnetometer (± 1.5 Gauss FS) were used for gathering kinematic data of pelvis, trunk, and head. More specifically, the pelvis-sensor was placed centered on the median sacral crest and just below the anterior sacral promontory. The trunk-sensor was placed under the suprasternal notch on the sternum body; while, the head-sensor was placed on the frontal bone over the superciliary arch. Each subject was instrumented by the same expert operator, to guarantee consistent sensor location on body segment.

IMUs were placed by means of suitable elastic belts to avoid relative movement between sensor and body. Sampling frequency was set at 40 Hz. Equipment was simultaneously triggered at both the beginning and the end of each acquisition. More specifically, the ad-hoc LabVIEW software was designed to simultaneously drive the RotoBit1D servo motor and provide an external trigger, i.e., a square signal ranging from 0 to +5 V, to the IMUs Awinda Station through an NI USB-6212 DAQ board. The Awinda was set to start and end the IMUs' acquisition on the rising and the falling edge of the trigger signal, respectively.

2.3. Experimental Procedure

The experimental protocol was conducted at the Department of Physical Medicine and Rehabilitation, Sapienza University of Rome, Italy. Before each session, all tested subjects were asked to perform a Functional Calibration procedure advised by an operator. The FC procedure provided sensor orientations with respect to body segment, to complete the body-to-sensor alignment procedure [31]. The FC procedure consisted of a standing and sitting task, each lasting 5 s. Afterwards, all subjects stood in a comfortable upright bipedal position with vertically hanging arms and externally rotated feet at a preferred angle with a symmetrical placement, on the top center of the robotic platform. All subjects were asked to wear heelless shoes. The experimental procedure included two different perturbation frequencies (lower and higher frequency) and two different visual conditions (eyes opened (EO) and eyes closed (EC)). More specifically, the participants' reactive postural responses were measured under four balance tasks: standing with EO during platform rotation at (i) lower (EO-L) and (ii) higher (EO-H) frequencies; and standing with EC considering both (iii) lower (EC-L) and (iv) higher (EC-H) platform frequency rotations. In the EO condition, subjects were asked to stare at a fixed point placed on the wall at 2 m from the platform.

Each task was performed three times randomizing the task order across subjects to avoid bias in results due to similar task sequences. In addition, subjects were not advised about frequency, to avoid habituation of postural responses or anticipatory strategies due to predictability [32].

2.4. Data Analysis

All data were analyzed off-line using MATLAB (v.2015b, MathWorks, Natick, MA) program. Angular rotations around the vertical axis, i.e., yaw angles, of the platform (pt), the head (h), the trunk (t), and the pelvis (p) segments were considered for the data analysis. Comparisons between rotation of the platform and body segment rotations in the transversal plane were obtained via the fast Fourier transform analysis, by considering Gain ratio (G) and phase shift (φ) indices, akin to [15]. G was computed as the ratio between the maximum amplitude value of the fundamental wave of the first signal and the amplitude value of the second signal at the same frequency. φ was obtained as the difference of the phase angles of the Fourier transform of the two signals at the frequencies having the maximum amplitude in the Fourier domain. Before transforming in Fourier domain, each body segment rotation in the transversal plane was demeaned. Among angular rotation around the vertical axis of the three body segments and platform, 6 pairs were considered for the analysis: platform-head, platform-trunk, platform-pelvis, head-trunk, trunk-pelvis, and head-pelvis. The following nomenclatures were chosen for G and φ indices to indicate the 6 pairs:

$$^{f}G_{s}, {}^{f}\varphi_{s} \tag{1}$$

where f and s represent the first and second sine waves considered, respectively. Perfect agreement in amplitude and timing between the first and the second element of the pair were observed considering a G value close to 100% and a φ value close to 0 °C. Instead, an anticipation/delay in phase angle of the second sine wave compared to the first one was defined as a positive/negative phase shift. For sake of clearness, ${}^{pt}G_h$, ${}^{pt}G_t$, ${}^{pt}G_p$, ${}^{pt}\varphi_h$, ${}^{pt}\varphi_t$, and ${}^{pt}\varphi_p$ were addressed as G-absolute and φ-absolute, because each of the yaw body angles (second sine wave) is referred to the yaw platform angle (first sine wave). While, ${}^{h}G_t$, ${}^{t}G_p$, ${}^{h}G_p$, ${}^{h}\varphi_t$, ${}^{t}\varphi_p$, and ${}^{h}\varphi_p$ were addressed as G-relative and φ-relative, because both the first and the second sine wave are referred to a body segment sine wave.

To assess inter-joint coordination on the transversal plane, the continuous relative phase (CRP) technique was computed on yaw body angles, i.e., body rotation on transversal plane [33]. The CRP analyzes the differences in phase angles of two body segments during a particular motion task. Differently from φ index, which provides the average phase shift of the specific sine wave above described, the CRP refers to all the frequency components of a signal, reporting coupling behaviors of two body segments over the entire motor task. Concerning the CRP analyses, the phase space usually consists of the time-dependent measured signal and its first derivative. Thus, the CRP for a particular task is obtained as the four-quadrant arctangent phase angle from the phase space. Several methods have been developed for the calculation of the phase angles based on phase portrait analysis or the Hilbert transform [33–35]. Among those methodologies, the Hilbert transform-based method has been proved to be more robust than the phase angles in performing the phase portrait, especially regarding none-purely sinusoidal signals [33]. The Hilbert transform allows the transformation of any real signal into complex, analytic signal according to:

$$\zeta(t) = x(t) + iH(t) \tag{2}$$

at time t_i, the phase angle can be computed by:

$$\theta(t_i) = \arctan\left(\frac{H(t_i)}{x(t_i)}\right) \tag{3}$$

The continuous relative phase CRP between two signals can be defined as the differences of the phase angles:

$$CRP(t_i) = \theta_1(t_i) - \theta_2(t_i) = \arctan\left(\frac{H_1(t_i)x_2(t_i) - H_2(t_i)x_1(t_i)}{x_1(t_i)x_2(t_i) + H_1(t_i)H_2(t_i)}\right) \tag{4}$$

where $H_1(t)$ and $H_2(t)$ are the Hilbert transform of signals of the proximal and distal segment, respectively. The CRP index can assume values between 0° and 180°. Values close to 0° indicate in-phase coupling of the two segments, while values close to 180° represent an out-phase coupling of signals.

To identify differences in inter-joint coordination between young and older adults, the mean absolute relative phase (MARP) was computed by averaging the absolute values of the curve points considering the overall trial duration [36], as in the following equation:

$$\text{MARP} = \frac{\sum_{i=1}^{p} \left|\overline{CRP_i}\right|}{p} \tag{5}$$

where p is the number of time points in each trial. Similar consideration done for the CRP can be adopted for the MARP value.

Furthermore, the Deviation Phase (DP) was analyzed in order to assess variability among trials in inter-joint coordination [36]. The DP can be assessed by averaging the standard deviation among trials of the CRP(t_i) over the trial duration, as in the following equation:

$$\text{DP} = \frac{\sum_{i=1}^{p} SDi}{p} \tag{6}$$

where SDi represents the standard deviation of the CRP among the three trials at the i-th time instant. The DP is a measure of the stability organization provided by the neuromuscular system [36]. DP values close to 0 °C attest less intra-subject variability of the inter-joint coordination.

In order to assess postural control of body motion in the anterior-posterior and medio-lateral directions in response to external balance perturbations, body displacements of the head, the trunk and the pelvis were estimated via a strap-down integration of the acceleration signal, similarly to [37]. Body displacement of pelvis sensor obtained through this method can be addressed as an estimation of the COM displacement, as authors reported [37]. However, differently from [37], in our study, we used the rotation matrix obtained from the quaternion output of the IMUs for the strapdown integration. In the original work [36] the authors rather used the gyro output for obtaining an estimation of the rotation matrix, as the system available for them did not provide the quaternion as an output.

To estimate the displacement, the acceleration signal was firstly rotated in the global coordinate frame to remove gravitational acceleration. After the gravitational acceleration removal, the acceleration signal of the inertial sensor in the global coordinate frame was straightforwardly integrated. Velocity was then high-pass filtered and the displacement was obtained through a second integration and filtering. The applied filters were zero-lag first-order Butterworth filters with a cut-off frequency of 0.2 Hz for the anterior-posterior (AP) and medio-lateral (ML) components and 0.5 Hz per the vertical (V) component.

Considering the body displacement, the following kinematic parameters were obtained for the statistical analysis: (i) the range of motion of the body segment displacement in the ML (RoM_{ML}) and AP (RoM_{AP}) directions expressed in mm; (ii) the total path length of the body displacement normalized to the task duration (PATH) and (iii) the maximum velocity of the displacement (MV) expressed in m/s. The before mentioned parameters were considered for the head, trunk, and pelvis displacements.

2.5. Statistical Analysis

All data were tested for normality by means of the Shapiro-Wilk test. Statistical analysis was performed with the SPSS package (IBM-SPSS Inc., Armonk, NY, USA). Two unpaired t-tests were conducted to assess differences in body mass and height between the two groups. In order to test differences in postural strategies induced by aging and different perturbation frequencies, a 2 × 2 two-way mixed ANOVA, with AGE as a between-subject factor (two levels: young adults and older adults) and FREQ (two levels: lower and higher frequency) as a within-subject factor was used separately for the EO and the EC conditions. When the assumption of sphericity was violated, the Greenhouse-Geisser correction was considered. A paired *t*-test within each group and an unpaired t-test between the two groups were performed when the interaction between the main effects was significant. The Bonferroni corrections were considered for all the statistical analysis. The significance level was set at 0.05 for all the statistical tests.

3. Results

All subjects were able to complete the experimental procedures without losing balance and/or experiencing fatigue.

No statistical differences were found in body mass and height between groups.

Considering all statistical analysis, the 2 × 2 two-way mixed ANOVA reported no significant interactions between main effects AGE and FREQ.

- G- and φ-absolute:

In Table 1, mean and standard deviation of G-absolute and φ-absolute values and *p*-values are reported.

With regards to AGE main effect in the EO task, older adults showed a significant lower G-absolute value of trunk and head than young adults. The pelvis body segment of young adults reached the highest mean values of G-absolute in both frequency conditions (57.04% and 48.00%), see Table 1 and Figure 1. Compared with younger population, older adults exhibited a smaller amount of the axial body motion. In the EC condition, a similar reduction of motion amplitude was found in G-absolute values of head and pelvis of older subjects. Considering the main effect FREQ, upper body G-absolute values statistically decreased as a function of rotation frequency increment, regardless of visual condition for all body segments.

Concerning AGE main effect, no differences were found in the EO condition related to φ-absolute. While the main effect of FREQ was found statistically significant for all the body segments. In the EC task, φ-absolute values of the pelvis were statistically different between young and older adults, according to AGE main effect. To face external yaw postural perturbations, older adults anticipated pelvis motion while younger adults adopted a delayed motion strategy. Head and pelvis phase shifts were found statistically significant in the main effect FREQ.

- G- and φ-relative:

In Table 2, mean and standard deviation of G-relative and φ-relative values and *p*-values are reported. In the EO task, the main effect AGE of the G-relative values were found to be statistically different for all the considered segment-couples. More specifically, young adults reported G-relative values close to 100% attesting a similar amplitude pattern between the rotation around the vertical axis of the proximal and the distal segment. By contrast, older adults exhibited lower values of G-relative attesting a reduction in amplitude pattern of the distal body compared to the proximal. As regard to the main effect FREQ, statically differences were found for all the considered body segment-couples. In the EC condition, similar trend of the EO condition was found for the main effect AGE. While regarding the main effect FREQ, statistical differences were found for the head-pelvis and trunk-pelvis couples, attesting that the amplitude value of the distal segment decreased as a function of frequency perturbation increment.

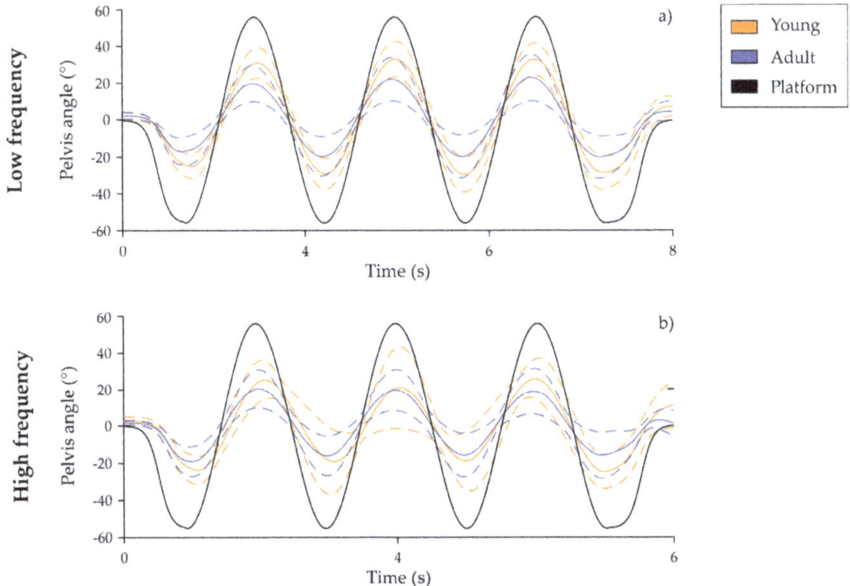

Figure 1. (**a**) Mean (solid line) and standard deviation (dashed lines) of pelvis angle in the transverse plane for the eyes opened condition during the low frequency task. (**b**) Mean (solid line) and standard deviation (dashed lines) of pelvis angle in the transverse plane for the eyes opened condition during the high frequency task. The orange curves refer to young group, the blue ones to the older adults while the black curve is platform trajectory.

Regarding φ-relative, significant differences were found in the head-pelvis couple between young adults and older ones, in EO condition. More specifically, the distal segment of older adults appeared delayed with respect to the proximal one, while in the young group, synchronization strategies were adopted between segment couples. Increasing frequency motion, a higher delay was observed in the distal segment, especially in older adults.

- Continuous relative phase: MARP and DP:

In Table 3, mean and standard deviation of MARP and DP indices expressed in [°] are reported. In the EO task, age-related differences were found for all the segment-couples. Higher MARP values were reached by older populations who preferred a more anti-phase motion strategy in response to balance rotational disturbance. An in-phase coupling strategy was observed in the young subjects. Increasing perturbations intensity, similar in-phase motion strategy was observed for all segment-couples. Considering the EC condition, a similar trend was found both regarding AGE as main effect. While considering FREQ as main effect, a more anti-phase strategy was observed in both groups when increasing perturbation intensity.

As regards coordination variability, older adults exhibited higher values of DP in all the segment-couples compared to young adults, regardless of the task condition (see Figure 2). Although the aforementioned results reported a wide amount of motor engagement, with the increase perturbation intensity, coordination variability of the trunk-pelvis couple increased as well in both visual conditions.

- Body displacement: RoM_{ML}, RoM_{AP}, PATH, and MV:

Table 1. Mean and standard deviation of G-absolute values expressed as a percentage [%] and φ-absolute values expressed in [°] concerning head, trunk, and pelvis, for both lower (L) and higher (H) frequencies and visual conditions, i.e., eyes opened (EO) and eyes closed (EC) in young and older adults. *p*-values of the main effects, AGE and FREQ, of the two-way mixed ANOVA are reported. Statistical differences ($p < 0.05$) in the main effects are single-starred and reported in bold.

		EO-L		EO-H		*p*-Values		EC-L		EC-H		*p*-Values	
		Young Adults	Older Adults	Young Adults	Older Adults	AGE	FREQ	Young Adults	Older Adults	Young Adults	Older Adults	AGE	FREQ
Head	G	53.3 ± 19.4	21.5 ± 14.2	39.4 ± 16.1	15.6 ± 13.7	<0.01 *	<0.01 *	59.9 ± 21.2	41.2 ± 16.2	48.3 ± 17.8	25.2 ± 17.6	0.02 *	<0.01 *
	Φ	−4.8 ± 15.9	−1.6 ± 23.3	−25.3 ± 22.4	−18.5 ± 45.8	0.45	0.04 *	-13.8 ± 23.7	−2.4 ± 9.2	−25.4 ± 17.0	−13.3 ± 25.9	0.14	0.04 *
Trunk	G	55.2 ± 19.5	36.2 ± 18.3	45.0 ± 16.8	29.00 ± 21.2	0.04 *	<0.01 *	59.9 ± 21.1	47.2 ± 16.6	48.5 ± 17.1	30.0 ± 19.7	0.07	<0.01 *
	Φ	−3.7 ± 13.9	8.3 ± 19.3	−22.3 ± 22.3	5.8 ± 43.3	0.11	0.02 *	−12.6 ± 23.7	0.5 ± 8.9	−22.9 ± 16.6	−15.5 ± 35.8	0.24	0.06
Pelvis	G	57.0 ± 15.2	42.1 ± 14.5	48.0 ± 14.0	37.7 ± 16.8	0.09	<0.01 *	56.3 ± 15.3	48.2 ± 11.9	48.9 ± 13.9	39.0 ± 14.5	0.14	<0.01 *
	Φ	0.9 ± 9.8	6.00 ± 6.4	−7.7 ± 11.7	1.1 ± 9.7	0.10	<0.01 *	−2.1 ± 14.6	4.6 ± 6.0	−8.8 ± 11.2	3.6 ± 6.9	0.03 *	0.03 *

Table 2. Mean and standard deviation of G-relative values expressed as a percentage [%] and φ-relative values expressed in [°] concerning head-pelvis, trunk-pelvis, and head-trunk, for both lower (L) and higher (H) frequencies and both visual conditions, i.e., eyes opened (EO) and eyes closed (EC) in young and older adults. *p*-values of the main effects, AGE and FREQ, of the two-way mixed ANOVA are reported. Statistical differences ($p < 0.05$) in the main effects are single-starred and reported in bold.

		EO-L		EO-H		*p*-Values		EC-L		EC-H		*p*-Values	
		Young Adults	Older Adults	Young Adults	Older Adults	AGE	FREQ	YOUNG ADULTS	Older Adults	Young Adults	Older Adults	AGE	FREQ
Head-Pelvis	G	93.7 ± 17.8	47.0 ± 19.6	82.0 ± 23.9	39.5 ± 18.6	<0.01 *	<0.01 *	103.5 ± 18.0	83.0 ± 18.6	95.8 ± 18.3	61.2 ± 20.3	<0.01 *	<0.01 *
	Φ	−6.5 ± 9.2	−13.6 ± 35.3	−7.4 ± 12.9	−62.1 ± 54.0	0.03	<0.01 *	−10.7 ± 14.6	−20.4 ± 35.3	−15.4 ± 11.1	−27.8 ± 50.1	0.42	0.24
Trunk-Pelvis	G	101.4 ± 14.0	81.8 ± 17.9	93.3 ± 17.4	75.3 ± 23.8	0.02 *	0.02 *	103.1 ± 18.4	96.8 ± 13.1	97.0 ± 14.8	74.4 ± 20.5	0.05	<0.01 *
	Φ	−5.0 ± 6.2	−22.3 ± 59.2	−10.2 ± 9.0	−23.6 ± 37.7	0.32	0.46	−9.3 ± 13.7	−17.8 ± 37.4	−13.3 ± 10.4	−27.0 ± 46.4	0.41	0.12
Head-Trunk	G	91.7 ± 7.5	56.1 ± 16.6	85.4 ± 16.8	51.5 ± 17.4	<0.01 *	0.04	100.5 ± 1.9	84.8 ± 14.8	98.0 ± 6.5	82.5 ± 16.8	<0.01 *	0.12
	Φ	−1.5 ± 3.7	8.7 ± 35.0	2.9 ± 14.4	−38.5 ± 45.6	0.30	0.03	−1.4 ± 1.6	−2.5 ± 3.2	−2.1 ± 2.8	−0.8 ± 8.9	0.40	0.95

Table 3. Mean and standard deviation of the mean absolute relative phase (MARP) and Deviation Phase (DP) index of the continuous relative phase (CRP) technique, both expressed in [°] concerning head-pelvis, trunk-pelvis and head-trunk, for both lower (L) and higher (H) frequencies and both visual conditions, i.e., eyes opened (EO) and eyes closed (EC) in young and older adults. *p*-values of the main effects, AGE and FREQ, of the two-way mixed ANOVA are reported. Statistical differences ($p < 0.05$) in the main effects are single-starred and reported in bold.

		EO-L		EO-H		*p*-Values		EC-L		EC-H		*p*-Values	
		Young Adults	Older Adults	Young Adults	Older Adults	AGE	FREQ	Young Adults	Older Adults	Young Adults	Older Adults	AGE	FREQ
Head-Pelvis	MARP	6.7 ± 2.2	27.7 ± 22.1	15.2 ± 9.7	36.9 ± 20.7	<0.01 *	0.12	6.1 ± 1.8	14.3 ± 7.8	14.6 ± 8.7	33.9 ± 26.2	0.02 *	0.02 *
	DP	3.2 ± 1.4	23.4 ± 22.6	6.8 ± 5.4	23.9 ± 13.2	<0.01 *	0.57	3.8 ± 3.6	17.1 ± 20.6	5.6 ± 3.6	17.8 ± 9.9	0.02 *	0.77
Trunk-Pelvis	MARP	4.9 ± 2.5	26.5 ± 31.1	12.8 ± 10.4	32.1 ± 25.2	0.01 *	0.43	5.4 ± 2.1	7.9 ± 4.0	11.2 ± 6.8	33.7 ± 35.2	0.08	0.04 *
	DP	1.8 ± 0.8	11.6 ± 10.4	5.9 ± 4.1	20.1 ± 14.4	0.01 *	0.01 *	3.4 ± 3.3	7.9 ± 12.2	5.5 ± 4.3	21.6 ± 15.4	0.03 *	0.03 *
Head-Trunk	MARP	3.1 ± 1.4	21.4 ± 26.5	4.0 ± 1.3	34.0 ± 37.4	0.05 *	0.08	1.7 ± 0.7	5.9 ± 4.3	3.3 ± 1.4	13.5 ± 10.8	<0.01 *	0.07
	DP	1.8 ± 0.8	17.4 ± 19.4	2.3 ± 1.0	21.4 ± 21.05	0.02 *	0.43	1.1 ± 0.6	10.4 ± 15.2	1.6 ± 0.4	10.4 ± 6.6	0.03	0.92

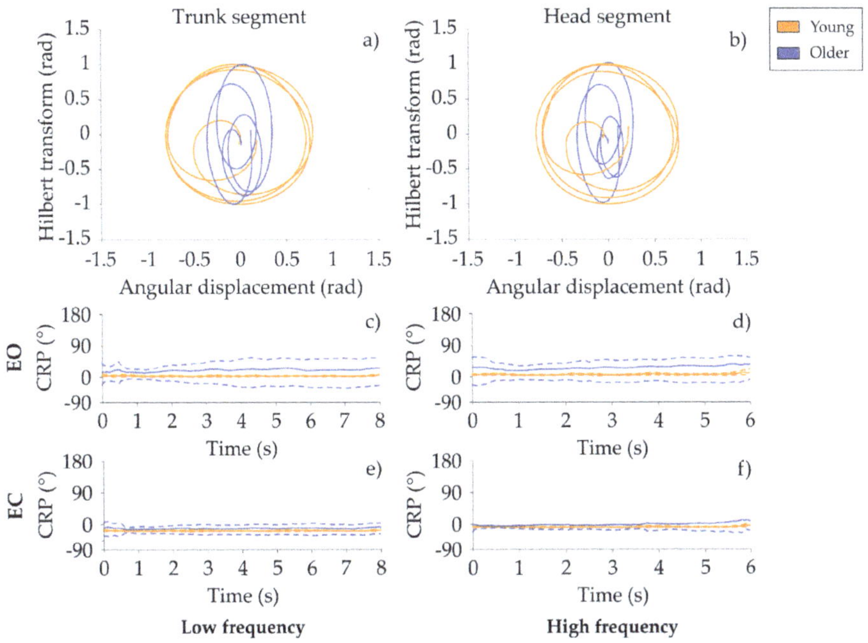

Figure 2. (**a**) and (**b**) are phase portraits of the trunk and head body segment of one healthy young subject and one older subject during EO condition at low frequency, respectively. (**c–f**) Mean (solid line) and standard deviation (dashed lines) of CRP of the head-trunk couple for the EO and EC condition for the low frequency task and EO and EC conditions for the high frequency task, respectively. The orange curves refer to the young group, the blue ones to older adults.

In Table 4, mean and standard deviation of upper body displacement measures and p-values of the mixed ANOVA are reported. In EO condition, RoM_{ML}, PATH, and MV of the head and pelvis displacements were statistically lower in older adults compared to young adults. Regarding FREQ as a main effect, statistical differences were found considering PATH, MV in all upper body segments. Additionally, frequency increment effect was also found in RoM_{ML} and RoM_{AP} in trunk displacement. A similar trend was reported in the EC task for both main effects.

Table 4. Mean and standard deviation of medio-lateral (ML) and anterior-posterior (AP) components of range of motion (RoM) expressed in [mm], the total path length of the body displacement divided by the task duration (PATH), and the maximum velocity of the displacement (MV) expressed in [m/s], all these displacement measures concerning head, trunk, and pelvis, for both lower (L) and higher (H) frequencies and both visual conditions, i.e., eyes opened (EO) and eyes closed (EC), in young and older adults. *p*-values of the main effects, AGE and FREQ, of the two-way mixed ANOVA are reported. Statistical differences ($p < 0.05$) in the main effects are single-starred and reported in bold.

		EO-L		EO-H		*p*-Values		EC-L		EC-H		*p*-Values	
		Young Adults	Older Adults	Young Adults	Older Adults	AGE	FREQ	Young Adults	Older Adults	Young Adults	Older Adults	AGE	FREQ
Head	RoM_{ML}	88.5 ± 33.8	43.4 ± 16.6	83.5 ± 33.1	45.2 ± 28.4	**0.01** *	0.08	109.5 ± 41.4	75.1 ± 34.0	117.8 ± 52.0	65.5 ± 37.2	**0.04** *	0.29
	RoM_{AP}	40.8 ± 15.8	24.1 ± 5.0	38.4 ± 17.9	23.9 ± 8.2	0.25	0.38	54.7 ± 22.6	36.0 ± 7.2	52.8 ± 24.1	40.1 ± 14.6	0.18	0.52
	PATH	0.4 ± 0.1	0.2 ± 0.1	0.4 ± 0.1	0.2 ± 0.1	**0.02** *	**<0.01** *	0.5 ± 0.2	0.3 ± 0.1	0.6 ± 0.3	0.3 ± 0.2	**0.02** *	**<0.01** *
	MV	0.3 ± 0.1	0.1 ± 0.1	0.3 ± 0.1	0.2 ± 0.1	**0.02** *	**<0.01** *	0.3 ± 0.2	0.2 ± 0.1	0.4 ± 0.1	0.2 ± 0.1	**0.02** *	**<0.01** *
Trunk	RoM_{ML}	31.9 ± 9.0	28.2 ± 8.7	40.1 ± 8.6	34.7 ± 13.8	0.27	**0.01** *	38.3 ± 12.6	34.2 ± 12.0	48.7 ± 18.8	41.5 ± 15.5	0.38	**0.01** *
	RoM_{AP}	20.9 ± 3.5	20.4 ± 5.3	23.6 ± 4.8	24.7 ± 10.4	0.88	**0.03** *	24.4 ± 4.1	24.5 ± 5.5	27.6 ± 7.2	29.3 ± 9.5	0.75	**0.01** *
	PATH	0.1 ± 0.1	0.1 ± 0.1	0.2 ± 0.1	0.2 ± 0.1	0.68	**<0.01** *	0.2 ± 0.1	0.2 ± 0.1	0.2 ± 0.1	0.2 ± 0.1	0.52	**<0.01** *
	MV	0.1 ± 0.1	0.1 ± 0.1	0.1 ± 0.1	0.1 ± 0.1	0.52	**<0.01** *	0.1 ± 0.0	0.1 ± 0.0	0.2 ± 0.1	0.2 ± 0.1	0.38	**<0.01** *
Pelvis	RoM_{ML}	81.2 ± 25.7	57.0 ± 17.3	87.2 ± 20.1	63.2 ± 18.5	**0.01** *	0.08	83.0 ± 15.4	65.2 ± 17.1	83.0 ± 25.4	47.4 ± 15.1	**0.04** *	0.52
	RoM_{AP}	52.1 ± 9.4	5.5 ± 3.0	52.9 ± 10.2	44.4 ± 15.0	0.25	0.38	52.6 ± 8.7	47.7 ± 14.9	57.5 ± 12.6	47.5 ± 15.1	0.18	0.29
	PATH	0.3 ± 0.1	0.2 ± 0.1	0.4 ± 0.1	0.3 ± 0.1	**0.02** *	**<0.01** *	0.4 ± 0.1	0.3 ± 0.1	0.4 ± 0.1	0.3 ± 0.1	**0.02** *	**<0.01** *
	MV	0.2 ± 0.1	0.2 ± 0.0	0.3 ± 0.1	0.2 ± 0.1	**0.02** *	**<0.01** *	0.2 ± 0.1	0.2 ± 0.0	0.3 ± 0.1	0.2 ± 0.1	**0.02** *	**<0.01** *

4. Discussion

In our study, age-related changes were found in balancing continuous yaw perturbations, regardless of visual condition and perturbation frequency. A more conservative and less destabilizing motion was observed as a postural strategy in the older population, suggesting that older adults compensate for their reduced physical capabilities by becoming more cautious while performing a postural task.

- G- and φ-absolute:

As the fast Fourier analysis on yaw segment angles reported, young adults showed a matching strategy between the body displacement and the platform rotation, taking into account the amplitude of the trajectories. Conversely, older adults exhibited a smaller amount of motion amplitude of the upper body to counteract an unbalanced rotational state. This reduction in amplitude was observed in all the upper body yaw angles in older adults, especially regarding the distal segments. As observed in our results, this stabilization strategy progressively decreases from proximal to distal segment, becoming noticeable especially in the head motion of the older adults when perturbation frequency increases. In this context, the different balance response of adults could be caused by different perceptions of the vestibular system. Perceiving risky head rotations, the vestibular system tends to modify body dynamics by minimizing head oscillations, regardless of visual information. Moreover, avoiding visual feedback, a delayed motor strategy of the proximal segment was found in comparison with the platform motion in the older population, attesting the incapability of older adults to manage perturbation with anticipatory adjustment of the axial body.

As reported in literature, continuous and predictable perturbations are more easily managed compared to discontinuous and impulsive perturbations [18]. However, contrarily to sudden perturbations, during continuous perturbation subjects blend two primarily postural mechanisms. Subjects focused on the adjustment of their body motion in accordance with the actual external perturbation [18]. Moreover, subjects anticipate the body motion predicting the mechanical effect of the reflexes triggered by the displacement itself. In this context, pursuing the external movement of the perturbation by complying balance response with the platform displacement was observed to be the most functional and less-expensive postural strategy adopted by healthy younger adults [14,18]. Similar postural strategies were observed in our results by healthy younger subjects in balancing continuous yaw perturbations.

- G- and φ-relative:

By considering the fast Fourier analysis on segment-couples, proximal and distal segments of the younger group were in perfect agreement regarding amplitude rotation as well as time shift. Conversely, older population motor response reported a reduction and a delay in distal segment rotation in comparison with the proximal one, especially considering the head-pelvis segment-couple. Basically, younger subjects were able to oscillate in accordance to platform motion. As a consequence, younger subjects adopted a strategy that minimized the active effort by generating a lower torque couple among body segment. Older adults instead reacted differently from young adults, adopting more complex motor control strategies targeted to head stabilization. These findings can be ascribed to the greater difficult of older subjects to manage head-trunk movements, as also demonstrated during walking tasks [38].

- Continuous relative phase: MARP and DP

As regards the coordination pattern, the CRP technique provides a measure of the coupling or the phase relationship between the actions of body segment-couples. Since coordination impairment represents a common sign among musculoskeletal [39] and neurological disease [40], the age-related differences in the coordinative compensatory strategy were investigated in this study. Segment-couple coordination patterns of the upper body were different in the two groups. During EO condition, compared to the older adults, the younger population responded coupling body segments across the trial and reporting a less variable intra-subject trial-to-trial relationship between the actions of the segment-couples. Older adults also tended to balance yaw perturbation moving joints in and out-of-phase opposite fashion with an increased variability among trials. These outcome can be justified by the well-known loss of spatio-temporal coupling of muscles during postural responses, as reported in [41].

Increased variability in coupling relationship due to the aging was already reported in Yen et al. [42] during the obstacle-crossing task. This age-related biomechanical modification is associated to a lower ability of older adults to maintain a stable body displacement when crossing obstacles with different heights. A similar trend was observed in our results when subjects shifted from eyes open to the eyes closed condition. The segment-couple coordination of the upper body assumed more in-phase motion behaviors. Avoiding out-phase segment movements might be a coordination strategy adopted in the process of mastering redundant degrees of freedom of the upper body in a more unsafe scenario.

- Body displacement: RoM_{ML}, RoM_{AP}, PATH, and MV:

In terms of displacement of body segments, the analysis of the upper body unveiled that subjects behave as a double inverted pendulum, mainly in the frontal plane. As Figure 3 shows, both groups tended to assume a more oscillating displacement in the medio-lateral fluctuation of the head and the pelvis body segment with respect to the trunk. In particular, when visual information was allowed, head and pelvis medio-lateral movements were found to be both in a wider range, while the trunk stood noticeably more stable, acting as a center pivot. The upright balance was continuously reach by actively counteracting the head body segment with respect to the pelvis one, as they appear in almost perfect phase opposition (see Figure 3). In Figure 3, results of the eyes opened condition during the high frequency task were reported. However, a similar trend was observed in all the experimental tasks. Despite subjects were challenged with a rotational perturbation in the transversal plane, results highlighted motor behaviors similar to those of the double inverted pendulum already observed in previous studies, in which subjects were exposed to purely translational perturbations. This aspect could be justified considering the effect of the centrifugal force. In our interpretation, the rotational perturbation generated a centrifugal force, which tended to displace the body center of mass in medio-lateral direction. Subjects counteracted by translating head in the opposite direction. Head trajectory, in fact, appeared to be perfectly in opposition with respect to the pelvis one (see Figure 3), similarly to what happens due to the hip strategy in the case of translational perturbations [6].

Despite the before-mentioned trend reported in both groups, a prominent age-dependent stiffening strategy was witnessed by a smaller amount of medio-lateral oscillation, a lower path length and a lower mean velocity of the pelvis and head displacements. As previously reported in the literature, in the upright balance, humans can be sketched as an Acrobot [43], consisting in a series of inverted pendulums related to feet, legs and torso. When a large deflection of the base of support occurs, a quick shift of the COM is required to maintain the upright stance. In a human model, a proper displacement at the hip level modulates the torque created by the gravitational force on the shifted body, allowing the subject to keep the COM over the base of support and the foot flat on the ground [44]. By approximating the human torso as a single link, the head displacement acted in the opposite direction, allowing the generation of the counteractive moment of the gravitational head force.

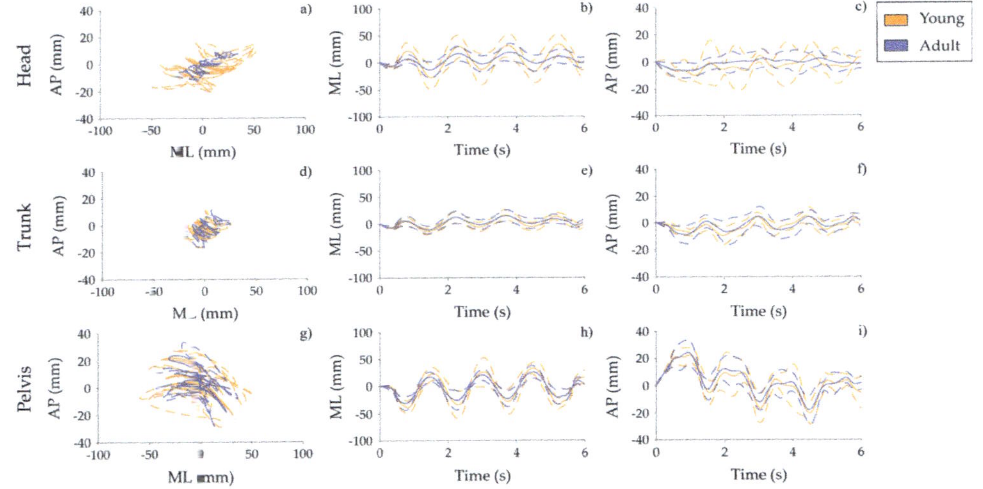

Figure 3. Mean (solid line) and standard deviation (dashed lines) of head, trunk, and pelvis displacement in the eyes opened condition during the high frequency task. The orange curves refer to the young group, the blue ones to older adults. In young group, medio-lateral sway of head is in phase opposition with respect to pelvis. Both have higher amplitude with respect to trunk, which acts as a pivot. This strategy is less evident in older subjects.

Interestingly, a different motor strategy was observed when vision was denied. The absence of visual inflow changed the body motion strategy adopted to keep the balance. Since the hip strategy was apparently preserved, the head displacement in the sagittal plane appeared more evident, overshooting pelvis motion in both groups. As a consequence of visual condition, the counteraction of body inertia was mainly in charge of the somatosensory and vestibular reference control, which reported a less effective role compared to the visual reference control in maintaining head stabilization. Similarly to findings in Nardone et al. [14], in which subjects were exposed to continuous anterior-posterior external perturbation, head oscillations were found more prominent during eyes-closed tasks. On the contrary to [14], older adults stimulated by continuous yaw perturbation presented a lower amount of head displacement compared to younger population, both regarding the eyes-open and the eyes-closed conditions.

By summarizing, our results highlighted age-dependent differences in rotational, translational, and coupling motor behaviors of the upper body of subjects elicited by external yaw perturbations. Future developments could be focused on the design of rehabilitation programs targeted to restore those impaired motor strategies. Those programs will be beneficial in reducing the risk of fall in older adults, especially those occurring during turning tasks, which still represent the 13% of all real-life falls [45].

Although this study provides insight into contributing factors to manage unbalanced conditions, some limitations should be taken into account. The small sample size and the large variability in the phase shift parameters could have affected the statistical analysis biasing results. In a future study, it will be necessary to increase the number of the participants to enforce statistical results regarding age-related differences, and design rehabilitation program aimed to enhance motor deficit of older population.

5. Conclusions

In this paper, we addressed the question of how aging affects postural control by examining the kinematic response under external yaw perturbation. The fundamental age-related postural change was mainly observed in the head stabilization strategy. Outcomes of older adults reported a decreased amount of the rotational and the translational body motion with a tendency of delayed behaviors and an out-phase and highly-varied coordinative compensatory strategy of the upper segment-couples. During low-frequency perturbation, postural strategies implied an easier following-strategy of the platform motion reporting a more complex reflex response than an anticipatory postural adjustment. During a more demanding frequency yaw rotation, a counteractive motor response was observed by the stiffening of the upper body motion and by strongly acting on the body inertial control.

Author Contributions: I.M., J.T., S.R., A.S. and E.P., designed the experiments; I.M., J.T., and E.P. performed the experiments; I.M. and E.P. processed kinematic data; I.M., J.T., S.R., Z.D.P., M.P., A.S. and E.P. analyzed data and interpreted results; I.M. wrote the first draft. I.M., J.T., S.R., Z.D.P., M.P., A.S. and E.P. reviewed and approved the manuscript. All authors have read and agreed to the published version of the manuscript.

Funding: This research received no external funding.

Conflicts of Interest: The authors declare no conflict of interest.

References

1. Fuller, G.F. Falls in the elderly. *Am. Fam. Physician* **2000**, *61*, 2159–2168, 2173–2174.
2. Boyé, N.D.A.; Van Lieshout, E.M.; Van Beeck, E.F.; Hartholt, K.A.; Van der Cammen, T.J.; Patka, P. The impact of falls in the elderly. *Trauma* **2013**, *15*, 29–35. [CrossRef]
3. Ganz, D.A.; Bao, Y.; Shekelle, P.G.; Rubenstein, L.Z. Will My Patient Fall? *JAMA* **2007**, *297*, 77–89. [CrossRef] [PubMed]
4. Runge, C.F.; Shupert, C.L.; Horak, F.B.; Zajac, F.E. Ankle and hip postural strategies defined by joint torques. *Gait Posture* **1999**, *10*, 161–170. [CrossRef]

5. Rogers, M.W.; Mille, M.-L. *Balance perturbations. In Handbook of Clinical Neurology*; Elsevier: London, UK, 2018; Volume 159, pp. 85–105.
6. Chaudhry, H.; Bukiet, B.; Ji, Z.; Findley, T. Measurement of balance in computer posturography: Comparison of methods-A brief review. *J. Bodyw. Mov. Ther.* **2011**, *15*, 82–91. [CrossRef]
7. Kharboutly, H.; Ma, J.; Benali, A.; Thoumie, P.; Pasqui, V.; Bouzit, M. Design of multiple axis robotic platform for postural stability analysis. *IEEE Trans. Neural Syst. Rehabilit. Eng.* **2015**, *23*, 93–103. [CrossRef]
8. Amori, V.; Petrarca, M.; Patané, F.; Castelli, E.; Cappa, P. Upper body balance control strategy during continuous 3D postural perturbation in young adults. *Gait Posture* **2015**, *41*, 19–25. [CrossRef]
9. Brown, L.A.; Jensen, J.L.; Korff, T.; Woollacott, M.H. The translating platform paradigm: Perturbation displacement waveform alters the postural response. *Gait Posture* **2001**, *14*, 256–263. [CrossRef]
10. Mileti, I.; Taborri, J.; Rossi, S.; Del Prete, Z.; Paoloni, M.; Suppa, A.; Palermo, E. Measuring age-related differences in kinematic postural strategies under yaw perturbation. In Proceedings of the 2018 IEEE International Symposium on Medical Measurements and Applications (MeMeA), Rome, Italy, 11–13 June 2018; pp. 1–6.
11. Taborri, J.; Mileti, I.; Del Prete, Z.; Rossi, S.; Palermo, E. Yaw Postural Perturbation Through Robotic Platform: Aging Effects on Muscle Synergies. In Proceedings of the 2018 7th IEEE International Conference on Biomedical Robotics and Biomechatronics (Biorob), Enschede, The Netherlands, 26–29 August 2018; pp. 916–921.
12. Kanekar, N.; Aruin, A.S. Aging and balance control in response to external perturbations: Role of anticipatory and compensatory postural mechanisms. *Age* **2014**, *36*, 9621. [CrossRef]
13. De Freitas, P.B.; Knight, C.A.; Barela, J.A. Postural reactions following forward platform perturbation in young, middle-age, and old adults. *J. Electromyogr. Kinesiol.* **2010**, *20*, 693–700. [CrossRef]
14. Nardone, A.; Grasso, M.; Tarantola, J.; Corna, S.; Schieppati, M. Postural coordination in elderly subjects standing on a periodically moving platform. *Arch. Phys. Med. Rehabil.* **2000**, *81*, 1217–1223. [CrossRef] [PubMed]
15. Cappa, P.; Patanè, F.; Rossi, S.; Petrarca, M.; Castelli, E.; Berthoz, A. Effect of changing visual condition and frequency of horizontal oscillations on postural balance of standing healthy subjects. *Gait Posture* **2008**, *28*, 615–626. [CrossRef] [PubMed]
16. Allum, J.H.J.; Carpenter, M.G.; Honegger, F.; Adkin, A.L.; Bloem, B.R. Age-dependent variations in the directional sensitivity of balance corrections and compensatory arm movements in man. *J. Physiol.* **2002**, *542*, 643–663. [CrossRef] [PubMed]
17. Cenciarini, M.; Loughlin, P.J.; Sparto, P.J.; Redfern, M.S. Stiffness and Damping in Postural Control Increase With Age. *IEEE Trans. Biomed. Eng.* **2010**, *57*, 267–275. [CrossRef] [PubMed]
18. Corna, S.; Tarantola, J.; Nardone, A.; Giordano, A.; Schieppati, M. Standing on a continuously moving platform: Is body inertia counteracted or exploited? *Exp. Brain Res.* **1999**, *124*, 331–341. [CrossRef] [PubMed]
19. De Nunzio, A.M.; Nardone, A.; Schieppati, M. Head stabilization on a continuously oscillating platform: The effect of a proprioceptive disturbance on the balancing strategy. *Exp. Brain Res.* **2005**, *165*, 261–272. [CrossRef]
20. Thelen, D.G.; Wojcik, L.A.; Schultz, A.B.; Ashton-Miller, J.A.; Alexander, N.B. Age differences in using a rapid step to regain balance during a forward fall. *J. Gerontol. A. Biol. Sci. Med. Sci.* **1997**, *52*, M8–M13. [CrossRef]
21. Madigan, M.L.; Lloyd, E.M. Age-Related Differences in Peak Joint Torques during the Support Phase of Single-Step Recovery from a Forward Fall. *J. Gerontol. Ser. A Biol. Sci. Med. Sci.* **2005**, *60*, 910–914. [CrossRef]
22. Lee, P.-Y.; Gadareh, K.; Bronstein, A.M. Forward–backward postural protective stepping responses in young and elderly adults. *Hum. Mov. Sci.* **2014**, *34*, 137–146. [CrossRef]
23. Lockhart, T.E.; Smith, J.L.; Woldstad, J.C. Effects of aging on the biomechanics of slips and falls. *Hum. Factors* **2005**, *47*, 708–729. [CrossRef]
24. Liaw, M.-Y.; Chen, C.-L.; Pei, Y.-C.; Leong, C.-P.; Lau, Y.-C. Comparison of the static and dynamic balance performance in young, middle-aged, and elderly healthy people. *Chang Gung Med. J.* **2009**, *32*, 297–304. [PubMed]
25. Leach, J.M.; Mellone, S.; Palumbo, P.; Bandinelli, S.; Chiari, L. Natural turn measures predict recurrent falls in community-dwelling older adults: A longitudinal cohort study. *Sci. Rep.* **2018**, *8*, 4316. [CrossRef] [PubMed]

26. Motta, C.; Palermo, E.; Studer, V.; Germanotta, M.; Germani, G.; Centonze, D.; Cappa, P.; Rossi, S.; Rossi, S. Disability and Fatigue Can Be Objectively Measured in Multiple Sclerosis. *PLoS ONE* **2016**, *11*, e0148997. [CrossRef] [PubMed]
27. Pacilli, A.; Mileti, I.; Germanotta, M.; Di Sipio, E.; Imbimbo, I.; Aprile, I.; Padua, L.; Rossi, S.; Palermo, E.; Cappa, P. A wearable setup for auditory cued gait analysis in patients with Parkinson's Disease. In Proceedings of the 2016 IEEE International Symposium on Medical Measurements and Applications (MeMeA), Benevento, Italy, 15–18 May 2016; pp. 1–6. [CrossRef]
28. Mileti, I.; Germanotta, M.; Alcaro, S.; Pacilli, A.; Imbimbo, I.; Petracca, M.; Erra, C.; Di Sipio, E.; Aprile, I.; Rossi, S.; et al. Gait partitioning methods in Parkinson's disease patients with motor fluctuations: A comparative analysis. In Proceedings of the 2017 12th IEEE International Symposium on Medical Measurements and Applications, (MeMeA), Rochester, NY, USA, 7–10 May 2017; pp. 402–407.
29. Mancini, M.; Salarian, A.; Carlson-Kuhta, P.; Zampieri, C.; King, L.; Chiari, L.; Horak, F.B. ISway: A sensitive, valid and reliable measure of postural control. *J. Neuroeng. Rehabil.* **2012**, *9*, 59. [CrossRef]
30. Ghislieri, M.; Gastaldi, L.; Pastorelli, S.; Tadano, S.; Agostini, V. Wearable Inertial Sensors to Assess Standing Balance: A Systematic Review. *Sensors* **2019**, *19*, 4075. [CrossRef]
31. Palermo, E.; Rossi, S.; Marini, F.; Patanè, F.; Cappa, P. Experimental evaluation of accuracy and repeatability of a novel body-to-sensor calibration procedure for inertial sensor-based gait analysis. *Measurement* **2014**, *52*, 145–155. [CrossRef]
32. Germanotta, M.; Taborri, J.; Rossi, S.; Frascarelli, F.; Palermo, E.; Cappa, P.; Castelli, E.; Petrarca, M. Spasticity Measurement Based on Tonic Stretch Reflex Threshold in Children with Cerebral Palsy Using the PediAnklebot. *Front. Hum. Neurosci.* **2017**, *11*, 277. [CrossRef]
33. Lamb, P.F.; Stöckl, M. On the use of continuous relative phase: Review of current approaches and outline for a new standard. *Clin. Biomech.* **2014**, *29*, 484–493. [CrossRef]
34. Burgess-Limerick, R.; Abernethy, B.; Neal, R.J. Relative phase quantifies interjoint coordination. *J. Biomech.* **1993**, *26*, 91–94. [CrossRef]
35. Miller, R.H.; Chang, R.; Baird, J.L.; Van Emmerik, R.E.A.; Hamill, J. Variability in kinematic coupling assessed by vector coding and continuous relative phase. *J. Biomech.* **2010**, *43*, 2554–2560. [CrossRef]
36. Ranavolo, A.; Donini, L.M.; Mari, S.; Serrao, M.; Silvetti, A.; Iavicoli, S.; Cava, E.; Asprino, R.; Pinto, A.; Draicchio, F. Lower-Limb Joint Coordination Pattern in Obese Subjects. *Biomed. Res. Int.* **2012**, *19*. [CrossRef] [PubMed]
37. Floor-Westerdijk, M.J.; Schepers, H.M.; Veltink, P.H.; Van Asseldonk, E.H.F.; Buurke, J.H. Use of inertial sensors for ambulatory assessment of center-of-mass displacements during walking. *IEEE Trans. Biomed. Eng.* **2012**, *59*, 2080–2084. [CrossRef] [PubMed]
38. Cromwell, R.L.; Newton, R.A.; Forrest, G. Head stability in older adults during walking with and without visual input. *J. Vestib. Res. Equilib. Orientat.* **2001**, *11*, 105–114.
39. Chiu, S.-L.; Lu, T.-W.; Chou, L.-S. Altered inter-joint coordination during walking in patients with total hip arthroplasty. *Gait Posture* **2010**, *32*, 656–660. [CrossRef]
40. Winogrodzka, A.; Wagenaar, R.C.; Booij, J.; Wolters, E.C. Rigidity and bradykinesia reduce interlimb coordination in Parkinsonian gait. *Arch. Phys. Med. Rehabil.* **2005**, *86*, 183–189. [CrossRef]
41. Greene, L.S.; Williams, H.G. Aging and coordination from the dynamic pattern perspective. *Adv. Psychol.* **1996**, *114*, 89–131.
42. Yen, H.; Chen, H.; Liu, M.; Liu, H.; Lu, T. Age effects on the inter-joint coordination during obstacle-crossing. *J. Biomech.* **2009**, *42*, 2501–2506. [CrossRef]
43. Spong, M.W. The swing up control problem for the Acrobot. *IEEE Control Syst.* **1995**, *15*, 49–55.
44. Horak, F.B.; Nashner, L.M. Central programming of postural movements: Adaptation to altered support-surface configurations. *J. Neurophysiol.* **1986**, *55*, 1369–1381. [CrossRef]
45. Robinovitch, S.N.; Feldman, F.; Yang, Y.; Schonnop, R.; Leung, P.M.; Sarraf, T.; Sims-Gould, J.; Loughin, M. Video capture of the circumstances of falls in elderly people residing in long-term care: An observational study. *Lancet* **2013**, *381*, 47–54. [CrossRef]

Article

Exploring Risk of Falls and Dynamic Unbalance in Cerebellar Ataxia by Inertial Sensor Assessment

Pietro Caliandro [1], Carmela Conte [2], Chiara Iacovelli [2,*], Antonella Tatarelli [3], Stefano Filippo Castiglia [4], Giuseppe Reale [5] and Mariano Serrao [4,6]

1 Unità Operativa Complessa Neurologia, Fondazione Policlinico Universitario A. Gemelli IRCCS, Largo A. Gemelli, 8, 00168 Rome, Italy; pietro.caliandro@policlinicogemelli.it

2 IRCCS Fondazione Don Carlo Gnocchi, Piazzale Morandi, 6, 20121 Milan, Italy; cconte@dongnocchi.it

3 Department of Occupational and Environmental Medicine, Epidemiology and Hygiene, INAIL, via Fontana Candida, 1, 00078 Monte Porzio Catone, Italy; antonellatatarelli@gmail.com

4 Department of Medical and Surgical Sciences and Biotechnologies, Sapienza University of Rome, Piazzale Aldo Moro, 5, 00185 Rome, Italy; stefanofilippo.castiglia@uniroma1.it (S.F.C.); mariano.serrao@uniroma1.it (M.S.)

5 Department of Neurosciences, Università Cattolica del Sacro Cuore, Largo F. Vito, 1, 00168 Rome, Italy; giureale@yahoo.it

6 Policlinico Italia, Movement Analysis Laboratory, Piazza del Campidano, 6, 00162 Rome, Italy

* Correspondence: ciacovelli@dongnocchi.it; Tel.: +39-0633086554

Received: 6 November 2019; Accepted: 12 December 2019; Published: 17 December 2019

Abstract: Background. Patients suffering from cerebellar ataxia have extremely variable gait kinematic features. We investigated whether and how wearable inertial sensors can describe the gait kinematic features among ataxic patients. Methods. We enrolled 17 patients and 16 matched control subjects. We acquired data by means of an inertial sensor attached to an ergonomic belt around pelvis, which was connected to a portable computer via Bluetooth. Recordings of all the patients were obtained during overground walking. From the accelerometric data, we obtained the harmonic ratio (HR), i.e., a measure of the acceleration patterns, smoothness and rhythm, and the step length coefficient of variation (CV), which evaluates the variability of the gait cycle. Results. Compared to controls, patients had a lower HR, meaning a less harmonic and rhythmic acceleration pattern of the trunk, and a higher step length CV, indicating a more variable step length. Both HR and step length CV showed a high effect size in distinguishing patients and controls ($p < 0.001$ and $p = 0.011$, respectively). A positive correlation was found between the step length CV and both the number of falls ($R = 0.672$; $p = 0.003$) and the clinical severity (ICARS: $R = 0.494$; $p = 0.044$; SARA: $R = 0.680$; $p = 0.003$). Conclusion. These findings demonstrate that the use of inertial sensors is effective in evaluating gait and balance impairment among ataxic patients.

Keywords: inertial sensors; cerebellar ataxia; movement analysis; gait analysis; balance; personalized medicine; rehabilitation

---

1. Introduction

Patients suffering from cerebellar ataxia exhibit peculiar spatiotemporal and kinematic features that contribute to an unstable gait [1–5]. The gait impairment typically worsens over time, in parallel with the functional decline associated to the neurodegenerative process [6,7]. While stable gait is characterized by repeatable walking patterns [8], steadiness in the case of perturbations [9–13], and effectiveness in maintaining upright balance [14,15], ataxic gait is extremely variable over gait cycles [1] and exhibits inefficient coordination between upper and lower segments of body, even in the absence of external perturbations [16]. Taking into account such conditions, it is reasonable to hypothesize that when perturbation occurs in ataxic patients, the consequent fall risk increases, and the gait pattern can be defined as unstable [6,7].

The evaluation of gait instability and fall risk is, therefore, pivotal in the study of ataxic gait to prevent further disabilities, and in order to maximize and optimize the information we gather from such evaluation, it should be performed in a real-life environment outside the motion analysis laboratory for a long period of time. In this context, wearable magnetic and inertial measurement units (MIMUs), consisting of a three-axial accelerometer, a gyroscope, and a magnetometer, represent a self-contained alternative to conventional laboratory-based motion capture systems [17–19]. This technology estimates the three-dimensional (3D) orientation of MIMUs with respect to a global coordinate system by specific sensor fusion algorithms, using angular velocity, gravity and magnetic field vectors.

A series of biomechanical stability measures based on MIMU evaluations have been proposed in several studies on neurological gait disorders with dynamic unbalance [3,20–22]. The maximum Lyapunov exponent (λmax) is an available method to evaluate gait instability [4] and fall risk [3] in ataxic patients, but the relationship between λmax and clinical severity has not been definitively established, since it has been demonstrated to be both positively [4] and negatively [3] correlated to International Cooperative Ataxia Rating Scale (ICARS) scores. A possible explanation could be found in the heterogeneous etiologies of the study samples, respectively acquired cerebellar lesions after tumor resection [4], and neurodegenerative ataxia [3]. Another important issue is that λmax properly explores the nonlinear dynamic local stability of the trunk during locomotion when at least 150 continuous strides are recorded [15]. However, such stride numbers are often not practically feasible in ataxic patients, and this could have influenced the correlation analysis between λmax and clinical severity.

To the best of our knowledge, no other studies in the literature have used additional indexes of stability, like harmonic ratio (HR) and the coefficients of variation (CV) based on MIMU data to detect the instability of ataxic patients. Therefore, the aim of this study is to evaluate these indexes of stability and, in particular, examine the ability of each index to detect the instability of ataxic patients compared to healthy controls and determine the fall risk. HR was chosen to evaluate the trunk acceleration patterns, a key feature in determining the severity of the ataxic gait [5,16], while CV was chosen to evaluate the variability of step length, an important compensatory mechanism in ataxic patients.

2. Materials and Methods

2.1. Participants

Seventeen patients affected by primary degenerative cerebellar ataxia were enrolled in the study. Table 1 summarizes the patients' clinical features and genotype.

The complete neurological assessment included (1) cognitive evaluation according to mini-mental state examination (MMSE) scale, (2) cranial nerve evaluation, (3) muscle tone evaluation, (4) muscle strength evaluation, (5) joint coordination evaluation, (6) sensory examination, (7) tendon reflex elicitation, and (8) disease severity measured by International Cooperative Ataxia Rating Scale (ICARS) and Scale for the Assessment and Rating of Ataxia (SARA) [23,24]. We excluded patients with gait impairment due to extracerebellar symptoms or orthopedic disorders. Regarding the extracerebellar disorders affecting gait, we excluded patients with spasticity, polyneuropathy, cognitive deficits, and extrapyramidal disorders. Of the recruited patients, no one presented with signs of spasticity, hyposthenia, hypoesthesia, and/or cognitive impairment (MMSE > 24). All patients were able to walk alone without any kind of assistance or aid, and were receiving physical therapy, including active and passive exercises for upper and lower limbs as well as balance and gait re-education. Furthermore, no patient had significant visual deficits according to the Snellen visual acuity test. Almost all of the patients had non-disabling oculomotor abnormalities, such as nystagmus or square wave jerks pursuit movements, because of the underlying disorder. A brain MRI showed that all patients had cerebellar atrophy. Regarding the fall risk assessment, all patients had to complete a specific questionnaire designed to evaluate the number of falls in the previous year, the characteristics of such falls (side, associated injury), and the circumstances in which they occurred. The number of falls in the last year was used for correlation analysis. Sixteen age-matched healthy adults (age, ataxic patients 53.53 ± 12.12 years, healthy controls, 50.94 ± 8.79 years, $p > 0.05$) were enrolled as the control group. We obtained informed consent from each patient and healthy subject, which complied with the Helsinki Declaration and was approved by the local ethics committee.

Table 1. Ataxic patients' clinical and anthropometric characteristics.

	Number/Total	%	Mean (SD)
Male	9/17	52.9	-
Female	8/17	47.1	-
Age (years)	-	-	53.53 (12.12)
Height (m)	-	-	1.65 (0.09)
Weight (kg)	-	-	71.03 (12.74)
ICARS	-	-	24.70 (10.80)
SARA	-	-	12.20 (4.25)
Disease duration (years)	-	-	12.11 (4.52)
Diagnosis			
SAOA	9/17	52.9	-
SCA1	2/17	11.8	-
SCA2	3/17	17.6	-
SCA3	1/17	5.9	-
SCA8	1/17	5.9	-
FRDA	1/17	5.9	-

SAOA: sporadic adult onset ataxia of unknown etiology; SCA: spinocerebellar ataxia; FRDA: Friedreich's ataxia.

2.2. Gait Analysis

We acquired data with an inertial sensor (BTS GWALK, BTS, Milan, Italy), attached to an ergonomic belt placed around the pelvis at the level of the L5 vertebra, connected to a portable computer via Bluetooth. The sampling rate was 100 Hz, and the sensor, endowed with a tri-axial accelerometer (16 bit/axes), a tri-axial magnetometer (13 bit), and a tri-axial gyroscope (16 bit/axes), measured the linear trunk accelerations and the trunk angular velocities in three space directions (i.e., AP: anterior-posterior; ML: mediolateral; VT: vertical direction).

2.3. Task Description

Before starting the experimental session, participants were asked to walk along a predetermined route in order to familiarize themselves with the procedure. Recordings of all the patients were obtained during overground walking. We asked participants to walk along a corridor (3 m wide and 20 m long) at their preferred speed. Control subjects were asked to walk at a low speed in order to match the two groups for speed (ataxic patients, 0.939 ± 0.195 m/s; controls, 0.924 ± 0.239 m/s; $p > 0.05$).

2.4. Inertial Sensor Data Processing

The 'walking protocol' of the inertial sensor (G-STUDIO, BTS, Milan, Italy) was used to detect: (1) trunk acceleration patterns, (2) right and left heel strikes, and (3) toe-off. The HR and the CV were calculated using MATLAB software (MATLAB 7.4.0, MAthWorks, Natick, MA, USA).

Harmonic ratio. The harmonic ratio (HR), initially described by Gage [25] and later modified by Smidt et al. [26], provides an indication of the acceleration patterns, smoothness, and rhythm. Since the unit of measurement from a continuous walking trial is a stride (two steps), a stable, rhythmic gait pattern should be characterized by multiples of two repeated acceleration patterns within any given stride. Accelerations patterns that do not repeat in multiples of two generate out of phase accelerations, reflecting irregular accelerations during a walking trial and, therefore, an unstable gait pattern. The harmonic content of the acceleration signals can be analyzed in each spatial direction using stride frequency as the fundamental frequency component. Based on each stride time, 20 harmonics were calculated. Trunk accelerations of each stride were broken down into individual sinusoidal waveforms using discrete Fourier transform (DFT).

Since a stable smooth gait pattern is characterized by acceleration signals in VT and AP directions that repeat in multiples of two during a single stride, HRs in the VT and AP directions were calculated as the ratio of the sum of the amplitudes of the first 10 even harmonics divided by the sum of the amplitudes of the first 10 odd harmonics. In the ML direction, acceleration signals were repeated once for any given stride, so HRs in the ML direction were calculated as the sum of the amplitudes of the odd harmonics divided by the sum of the amplitudes of the even harmonics. We used a high-pass filter with cutoff at 20 Hz to eliminate noise signals.

HRs per stride were determined and averaged across a steady walk, resulting in a mean HR. HR in AP and VT, and in the ML direction, were calculated as below [19]:

HR in anterior–posterior and vertical directions

$$HR = \frac{\sum_{i=1}^{10} A_{2i}}{\sum_{i=1}^{10} A_{2i-1}}$$

HR in the medio-lateral direction

$$HR = \frac{\sum_{i=1}^{10} A_{2i-1}}{\sum_{i=1}^{10} A_{2i}}$$

where A_{2i} denotes the amplitude of the first 20 even harmonics and A_{2i-1} indicates the amplitude of the first 20 odd harmonics. The higher the HR value, the smoother the walking pattern.

Coefficient of variation. In order to compute the step length CV, the step length was estimated using the upward and downward movements of the trunk, as proposed by Zijlstra and Hof [27]. Assuming a compass gait type, the body's center of mass (CoM) movements in the sagittal plane follow a circular trajectory during each single support phase. In this inverted pendulum model, changes in height of CoM depend on step length [27]. Thus, step length can be deduced by known height changes and predicted from geometrical characteristics as follows: step length = $2\sqrt{2lh - h^2}$.

In this equation, h is equal to the change in height of the CoM, and l represents the pendulum length. Changes in vertical position were calculated by a double integration of the vertical acceleration. A high-pass filter (fourth-order zero-lag Butterworth filter at 0.1 Hz) was used in order to avoid integration drift. The difference between highest and lowest position during a step cycle was used to determine the amplitude of changes in the vertical position (h). Leg length was considered as pendulum length (l). Step length was calculated as the mean of step lengths observed during seven subsequent steps of each subject.

Then, the step length coefficient of variation (CV) was computed as follows: $CV = 100\frac{SD}{mean}$ where mean is the mean step length and SD is the standard deviation over the entire step length for each subject [1]. The CV is a measure of the variability of a data set; the closer to 0 the CV is, the less variable the data are.

2.5. Statistical Analysis

We used the SPSS 17.0 software (SPSS Inc. Chicago, IL, USA) for statistical analysis. All data were expressed as mean ± standard deviation; $p < 0.05$ was considered statistically significant. We assessed the normality of distributions using the Shapiro-Wilk test.

Mean and standard deviation within subjects were computed for speed and stability indexes. We used the independent-samples t test to look for differences between the stability indexes of ataxic patients vs. controls. Cohen's d index was used to assess the effect size of the stability indexes in the three spatial directions [28,29]. We used the Pearson's test to investigate any correlation We used the Pearson test to investigate any correlation of acceleration HR and step length CV with (1) age, (2) height, (3) weight, (4) disease duration, (5) total ICARS and SARA scores and (6) number of falls in the last year.

3. Results

Looking at the low scores of ICARS and SARA, the recruited patients mainly showed cerebellar symptoms (see Table 1).

HR in all three directions and step length CV were all significantly different when compared to the controls (Table 2). Briefly, the HR of patients was lower than the HR of healthy subjects, meaning a less harmonic and rhythmic acceleration pattern of the trunk, while the CV of step length was greater in patients than in the controls, indicating a more variable step length in ataxic patients. Both HR and CV of step length showed a high effect size in distinguishing patients and controls, but HR in all three directions showed a higher effect size score when compared to the CV (Table 2).

Table 2. Comparisons of the stability indexes between 17 ataxic patients and 16 controls at matched gait speed.

Parameter	Patients	Controls	t	p	Cohen's d
HR-AP	1.665 ± 0.300	2.414 ± 0.540	4.964	<0.001	1.714
HR-ML	1.639 ± 0.282	2.347 ± 0.559	4.631	<0.001	1.599
HR-VT	1.694 ± 0.304	2.549 ± 0.715	4.519	<0.001	1.556
Step length CV (%)	21.249 ± 10.293	13.205 ± 6.004	−2.720	0.011	0.955
Step length (m)	0.499 ± 0.087	0.569 ± 0.067	−2.382	0.024	0.112
Speed (m/s)	0.939 ± 0.195	0.924 ± 0.239	−0.207	0.838	0.069

Mean ± standard deviation values, the results of the independent samples t-test and Cohen's d are reported. Values of p lower than 0.05 were considered statistically significant. HR-AP: harmonic ratio in the anterior–posterior direction; HR-ML: harmonic ratio in the mediolateral direction; and HR-VT: harmonic ratio in the vertical direction.

Surprisingly, no correlation was found between HR in all directions, falls/year, and clinical severity (ICARS and SARA scores) (Table 3), while a significant positive correlation was found between the CV of step length and the falls/years and ICARS and SARA scores (Figure 1).

Table 3. Correlation analysis between HR in all directions and ICARS, SARA, and falls/year.

Parameter	ICARS (R, p)	SARA (R, p)	falls/year (R, p)
HR-AP	−0.35, 0.24	−0.35, 0.13	−0.10, 0.66
HR-ML	−0.47, 0.10	−0.36, 0.11	0.02, 0.92
HR-VT	−0.41, 0.88	−0.43, 0.06	−0.01, 0.99

The reported values represent Pearson correlation value (R) and statistical significance value (p). HR-AP: harmonic ratio in the anterior–posterior direction; HR-ML: harmonic ratio in the mediolateral direction; and HR-VT: harmonic ratio in the vertical direction.

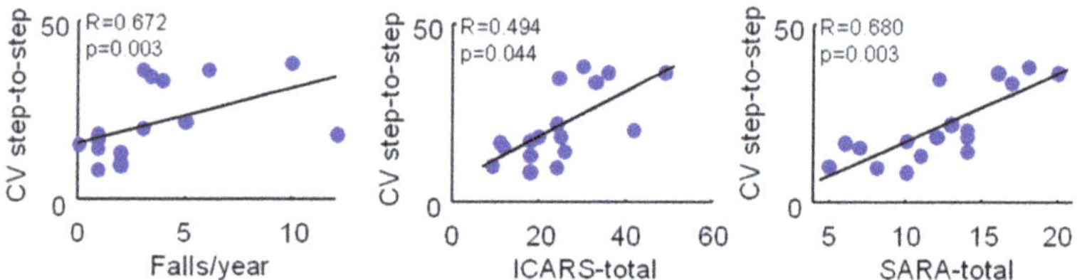

Figure 1. Correlations between the maximum step-to-step coefficient of variation and the falls/year, ICARS-total, and SARA-total scores in 17 ataxic patients. Pearson's R coefficient (R) and significance (p) are reported.

4. Discussion

In the present study, we found that trunk acceleration smoothness, as described by HR values, and the variability of step length, as described by the CV, may provide insights about gait stability in subjects with degenerative ataxia. Furthermore, the variability of step length correlated with both clinical severity and fall risk.

Regarding the acceleration patterns of the trunk, the HR of patients significantly differed from that of healthy controls in all three spatial planes. Moreover, it showed a high effect size, according to Cohen's d index (Table 2). This means that ataxic patients, compared to healthy subjects, exhibit a substantial reduction of trunk movement smoothness. When discussing these findings, we should bear in mind that the trunk has a great functional importance in minimizing the magnitude of linear and angular displacement of the head, ensuring clear vision [30,31], facilitating the integration of vestibular information [32], contributing to the maintenance of balance [5,6,16,33,34], and acting as a driving force for locomotion [35]. Consequently, investigating upper body stability in patients with degenerative cerebellar ataxia is essential, since the lack of motor control [5] and coordination [16] makes the trunk itself generate perturbations in a sort of vicious circle in parallel to the clinical decline [2]. In this context, trunk acceleration smoothness, as described by the HR values, provides a deeper insight into gait disturbances [14,36]. From the literature, we know that trunk acceleration smoothness during walking is predictive of gait dysfunction [37,38] and fall risk in older people [14,39]. Moreover, HR has already been found to be abnormal in patients who have suffered a stroke, Parkinson's disease, or multiple sclerosis [19,20,22,40].

Overall, these findings suggest that HR can substantially describe trunk accelerative behavior abnormalities among patients with degenerative ataxia [41]. On the other hand, we did not find any correlation between HR, the number of falls, and clinical severity. This last result is apparently in contrast with previous studies that found a relationship between clinical severity, increased range of motion of trunk [5] and trunk–thigh coordination deficit [16]. Considering the small sample size of our study, we cannot exclude a type II error. Nevertheless, another possible explanation might come from the different implemented technologies and protocols. In fact, previous studies assessed the kinematic patterns of the upper segment of the head and the trunk via optoelectronic systems [5,16]. This means that the body markers were located on body segments (i.e., the head and upper trunk) whose range of movements was wider than the lumbar one, as investigated by a BTS GWALK device located on L5 vertebra. Further studies will assess such differences, evaluating the role of the ergonomic belt placed around the thorax just underneath the axilla, and will validate inertial sensor findings against optoelectronic systems.

The other parameter we considered was the CV of step length, which has been reported to be significantly different in subjects with cerebellar ataxia when compared to healthy subjects [42]. During the progression of the disease (i.e., >4 years from the onset, as in our sample), subjects with degenerative ataxia tend to lose the ability to both enlarge their step width and fasten their walking speed and—maintaining the same step width and speed—they shorten their step length in order to reduce their single support time [43], with a significant increase in step length CV that can lead to an increased risk of falls. In fact, we found that the CV of step length was higher in patients with ataxia than the controls and, unlike HR, the CV of step length significantly correlated with the ICARS and SARA scores and with the number of falls per year. These findings differ from those from a previous study, where a correlation between the CV of step length and clinical severity was not detected [1]. This difference might be due to both the use of different movement analysis technologies (inertial sensor vs. optoelectronic system) and different investigated samples (sporadic adult onset ataxia of unknown etiology (SAOA)/ spinocerebellar ataxia (SCA) vs. SCA/SAOA/Friedreich's ataxia (FRDA)). Since a camera-based optoelectronic system can capture a smaller change of gait than MIMUs, our data should be interpreted with caution. However, the investigation of a large number of patients with FRDA in Serrao et al. [1] might explain, at least in part, such discrepant results. In this view, patients with FRDA and those with SCA and SAOA may show a different relationship between clinical features and gait stability; further studies are needed to explore this issue. However, our aim was not to obtain an alternative measure of step length, but to detect the relationship between the multifactorial gait impairment [5,6,16,33,34], clinical severity, and the fall risk. Because the MIMU-measured CV of step length is influenced by movements of the trunk [27], and trunk–thigh coordination is impaired in ataxic patients [16], our MIMU-measured CV might reflect trunk–thigh coordination variability. In this respect, the aforementioned limitation might come in handy, being such a multifactorial parameter able to summarize factors that, put together, explain gait instability in ataxic patients.

Finally, our results cannot be generalized as representative of the ataxic population because they refer to patients with a disease duration of at least 8 years, preserved walking ability, and without extracerebellar symptoms as disabling oculomotor abnormalities. Moreover, our findings highlight the need to investigate the relationship between each MIMU-measured index and the corresponding ones measured by traditional optoelectronic systems in order to have proper validation.

5. Conclusions

In conclusion, the present study highlighted that both HR and CV differed between ataxic patients and healthy subjects. However, when considering the correlation with clinical severity and fall risk, only MIMU-measured CV of step length was able to describe the burden of ataxic symptoms and to draw clinical attention towards a possible increased fall risk. These MIMU-based parameters might provide real-world information on patients' disabilities and falls, since they are obtained through wearable and comfortable devices.

Author Contributions: Conceptualization, P.C., M.S., C.I., C.C. and A.T. Methodology, P.C., C.C., M.S. and C.I.; Investigation, C.C., S.F.C., C.I.; Data Curation, C.I., C.C. and S.F.C. Writing—Original Draft Preparation, P.C., C.I., C.C. and S.F.C. Writing—Review & Editing: M.S. and G.R. Supervision: M.S.

Funding: This research received no external funding.

Acknowledgments: Thanks to Matteo Gratta for language editing.

Conflicts of Interest: The authors declare no conflict of interest.

References

1. Serrao, M.; Pierelli, F.; Ranavolo, A.; Draicchio, F.; Conte, C.; Don, R.; Di Fabio, R.; LeRose, M.; Padua, L.; Sandrini, G.; et al. Gait pattern in inherited cerebellar ataxias. *Cerebellum Lond. Engl.* **2012**, *11*, 194–211. [CrossRef] [PubMed]
2. Serrao, M.; Ranavolo, A.; Casali, C. Neurophysiology of gait. *Handb. Clin. Neurol.* **2018**, *154*, 299–303. [PubMed]
3. Chini, G.; Ranavolo, A.; Draicchio, F.; Casali, C.; Conte, C.; Martino, G.; Leonardi, L.; Padua, L.; Coppola, G.; Pierelli, F.; et al. Local Stability of the Trunk in Patients with Degenerative Cerebellar Ataxia During Walking. *Cerebellum Lond. Engl.* **2017**, *16*, 26–33. [CrossRef] [PubMed]
4. Hoogkamer, W.; Bruijn, S.M.; Sunaert, S.; Swinnen, S.P.; Van Calenbergh, F.; Duysens, J. Toward new sensitive measures to evaluate gait stability in focal cerebellar lesion patients. *Gait Posture* **2015**, *41*, 592–596. [CrossRef]

5. Conte, C.; Pierelli, F.; Casali, C.; Ranavolo, A.; Draicchio, F.; Martino, G.; Harfoush, M.; Padua, L.; Coppola, G.; Sandrini, G.; et al. Upper body kinematics in patients with cerebellar ataxia. *Cerebellum Lond. Engl.* **2014**, *13*, 689–697. [CrossRef]
6. Schniepp, R.; Wuehr, M.; Schlick, C.; Huth, S.; Pradhan, C.; Dieterich, M.; Brandt, T.; Jahn, K. Increased gait variability is associated with the history of falls in patients with cerebellar ataxia. *J. Neurol.* **2014**, *261*, 213–223. [CrossRef]
7. Schniepp, R.; Schlick, C.; Pradhan, C.; Dieterich, M.; Brandt, T.; Jahn, K.; Wuehr, M. The interrelationship between disease severity, dynamic stability, and falls in cerebellar ataxia. *J. Neurol.* **2016**, *263*, 1409–1417. [CrossRef]
8. Dingwell, J.B.; Marin, L.C. Kinematic variability and local dynamic stability of upper body motions when walking at different speeds. *J. Biomech.* **2006**, *39*, 444–452. [CrossRef]
9. Terrier, P.; Dériaz, O. Kinematic variability, fractal dynamics and local dynamic stability of treadmill walking. *J. Neuroeng. Rehabil.* **2011**, *8*, 12. [CrossRef]
10. England, S.A.; Granata, K.P. The influence of gait speed on local dynamic stability of walking. *Gait Posture* **2007**, *25*, 172–178. [CrossRef]
11. Kobayashi, M.; Nomura, T.; Sato, S. Phase-dependent response during human locomotion to impulsive perturbation and its interpretation based on neural mechanism. *Jpn. J. Med. Electron. Biol. Eng.* **2000**, *38*, 20–32.
12. Nessler, J.A.; Spargo, T.; Craig-Jones, A.; Milton, J.G. Phase resetting behavior in human gait is influenced by treadmill walking speed. *Gait Posture* **2016**, *43*, 187–191. [CrossRef] [PubMed]
13. Nomura, T.; Kawa, K.; Suzuki, Y.; Nakanishi, M.; Yamasaki, T. Dynamic stability and phase resetting during biped gait. *Chaos Woodbury N* **2009**, *19*, 026103. [CrossRef] [PubMed]
14. Menz, H.B.; Lord, S.R.; Fitzpatrick, R.C. Acceleration patterns of the head and pelvis when walking on level and irregular surfaces. *Gait Posture* **2003**, *18*, 35–46. [CrossRef]
15. Bruijn, S.M.; Meijer, O.G.; van Dieën, J.H.; Kingma, I.; Lamoth, C.J.C. Coordination of leg swing, thorax rotations, and pelvis rotations during gait: The organisation of total body angular momentum. *Gait Posture* **2008**, *27*, 455–462. [CrossRef]
16. Caliandro, P.; Iacovelli, C.; Conte, C.; Simbolotti, C.; Rossini, P.M.; Padua, L.; Casali, C.; Pierelli, F.; Reale, G.; Serrao, M. Trunk-lower limb coordination pattern during gait in patients with ataxia. *Gait Posture* **2017**, *57*, 252–257. [CrossRef]
17. Filippeschi, A.; Schmitz, N.; Miezal, M.; Bleser, G.; Ruffaldi, E.; Stricker, D. Survey of Motion Tracking Methods Based on Inertial Sensors: A Focus on Upper Limb Human Motion. *Sensors* **2017**, *17*, 1257. [CrossRef]
18. Picerno, P. 25 years of lower limb joint kinematics by using inertial and magnetic sensors: A review of methodological approaches. *Gait Posture* **2017**, *51*, 239–246. [CrossRef]
19. Iosa, M.; Picerno, P.; Paolucci, S.; Morone, G. Wearable inertial sensors for human movement analysis. *Expert Rev. Med. Devices* **2016**, *13*, 641–659. [CrossRef]
20. Buckley, C.; Galna, B.; Rochester, L.; Mazzà, C. Upper body accelerations as a biomarker of gait impairment in the early stages of Parkinson's disease. *Gait Posture* **2019**, *71*, 289–295. [CrossRef]
21. Beck, Y.; Herman, T.; Brozgol, M.; Giladi, N.; Mirelman, A.; Hausdorff, J.M. SPARC: A new approach to quantifying gait smoothness in patients with Parkinson's disease. *J. Neuroeng. Rehabil.* **2018**, *15*, 49. [CrossRef] [PubMed]
22. Pau, M.; Mandaresu, S.; Pilloni, G.; Porta, M.; Coghe, G.; Marrosu, M.G.; Cocco, E. Smoothness of gait detects early alterations of walking in persons with multiple sclerosis without disability. *Gait Posture* **2017**, *58*, 307–309. [CrossRef] [PubMed]
23. Schmitz-Hübsch, T.; du Montcel, S.T.; Baliko, L.; Berciano, J.; Boesch, S.; Depondt, C.; Giunti, P.; Globas, C.; Infante, J.; Kang, J.-S.; et al. Scale for the assessment and rating of ataxia: Development of a new clinical scale. *Neurology* **2006**, *66*, 1717–1720. [CrossRef] [PubMed]
24. Trouillas, P.; Takayanagi, T.; Hallett, M.; Currier, R.D.; Subramony, S.H.; Wessel, K.; Bryer, A.; Diener, H.C.; Massaquoi, S.; Gomez, C.M.; et al. International Cooperative Ataxia Rating Scale for pharmacological assessment of the cerebellar syndrome. The Ataxia Neuropharmacology Committee of the World Federation of Neurology. *J. Neurol. Sci.* **1997**, *145*, 205–211. [CrossRef]

25. Gage, H. *Accelerographic Analysis of Human Gait*; American Society for Mechanical Engineers: Washington, DC, USA, 1964.
26. Smidt, G.L. Methods of studying gait. *Phys. Ther.* **1974**, *54*, 13–17. [CrossRef] [PubMed]
27. Zijlstra, W. Assessment of spatio-temporal parameters during unconstrained walking. *Eur. J. Appl. Physiol.* **2004**, *92*, 39–44. [CrossRef] [PubMed]
28. Lang, T. Statistical Analyses and Methods in the Published Literature: The SAMPL Guidelines. In *Guidelines for Reporting Health Research: A Users' Manual*, 1st ed.; Moher, D., Altman, D., Schulz, K., Simera, I., Wager, L., Eds.; John Wiley & Sons, Ltd.: The Atrium, Southern Gate, Chichester, PO19 8SQ, UK, 2014.
29. Sullivan, G.M.; Feinn, R. Using Effect Size-or Why the P Value Is Not Enough. *J. Grad. Med. Educ.* **2012**, *4*, 279–282. [CrossRef]
30. Grossman, G.E.; Leigh, R.J.; Abel, L.A.; Lanska, D.J.; Thurston, S.E. Frequency and velocity of rotational head perturbations during locomotion. *Exp. Brain Res.* **1988**, *70*, 470–476. [CrossRef]
31. Hirasaki, E.; Moore, S.T.; Raphan, T.; Cohen, B. Effects of walking velocity on vertical head and body movements during locomotion. *Exp. Brain Res.* **1999**, *127*, 117–130. [CrossRef]
32. Pozzo, T.; Berthoz, A.; Lefort, L. Head stabilization during various locomotor tasks in humans. *Exp. Brain Res.* **1990**, *82*, 97–106. [CrossRef]
33. Prince, F.; Winter, D.; Stergiou, P.; Walt, S. Anticipatory control of upper body balance during human locomotion. *Gait Posture* **1994**, *2*, 19–25. [CrossRef]
34. Winter, D.A.; Mcfadyen, B.J.; Dickey, J.P. Adaptability of the CNS in Human Walking. *Adv. Psychol.* **1991**, *78*, 127–144.
35. Gracovetsky, S. An hypothesis for the role of the spine in human locomotion: A challenge to current thinking. *J. Biomed. Eng.* **1985**, *7*, 205–216. [CrossRef]
36. Lowry, K.A.; Lokenvitz, N.; Smiley-Oyen, A.L. Age- and speed-related differences in harmonic ratios during walking. *Gait Posture* **2012**, *35*, 272–276. [CrossRef]
37. Latt, M.D.; Menz, H.B.; Fung, V.S.; Lord, S.R. Walking speed, cadence and step length are selected to optimize the stability of head and pelvis accelerations. *Exp. Brain Res.* **2008**, *184*, 201–209. [CrossRef]
38. Latt, M.D.; Menz, H.B.; Fung, V.S.; Lord, S.R. Acceleration patterns of the head and pelvis during gait in older people with Parkinson's disease: A comparison of fallers and nonfallers. *J. Gerontol. A Biol. Sci. Med. Sci.* **2009**, *64*, 700–706. [CrossRef]
39. Doi, T.; Hirata, S.; Ono, R.; Tsutsumimoto, K.; Misu, S.; Ando, H. The harmonic ratio of trunk acceleration predicts falling among older people: Results of a 1-year prospective study. *J. Neuroeng. Rehabil.* **2013**, *10*, 7. [CrossRef]
40. Conway, Z.J.; Blackmore, T.; Silburn, P.A.; Cole, M.H. Dynamic balance control during stair negotiation for older adults and people with Parkinson disease. *Hum. Mov. Sci.* **2018**, *59*, 30–36. [CrossRef]
41. Kelley, K.; Preacher, K.J. On effect size. *Psychol. Methods* **2012**, *17*, 137–152. [CrossRef]
42. Mari, S.; Serrao, M.; Casali, C.; Conte, C.; Martino, G.; Ranavolo, A.; Coppola, G.; Draicchio, F.; Padua, L.; Sandrini, G.; et al. Lower limb antagonist muscle co-activation and its relationship with gait parameters in cerebellar ataxia. *Cerebellum Lond. Engl.* **2014**, *13*, 226–236. [CrossRef]
43. Serrao, M.; Chini, G.; Casali, C.; Conte, C.; Rinaldi, M.; Ranavolo, A.; Marcotulli, C.; Leonardi, L.; Fragiotta, G.; Bini, F.; et al. Progression of Gait Ataxia in Patients with Degenerative Cerebellar Disorders: A 4-Year Follow-Up Study. *Cerebellum Lond. Engl.* **2017**, *16*, 629–637. [CrossRef] [PubMed]

Article

High-Specificity Digital Architecture for Real-Time Recognition of Loss of Balance Inducing Fall

Daniela De Venuto and Giovanni Mezzina *

Department of Electrical and Information Engineering, Politecnico di Bari, 70125 Bari, Italy; daniela.devenuto@poliba.it

* Correspondence: giovanni.mezzina@poliba.it; Tel.: +39-0805963562

Received: 1 November 2019; Accepted: 30 January 2020; Published: 31 January 2020

Abstract: Falls are a significant cause of loss of independence, disability and reduced quality of life in people with Parkinson's disease (PD). Intervening quickly and accurately on the postural instability could strongly reduce the consequences of falls. In this context, the paper proposes and validates a novel architecture for the reliable recognition of losses of balance situations. The proposed system addresses some challenges related to the daily life applicability of near-fall recognition systems: the high specificity and system robustness against the Activities of Daily Life (ADL). In this respect, the proposed algorithm has been tested on five different tasks: walking steps, sudden curves, chair transfers via the timed up and go (TUG) test, balance-challenging obstacle avoidance and slip-induced loss of balance. The system analyzes data from wireless acquisition devices that capture electroencephalography (EEG) and electromyography (EMG) signals. The collected data are sent to two main units: the muscular unit and the cortical one. The first realizes a binary ON/OFF pattern from muscular activity (10 EMGs) and triggers the cortical unit. This latter unit evaluates the rate of variation in the EEG power spectrum density (PSD), considering five bands of interest. The neuromuscular features are then sent to a logical network for the final classification, which distinguishes among falls and ADL. In this preliminary study, we tested the proposed model on 9 healthy subjects (aged 26.3 ± 2.4 years), even if the study on PD patients is under investigation. Experimental validation on healthy subjects showed that the system reacts in 370.62 ± 60.85 ms with a sensitivity of 93.33 ± 5.16%. During the ADL tests the system showed a specificity of 98.91 ± 0.44% in steady walking steps recognition, 99.61 ± 0.66% in sudden curves detection, 98.95 ± 1.27% in contractions related to TUG tests and 98.42 ± 0.90% in the obstacle avoidance protocol.

Keywords: near falls, loss of balance, pre impact fall detection; activities of daily life; bio signals; EEG; EMG

---

1. Introduction

Recently, freezing of gait (FOG) and falls received increasing recognition as strongly debilitating features of Parkinson's disease (PD) [1,2]. By contrast with the tremor, which dominates the early stage of PD [2], falls and FOG are most common in advanced PD stages. The two phenomena seem related to each other, according to the study in [2], in which it is shown how sudden FOG can disturb the balance and, thereby, represents a common cause of falls in PD. Epidemiologic prospective studies conducted with 1-year, or 6-months, follow up [1–6], showed that the 45–68% of people with PD experience at least one fall per year, with a large portion (50–86%) falling recurrently [3–5]. It is not surprising that including the "near falls" this rate increases up to ~90% [6]. In this context, a near fall is a situation in which, despite a loss of balance, the body-ground impact could be avoided by grasping a support [6].

The clinical presentation in [2] shows that in PD patients most falls result from sudden changes in posture (in particular, turning movements of the trunk), rapid changes in the walking tasks (curve, transfers from the bed or the chair, etc.) or because they try to perform more than one activity simultaneously with walking or balancing.

In this context, the advances in wireless sensors networks, wearable acquisition devices, and new and more reliable digital signal-processing approaches for kinematic and biosignals analysis prompted the scientific community to develop technological solutions for early fall detection (FD).

The systematic review of FD solutions in [7] showed that body-worn accelerometers can be used to detect impacts and changes in orientation associated with falls. In the same context, the authors in [7] conclude that the accuracy of these FD systems may be improved by jointly using multiple sensors, e.g., signals from smartphone gyroscopes or barometers to define the height changes associated with falls [7–9]. These technologies aim to provide fast detection of falls but, at the moment, they are still not able to fully prevent injuries resulting from falls (e.g., hip fractures and traumatic brain injury) [8,9]. For this purpose, the focus of research contextually moved on fall risk assessment. This area of interest oversees the identification of the people's risk of falling, facilitating in this way early interventions via FD systems. Currently, fall risk-assessment procedures take into account the clinical evaluation of multiple domains such as balance control, mobility, physiology (strength, vision), psychology (fear of falling), cognition and environmental risk [7].

In this context, detecting near-falls (or recoverable imbalances) provides new opportunities to identify people with a high risk of falling before an actual fall occurs [10]. Near falls are defined as loss of balance that does not result in a fall because corrective action is taken to recover balance. They typically consist of slips, trips, and missteps. Moreover, since older people who frequently experience near falls are at increased risk of future falls [7,10], remote monitoring of these events during daily life could provide useful information to target falls and related circumstances as part of fall prevention initiatives [7].

Ultimately, an accurate algorithm for the detection of near falls could enhance the quality of existing fall detection systems by reducing false alarms [7].

In this respect, Table 1 summarizes state-of-the-art solutions [11–15] declared to be able to recognize near falls. The table reports the architectures in terms of used acquisition equipment, fall indicators (i.e., the feature(s) to be monitored and classified) and chosen classification method. Table 1 also dedicates a field to the Activities of Daily Life (ADL) and near-fall scenarios included in the discrimination. Finally, the last two rows summarize the declared system performance (i.e., accuracy and efficiency) and the applicability of proposed systems to daily-life and/or ambulatory contexts, as well as their suitability in the context of real-time near falls detection and fall prevention. All the studies selected for the comparison analyze unexpected slippages, classifying them as near falls because all the perturbations analyzed in [11–15] led to balance recovery.

Table 1 shows that the most used technologies in the loss of balance detection are motion capture systems (MCS) [12,13,15] and inertial measurement units (IMU) [11,14]. Solutions based on MCS are classified as context-aware and typically consists of a set of reflective markers and fixed cameras. For this reason, MCS-based fall detection systems present two limits: they are expensive and only suitable for ambulatory applications [12,13]. Pointing at daily-life applicability, authors in [11,14] propose wearable solutions mostly based on IMU sensors. More in detail, the authors in [11] analyze acceleration and angular velocity from 7 IMU sensors via a machine-learning (ML) approach. In a similar way, the authors in [14] exploit acceleration data from a single device, placed on the waist, to record vertical velocity from trials belonging to the chosen classification clusters (i.e., ADL vs. near falls). In terms of adopted classification methods, the result is that the most used approaches are still based on thresholds, in order to preserve a good speed in system response [12,14,15].

Table 1. State-of-the-art overview: features, performance (accuracy and efficiency) and applicability.

Reference		[11]	[12]	[13]	[14]	[15]
Acquisition Devices [1]		n.7 IMU 6DoF sensor (GYR + ACC)	MCS: total body	MCS: total body	n.1 IMU 6DoF sensor (GYR + ACC) on waist (close CoG)	Hips Encoder on APO + MCS: validation on lower limbs
Fall Indicators [2]		Means and variances of X, Y, Z for each ACC and GYR	Acceleration and VV of upper arms, trunk, tibia and head	Acceleration of all the monitored body segments	VV by integrating ACC data	ErF between Current Hips angle and predicted one
Classification Approach [3]		**ML:** RBF kernel SVM	**Multiple Thr + Statistics:** threshold based on an ARIMA model	**ML:** (1) accelerations analysis by ICA (2) ANN	**Single Thr:** user-specific pre-impact VV threshold	**Single Thr:** increment of the ErF
Recognized Classes [4]	**ADL**	Walking, Standing, correct chair transition, lying, picking up objects, ascending and descending stairs	Walking, Standing	Walking, Standing	Walking, Standing, correct chair transition, lying, picking up objects, ascending and descending stairs	Walking, Standing
	Near Fall Scenarios	Slip, Trip, Incorrect chair transfer, misstep (*recovery*)	Slip (*recovery*)	Slip (*recovery*)	Slip, Trip, Incorrect chair transfer, misstep (*recovery*)	Slip (*recovery*)
System Performance	**Se (%)**	80–96	88.5	92.7	95.2	92.7
	Sp (%)	90.8–99.2	92.9	98.0	97.6	98.0
	DT (ms)	offline	Mean: 680.00	Mean: 351.00	Mean: 469.00	Mean: 403.0
Applicability [5]	**OL**	✘	✘	✘	✔	✘
	Clin.	✔	✔	✔	✔	✔
	FD	✘	✘	✔	✔	✔

[1] **Acquisition Devices Acronyms:** IMU: inertial measurement unit, DoF: degree of freedom, GYR: gyroscope, ACC: accelerometer, MCS: motion capture system, CoG: center of gravity, APO: Active Pelvis Orthosis. [2] **Fall Indicators Acronyms:** VV: vertical velocity, ErF: error function. [3] **Classification Acronyms:** ML: machine learning, RBF: Radial Basis Function, Thr: threshold, SVM: support vector machine, ARIMA: autoregressive integrated moving average, ICA: independent component analysis, AR: Autoregression. [4] **Classes Acronyms:** ADL: Activities of Daily Life [5] **Applicability Acronyms:** OL: ordinary (or daily) life applicability, Clin.: Ambulatory applicability, FD: Fall-detection system suitability.

Machine-learning based solutions have been also investigated by authors in [11,13]. In this respect, noteworthy is the approach in [13], where the authors use an artificial neural network (ANN) to classify acceleration independent components, providing an interesting tradeoff between overall accuracy and fall-detection timing.

Table 1 reports the most investigated performance in FD and near-fall detection applications: the fall/imbalance recognition accuracy and the detection time [16]. The system accuracy parameter is composed of the sensitivity and the specificity. For the sake of comparison, all the analyzed works [11–15] share the same sensitivity and specificity definition. The first (i.e., sensitivity, Se (%) in Table 1) is defined as the ratio between the number of successfully detected falls/losses of balance over the total number of recorded perturbations. While, the specificity (Sp (%) in Table 1) is determined by the ratio between the number of successfully detected ADL over the total number of ADL-related trials. Finally, the detection time (DT in Table 1) characterizes the system efficiency. It is defined as the time difference between the fall initiation (that can be uniquely defined) and fall detection. This parameter gives an idea on how rapid the fall-detection system responds to a fall.

Concerning the applicability field of Table 1, the device's wearability and the proper specificity characterization relate to the suitability of the application to ordinary-life contexts. While, the detection time under a balance recovery limit (i.e., ~550 ms) [17] determines the system suitability for pre-impact FD strategy improvement.

In this paper, we propose and preliminarily validate a digital architecture for the loss of balance recognition during unexpected slippages potentially inducing fall. The main contributions of the paper concern:

- Fall Indicators. The architecture exploits a novel joint analysis of bio signals: electromyography (EMG) and electroencephalography (EEG).
- High Sensitivity and Specificity. The algorithm robustness is tested both in presence of unexpected slippages (near-fall scenarios) and during four ADL-like tasks: (i) steady walking, (ii) sudden curves, (iii) chair transfers via timed up and go (TUG) test and (iv) balance-challenging obstacle avoidance.
- Quick Loss of Balance Recognition. The system detection time reached by the proposed architecture is conservatively below the maximum intervention time limit for the countermeasures implementation [17].
- Wearability. The proposed architecture is fully based on wireless and wearable sensors, ensuring—together with the high-specificity constraint—the suitability in ordinary life applications.

The architecture proposed here exploits medical evidence from recent studies [18–23], according to which the cerebral cortex can regulate the postural stability according to environmental demands [18]. Specifically, the authors in [19,20] proved that low-frequency cortical rhythms ($f < 13$ Hz) are related to perception and cognitive control. In the loss of balance context, the modulation of the bands θ (4–7 Hz) and α (8–12 Hz) seems to be related to the visual field stabilization and active decoding of data from the vestibular system. Contextually, authors in [21–23] concluded that high-frequency cortical rhythms ($f > 13$ Hz) are commonly related to highly specific motor functions. Specifically, the β bands (i.e., β I, β II, β III), play a main role in muscle firing operations to compensate balance.

Besides the cortical dynamics' characterization, the muscular behavior could also be uniquely characterized. In this respect, the authors in [22–25] demonstrated that for accelerations or decelerations of the supporting surface (e.g., slippage) a low latency response (70–300 ms) occurs in the muscles near the ankles. It results in a muscular pattern characterized by co-contractions between agonist and antagonist muscle bundles [24].

Keeping this evidence in mind, the novel architecture exploits electrophysiological measurements from 10 EMG electrodes, to assess the muscular activity, and 13 EEG channels, to analyze the subject's cortical involvement during reactive response or normal motor planning.

Electrophysiological signals (i.e., EEG/EMG) are synchronously acquired via a central gateway. The gateway streams data to two computational units that distinctly analyze muscular and cortical activity. The unit dedicated to the muscular characterization has two main roles: realizing a binary ON/OFF pattern from muscular activity and triggering the cortical analysis unit. Once triggered, this latter unit quantifies the cortical involvements as the rate of variation in the EEG power spectrum density (PSD), considering the five bands of interest identified by authors in [18–23]. The parameters extracted from these units define some neuromuscular features of the subject under monitoring. As a final step, these features are sent to a logical network, which embeds a set of dynamic thresholds from the system calibration phase. In this application, the system calibration progressively builds a conservative range in which the neuromuscular features can be considered as "standard" and, thus, safe for the balance. The expectancy is that the ADL do not strongly affect the cortico-muscular parameters as, instead, happens during a loss of balance.

The paper is structured as follows. Section 2 outlines the experimental protocols, the setup and the implemented algorithm. Section 3 is dedicated to experimental results. Section 4 proposes a discussion about the system outcomes and Section 5 concludes the paper, presenting future perspectives.

2. Materials and Methods

2.1. Participants

Nine young and healthy volunteers (8 males, 1 female, 26.3 ± 2.4 years old, 64.5 ± 9.8 kg, 1.71 ± 0.06 m) were enrolled for this study. Six of them contributed to a near-fall scenarios test, while three subjects were actively involved in the system robustness test via ADL-like tasks. Before starting the experimental sessions, all the participants signed the informed consent. Research procedures were in accordance with the Declaration of Helsinki and was approved by the Local Ethical Committee (Protocol n. 2019_0025904).

2.2. Architecture Overview

Figure 1 shows a block diagram of the proposed loss of balance detection architecture. According to the figure, the system can be divided in four main sections: the acquisition unit, the muscular and cortical units and, finally, the classification block. As depicted in Figure 1, the proposed digital architecture synchronously operates on a STM32L4x microcontroller for the muscular analysis, and by means of Simulink real-time modeling to assess the cortical involvement. The Simulink model has been fully realized by blocks from the Digital Signal Processing (DSP) library in order to be implemented on a microcontroller.

The system working principle is inspired by our previous works [26,27], which laid the methodological bases for the joint analysis of EEG and EMG signals in the fields of gait analysis and involuntary movements detection. The overall processing chain is detailed in Figure 1.

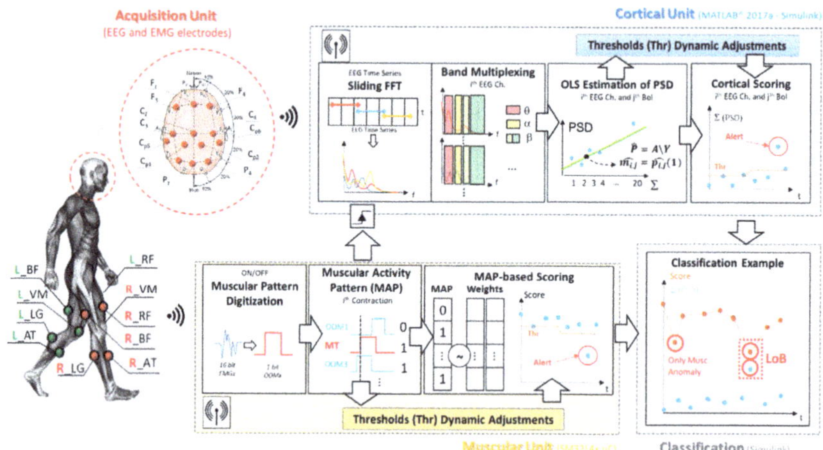

Figure 1. Proposed architecture block diagram. The figure shows the electroencephalography/ electromyography (EEG/EMG) experimental setup, as well as the graphical representation of the working flow of each involved block.

2.2.1. Acquisition Unit

The acquisition unit consists of a multi-sensing interface that jointly collects data from 10 surface EMGs and an EEG headset. The acquisition equipment has been selected to be fully wireless and wearable, allowing the subject complete freedom of movement.

In more detail, during the test and data collection phases, subjects wore a 32-channels EEG wireless headset (g.Nautilus Research by g.Tec [28]) and a set of 10 wireless surface EMG electrodes (Cometa WavePlus by Cometa srl [29]). According to the experimental measurement setup sketch in Figure 1, thirteen EEG sites have been monitored: F3, Fz, F4, C3, Cz, C4, Cp5, Cp1 Cp2, Cp6, P3, Pz, P4, according to the international 10–20 system. The O2 electrode was used for noise suppression, AFz as ground and the A2 (right earlobe) as the reference electrode. The EEG data were sampled at 500 Hz with 24-bit resolution.

On the muscular side, 10 surface EMG channels were monitored from following bilateral muscle groups: Anterior tibialis (AT), Lateral gastrocnemius (LG), Vastus medialis (VM), Rectus femoris (RF), and Biceps femoris (BF). The EMG signals were recorded with a sample rate of 2048 Hz and down sampled to 500 Hz (@16-bit resolution) before the transmission.

Data from the 10 EMG nodes are wirelessly streamed to a dedicated gateway, which is mounted on a Nucleo STM32L476RG board via a dedicated Printed Circuit Board (PCB) shield. Then the Muscular Unit algorithm runs on the microcontroller, analyzing the signals sample-by-sample.

Data from the EEG headset are sent to a base station connected via USB to a central computation unit that runs the Simulink model. The base station is also equipped with a 26-pin D-SUB connector used for the parallel reception of 8 digital input pins (DIN). These DINs will be used to receive data and triggers from the microcontroller. On the Simulink model side of the cortical unit, data from the monitored channels are continuously sent to n_{ch} = 13 circular registers, waiting for the enable signal from the muscular block. In this application, the central computation unit that runs the Simulink model consist of a HP Y5L00AE computer embedding an AMD A10-9600P processor (Hewlett-Packard—Palo Alto, CA, USA).

Pre-processing. The EEGs were progressively band-filtered between 1 Hz and 40 Hz by using a built-in 8th order Butterworth filter before the transmission [30]. The EMG node band-pass filters the signal between 15 Hz and 250 Hz before to be sent data to microcontroller [31]. Finally, a numeric notch filter 48–52 Hz has been implemented via the Simulink model for both EEG and EMG signals.

2.2.2. Experimental Protocols

To test the robustness of the algorithm proposed here and to ensure system suitability for daily-life contexts, the system response was assessed during four different ADL-like tasks. Figure 2 shows, through a snapshot grid, the experimental protocols carried out by the participants. Each row in the figure is composed of 6 frames, realizing a demonstrative sequence of the four experimental tasks:

1. **Steady walking to near fall (slip).** During this protocol, already presented in [26], the participants were asked to manage a slippage, unexpectedly provided during the steady walking by a mechatronic platform, called SENLY [32]. Specifically, the involved subjects underwent a series of 10 consecutive trials where their steady walking was unexpectedly perturbed by a slipping-like perturbation delivered in a pseudo-randomized order. Slippages consisted of a sudden and unexpected movement of one belt toward the antero-posterior (AP) direction. A demo of the protocols is shown in Figure 2a, panels (1) to (6).
2. **Steady walking with sudden curves.** In this protocol, the participants were asked to manage a tight turn around a preset delimiter by keeping the walking speed as constant as possible. The panels (3) and (4) of Figure 2b provides an idea of the protocol described. To evaluate the system specificity against the ADL-like task response, only the contractions related to the sudden curves were collected.
3. **Chair transfer via timed up and go test.** During the TUG test, the participants were asked to stand-up from a chair, walk toward a delimiter, carry out a tight turn around it and go back to the chair to sit- down again. The Figure 2b summarizes in 6 frames the TUG protocol. In this case, the contractions related to the sudden curves are kept in the sudden curves specificity database, while sit-down and stand-up contractions are collected in the dedicated TUG database.
4. **Balance-challenging obstacle avoidance.** This protocol is shown via the 6-frame sequence in Figure 2c. In this protocol, the participants were asked to manage a sequence of obstacle avoidances, by alternating the support foot for every trial. Obstacle avoidance-related contractions have been collected in the dedicated database for the system specificity computation.

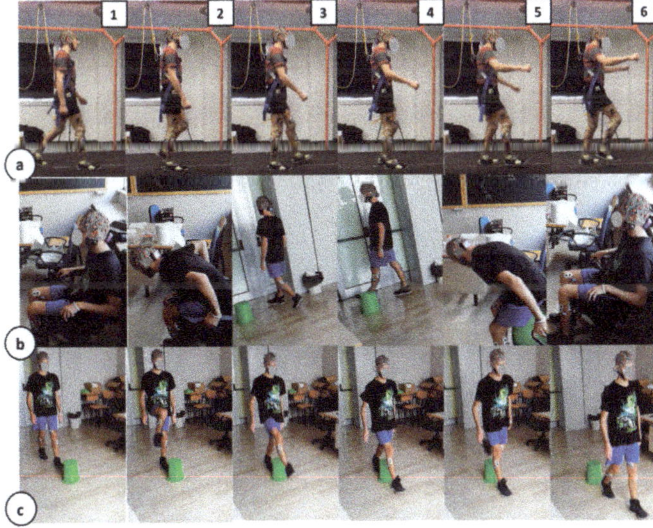

Figure 2. Experimental protocols grid. Each row in the panel represents a 6-frame demo sequence of the experimental protocol carried out by the participants. (**a**) Steady walking to near fall (slip) protocol; (**b**) Chair transfer via timed up and go test; (**c**) Balance-challenging obstacle avoidance.

2.2.3. ON/OFF Muscular Pattern Extraction

The muscular unit operates on the collected EMGs, generating an ON/OFF binary pattern of muscular activation (OOM—Figure 1) starting from the electrophysiological signal of each monitored muscle. Briefly, the implemented algorithm set OOM = 1 when the muscle is contracted, otherwise reset OOM = 0. This binarization procedure is entrusted to a moving threshold approach detailed in [26,33], because it demonstrated to be able in following the muscle tone changes (e.g., due to fatigue).

The ON/OFF muscular pattern extraction routine implemented on the STM32L4 μC can be briefly summarized by the following steps: the system progressively stores, for each muscle, a time-window containing the last M = 250 samples received.

It then extracts two data blocks: the first one containing the full EEG time-window (M = 250 samples, i.e., ~500 ms) and the second one that includes only the last N = 125 samples (i.e., ~250 ms). The algorithm squares these two vectors, averaging their elements. The resulting EMG power value for the longer time window (PM) acts as adaptive threshold, while the same parameter for the shorter time window (PN) as the instantaneous power. Finally, the two values are compared: if PN > PM, the system set OOM = 1, otherwise OOM = 0. The PM and PN values refresh and progressively adapt to each sample.

This ON/OFF muscular pattern digitization step (Figure 1) generates 10 parallel OOMs (one per muscle), which are sent to the muscular activity pattern (MAP) step according to the block diagram in Figure 1.

Two OOMs from both the Gastrocnemii are selected to trigger the Cortical Unit. These OOMs will be named master trigger (MT, Figure 1) hereafter.

To exclude, from the computation, the cortical activity that it is not strictly related to the specific movement, protecting from false alarms in the EEG unit, we selected as MT the gastrocnemius because it uniquely intervenes during the midstance gait phase.

2.2.4. Cortical Involvement Assessment

Once enabled via MT (side independent), the cortical unit extracts from the circular buffers 13 time-windowed EEGs of 400 samples (~800 ms) preceding the MT onset.

As a first step, these subset of EEG data undergo the on-line Riemannian artifact subspace reconstruction (rASR) [34]. The rASR is an online/offline artifacts attenuation method for mobile EEG data based on an ASR with Riemannian geometry.

The cortical unit analyzes these artifacts-free brain signals, quantifying the rate of variation in the EEGs power within the five bands of interest identified by authors in [18–23]: θ (4–7 Hz), α (8–12 Hz), β I, β II, β III (13–15, 16–20, 21–40 Hz). In more detail, the power spectrum density measurements are done by applying a sliding-window fast Fourier transform (FFT) on the considered EEG subset. For the purpose, the artifacts-free EEG subset is split in 20 overlapped windows long 200-samples with a step of 10 samples, covering the entire length of the subset.

Considering a single EEG window, the application of the FFT leads to a spectral resolution of 2.5 Hz (considering fs_{EEG} = 500 sps and L_{win} = 200 samples), which is suitable for the band multiplexing [26,35].

For each evaluated window, the system extracts a matrix named $\mathbf{S_{BoI}} \in \mathbf{R}^{nch,\, nBoI}$, with n_{ch} = 13 and nBoI = 5 number of bands involved in the multiplexing.

Each S_{BoI} element is the sum of the spectral contents falling within the selected j-th band due to the multiplexing:

$$S_{BoI}(i,j) = \frac{\sum_{k=(\text{jth band})} (S(k))\big|_{dB}}{Rg} \qquad i = 1 : n_{ch},\ j = 1 : n_{BoI},\ k = 1 : Rg \tag{1}$$

where the j-th band can mean the θ (k = 2:3), α (k = 3:5), β I (k = 6:7), β II (k = 8:10), β III (k = 11:16) band range, while Rg is the maximum k index (i.e., length of the j-th band).

The $\mathbf{S_{BoI}}$ is then extended to the 20 overlapped windows, generating a 3D matrix: $\mathbf{Y} \in \mathbf{R}^{\text{nch, nBoI, nW}}$ with nW = 20 number of measurements.

For the sake of clarity, considering a single band of interest (e.g., α band), data from the 20 FFT steps undergo a linear fitting via an ordinary least squares (OLS) estimator according to the equation:

$$\hat{\mathbf{p}}(\mathrm{i})|_{\alpha} = \mathbf{A} \backslash \mathbf{Y}(\mathrm{i}, \alpha, 1 : \mathrm{nW}) \tag{2}$$

where $\hat{\mathbf{p}}(\mathbf{i})\big|_{\alpha}$ is the OLS-based parameter vector for the i-th channel on the α band. It contains, in the order, the estimated linear model intercept $\hat{q} = \hat{\mathrm{p}}(\mathrm{i})|_{\alpha}[1]$ and the estimated straight-line slope $\hat{m} = \hat{\mathrm{p}}(\mathrm{i})|_{\alpha}[2]$. In the same equation (i.e., Equation (2)), **A** is the matrix of the basic functions containing a column of 1 and column of time vector (t = 20:800 ms, step 20 ms). Finally, $\mathbf{Y}(\mathrm{i}, \alpha, 1 : \mathrm{nW})$ is the vector that contains the FFT measurements on the i-th channel and the α band. The resulting linear models (OLS estimation—Figure 2) permit to approximate the cortical involvement parameter as the straight-line slope, $\hat{m}$. More details about the EEG computation branch implementation has been provided in our previous works [26,33]. The OLS-based estimation procedure is contextually applied to 13 channels and 5 band of interests, generating 65 $\hat{m}$ values.

2.2.5. Muscular Activity Pattern Extraction

The muscular unit hosted by the microcontroller operates in parallel with the cortical involvement analysis. In this frame, the muscular unit analyzes the 10 parallel OOMs via the MAP extraction routine. This stage aims to analyze the contraction status of each analyzed muscle "in correspondence" of the MT rising edge. Specifically, is time windows of 20 ms (11 samples), 10 ms before and 10 ms after the MT rising edge, is considered. The resulting OOM observation is named $\mathbf{wOOM} \in \mathbf{R}^{\text{nEMG, Lw}}$, where nEMG is the number of monitored EMG nodes and Lw is the number of samples composing the subset. In this application nEMG is 10, while Lw is 11. Also, the element wOOM(i,j) corresponds to the j-th sample of the i-th OOM observation window. In view of this, the **MAP** vector could be mathematically extracted as follows:

$$\mathbf{MAP}(\mathrm{i}) = \frac{\sum_{j=1}^{Lw} \mathbf{wOOM}(\mathrm{i},\mathrm{j})}{Lw} = \begin{cases} 1 & \mathbf{MAP}(\mathrm{i}) > 0.5 \\ 0 & \text{otherwise} \end{cases} \tag{3}$$

According with Equation (3), the outcome of this computation block consists of a 10-element vector (i.e., **MAP**). Each vector element corresponds to a muscle and it is 1 if the considered muscle is active (contracted) for more than half of the observation time, otherwise 0 (i.e., time predominance rule).

All the MAPs collected during a first brief stage of unperturbed gait or other ADL allow the system to build a first muscular behavior statistic. Specifically, they are used to extract a set of weights. These weights are based on the occurrence of a specific muscle contraction in correspondence of the MT contractions comprising the database. Two weights vectors are derived, one for the right leg (RL) movements and one from the left leg ones (to avoid asymmetry issues). In this way, it is possible to extract the most probable muscular pattern and, thus, a scoring method able to provide a high score if the incoming MAP is similar to the standard pattern, otherwise a low score (in presence of anomaly such as a perturbation). The weights vectors are continuously updated when requested by classification block, according to changes in user rhythms.

2.2.6. Muscular Activity Pattern (MAP)-Based Scoring Section

In a real-time application context, the MAP-based scoring block (Figure 1) analyzes the incoming MAP binary vector by dot-multiplying it by the related weight vector. For instance, MAP coming from the right Gastrocnemius contraction is dot-multiplied by the right leg-related weight vector and, finally, normalized. The score assignment outcome tends to 1 if the incoming MAP is similar to the muscular standard, otherwise, it tends to 0.

Naming $\mathbf{W_R} \in \mathbf{R}^{1,\,\mathrm{nEMG}}$ the weight vector from right leg movements, and $\mathbf{MAP_R}$ the resulting vector from Equation (3) when the MT is the right Gastrocnemius, the contraction score can be mathematically derived as:

$$\mathrm{Score\ RL} = \frac{\sum_{\mathrm{i}=1}^{\mathrm{nEMG}} \mathbf{MAP_R}(\mathrm{i})\mathbf{W_R}(\mathrm{i})}{\sum_{\mathrm{i}=1}^{\mathrm{nEMG}} \mathbf{W_R}(\mathrm{i})} \tag{4}$$

where score RL is the score related to a generic MT contraction from the right leg. Equation (4) can be easily extended to a MT contraction of the left leg MT, by changing the subscript R with L. In this latter case, the score is named score LL (LL, left leg) and it is derived via Equation (4) by considering the proper weight vector $\mathbf{W_L}$ and MAPs from the left leg, $\mathbf{MAP_L}$. A demonstrative example of the general muscular score during experimental walking to slip test is shown in Figure 3. The general score includes scores from right leg contractions (score RL) and, also, from left ones (score LL). The figure also shows a preview of the dynamic threshold extracted during the Calibration phase, detailed in Section 2.2.8 (red dotted line).

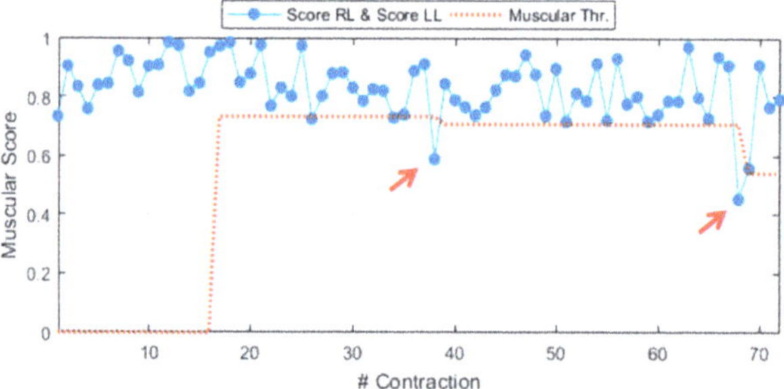

Figure 3. General muscular score (score right leg (RL) ∧ score left leg (LL)) during experimental walking to slip test (Sub. 3 – Trial 4).

2.2.7. Cortical Scoring Section

The score assignment embedded in the cortical unit passes through two main steps: the generalization and the lateralization assessment. The generalization step aims to reduce the data to be analyzed (65 vectors of $\hat{m}$ values from 13 channels and 5 bands of interest), providing a qualitative control about the subject's general cortical involvement. In this respect, the generalization step considers the $\hat{m}$ values on four cortical groups, which roughly identify functional macro areas:

- Supplementary motor area (SMA): F3, Fz, F4;
- Motor area (M1): C3, Cz, C4;
- Sensory-motor area (S1): Cp5, Cp1, Cp2, Cp6;
- Parietal area (PPC): P3, Pz, P4.

This means that, considering an incoming i-th contraction, the system extracts 20 $\hat{m}$ values (one per each functional group extended to 5 bands of interests). To clarify the concept, let us consider the α band involvement on the SMA. The generalized $\hat{m}$ value on the functional group SMA, considering the α band on the i-th contraction can be derived by the following equation:

$$\hat{m}_{SMA,\alpha}(i) = (\hat{m}_{C3,\alpha}(i) + \hat{m}_{C4,\alpha}(i) + \hat{m}_{Cz,\alpha}(i)) \tag{5}$$

The notation can be easily extended to the other formula parameters.

By contrast with the generalization step, the lateralization one evaluates the incidence of the power increment on a specific side (i.e., left or right) by analyzing the ratio between two specific macro areas: the left side containing {F, C, P}3 and the averaged {Cp1, Cp5}, and the right side that involves {F, C, P}4 and the averaged {Cp2, Cp6}.

The double-check implementation (i.e., generalization and lateralization) is justified by literature findings [19–23], which demonstrated that a reactive response leads to a widespread cortical involvement, while during unperturbed steps or non-challenging ADL, the cortical response remains more lateralized according to the limbs involved in the movement.

The Cortical Scoring block provides 20 values from the generalization step (one per each functional group on 5 bands of interests) and 5 values from the lateralization one (ratio between left and right-side involvement).

2.2.8. Logic Network-Based Classification

The logic network-based classification block concludes the system workflow according to Figure 1. It consists of two phases: the system adaptive calibration and the logic-network based classifier.

The system adaptive calibration oversees extracting dynamic thresholds (Thr—Figures 1 and 3) for every neuromuscular parameter involved in the classification, i.e., muscular score and the 25 values from the cortical generalization and lateralization steps.

Since the proposed architecture does not embed a learning phase, providing an auto-adaptive turnkey solution, these thresholds are continuously refreshed, contraction by contraction, by means of a sliding observation time window. In this window, the system checks the presence of thresholds lowering via statistical methods (some ADL can drag down the thresholds more than others). If a lowering is recorded, the thresholds are automatically adapted to the next value.

The role of these thresholds is to make the neuromuscular values as handleable as possible, for example, by associating a binary alert to each unexpected behavior. For instance, if the muscular score is below its dedicated threshold (red arrows in Figure 3) the muscular alert goes ON.

In a similar way, the procedure can be applied to the resulting $\hat{m}$ values.

The main goal of the implemented classifier is to cross-relate, among each other, these binary alerts from muscular and cortical sides. Specifically, the classification stage implements a logic network developed on 3 levels as shown by Figure 4.

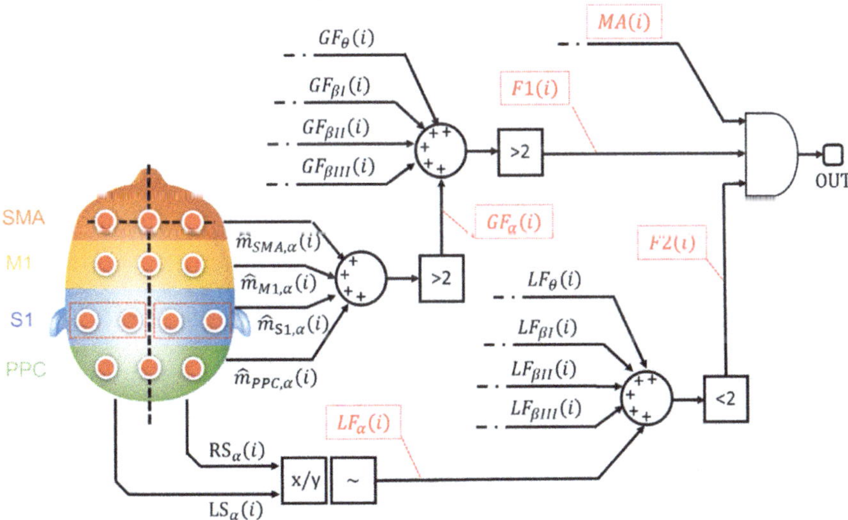

Figure 4. Logic network-based classifier. The flags that contribute to the classification stage are reported in red dotted boxes.

The 1st level considers the binary alerts from the 4 macro cortical areas (e.g., $\hat{m}_{SMA,\alpha}(i)$- Figure 4)), all over the 5 bands of interest. The system verifies the presence of a widespread increase in brain signal power. If more (>) than 2 cortical areas, for each evaluated band, are involved in the power increment, the architecture sets a generalization flag (GF_α—Figure 4) to 1, otherwise 0.

Generalization flags (GFs) from all the evaluated bands are further analyzed. If more (>) than 2 bands are involved in the brain power increase, the 1st level flag, F1 (i) in Figure 4, goes to 1.

The 2nd level analyzes the ratio between the left and the right cortical side (x/y—Figure 4) as described in Section 2.2.7. If the ratio is higher than 1+ε or lower than 1-ε, with ε specific tolerance (~), a lateralized increment is formally recorded. Similarly, the system generates a binary flag, named LF_α in Figure 4, which is equals to 1 if a lateralized brain activity is recognized. The system checks the number of lateralization flag as shown in Figure 4: if less (<) than 2 lateralization flags are active, the 2nd-level outcome (F2 (i) in Figure 4) goes to 1.

According to Figure 4, both the 1st-level output (i.e., F1 (i)), from generalization assessment, and the 2nd-level output (i.e., F2 (i)), from lateralization check, are sent to a final AND gate. If both the flags are ON, it means that the system recognized a not-lateralized increment of the cortical involvement. Finally, the classifier runs the 3rd level. This level considers the outcome from the AND gate, toggling the presence of a muscular alert (MA(i)—Figure 4) from the MAP-based scoring block (Figure 1).

Ultimately, if a not standard muscular behavior, jointly with a widespread and not lateralized cortical behavior, is found, the system classifies the i-th contraction as a potential loss of balance.

The classification output of this logical network can be used to enable a fall-prevention strategy (e.g., through wearable robotics and exoskeletons).

3. Results

The proposed system has been validated in near-fall scenarios and ADL-like tasks. During the walking-to-slip test protocol all the participants were secured by a safety harness attached to an overhead as shown in Figure 2a and no falls were reported during the trial. Participants were able to perform multistep recovery reaction to find back their balance.

Before starting the experimental sessions, all participants signed informed consent. Research procedures were in accordance with the Declaration of Helsinki and were approved by the Local Ethical Committee (Prot. no. 0028266/2019).

Section 3.1 briefly recaps general performance: sensitivity, detection time and specificity concerning steady walking steps versus near fall scenarios. Section 3.2 focuses on the daily life suitability of the system, discussing the method robustness against ADL. Section 3.3 briefly outlines the acquisition equipment features.

3.1. *Architecture Performance: Loss of Balance versus Steady Walking*

As already stated in the state-of-the-art comparison, the performance of a near-fall detection strategy is usually quantified in terms of accuracy and efficiency. According to all the evaluated works [11–15], the accuracy can be evaluated by considering the sensitivity and specificity parameters. Mathematically, the sensitivity can be defined as:

$$\text{Se}\ (\%) = (\#(\text{TrNF})/\text{N}_{\text{LoB}})\cdot 100 \tag{6}$$

where #(TrNF) is the number of correctly detected near fall events (i.e., induced slippages) and N_{LoB} represents the total number of evaluated loss of balance situations.

In a complementary way, the specificity is identified as:

$$\text{Sp}\ (\%) = (\#(\text{TrADL})/\text{N}_{\text{ADL}})\cdot 100 \tag{7}$$

where #(TrADL) is the amount of successfully detected ADL-like actions (i.e., walking steps, sudden curves, TUG and obstacle avoidance) and N_{ADL} is the total number of the evaluated ADL related trials.

In this section, we analyze results recorded during a real-time application of the walking-to-slip protocol. Two data pools were built: the first dataset is composed of 60 contractions from near-fall scenarios (10 perturbations per 6 subjects), while the second dataset includes 2091 contractions from the steady walking of all the subjects.

The experimentally extracted sensitivity, specificity (related to walking steps) and the detection time values are summarized in Table 2. The proposed multi-sensor architecture shows a sensitivity of 93.33 ± 5.16% and a walking steps vs. loss of balance (slip) specificity of 98.91 ± 0.44%.

Table 2. Proposed architecture accuracy and efficiency: near falls vs. steady walking.

Subject	Se (%)	SpWS [1](%)	DT (ms)	
			μ±σ	Max|Min
1	90.00 (9/10)	99.22 (386/389)	369.83 ± 97.49	536.11 | 202.02
2	100.00 (10/10)	98.32 (292/297)	436.72 ± 86.66	634.21 | 371.15
3	90.00 (9/10)	98.71 (308/312)	299.76 ± 107.99	432.00 | 194.60
4	90.00 (9/10)	98.55 (339/344)	355.85 ± 151.38	581.35 | 198.73
5	90.00 (9/10)	99.46 (370/372)	446.72 ± 112.89	626.45 | 374.36
6	100.00 (10/10)	99.20 (374/377)	314.82 ± 105.34	501.23 | 160.42
Avg [2]	**93.33 ± 5.16**	**98.91 ± 0.44**	**370.62 ± 60.85**	**634.21 |160.42** [3]

[1] **SpWS**: specificity strictly related to walking steps as not loss of balance actions; [2] **Avg**: averages among all the analyzed subjects (Sub. 1-6); [3] Values refer to the highest maximum and lowest minimum values among the reported data.

The system detection time is about 370.62 ± 60.85 ms, of which—on average—only 21.75 ms are dedicated to the overall computation chain for muscular and cortical units. The computation time comprises: (i) muscle ON/OFF pattern extraction (ii) sliding window FFT, (iii) band multiplexing, (iv) generalization and lateralization step (v) logic network-based classification and (vi) re-calibration of thresholds. Table 2 also shows that in the worst case (i.e., Sub 2, Trial 5) the system demands about 631 ms to intervene, while in the best case (i.e., Sub 6, Trial 4), the system recognizes the loss of balance in about 160 ms.

3.2. Architecture Performance: System Robustness against Activities of Daily Life (ADL)

To ensure the daily-life applicability of the proposed architecture, the wearability of the device is not the only constraint. Another important applicability limit lies in the system's robustness against movements that usually a subject does in his/her domestic environment. These are generally named Activities of Daily Life (ADL). In this respect, this section focuses on the system specificity characterization considering three ADL-like actions as: (i) sudden (and tight) curves, (ii) chair transfers (via TUG test) and (iii) obstacle avoidance.

For the sake of completeness, distinct datasets have been created starting from a real-time application of the system to the three tasks. Each test trial shown in the following consisted of a mixed pattern of these three tasks: walking with sudden curve, TUG and obstacle avoidance. Overall, the offline extraction of specific contractions resulted in a first dataset of 331 contractions (3 subjects) related to tight curves, a second dataset of 512 contractions (3 subjects) from the TUG test and, finally, a dataset that includes 352 contractions (3 subjects) related to the obstacle avoidance.

The steady walking specificity has been evaluated in the previous section by using a dataset of 2091 unperturbed steps (no recovery and near perturbed steps), leading to a value of 98.91 ± 0.44% that we assume as a final characterization parameter for the sake of readability.

Figure 5 provides a graphical characterization of the system robustness against the ADL.

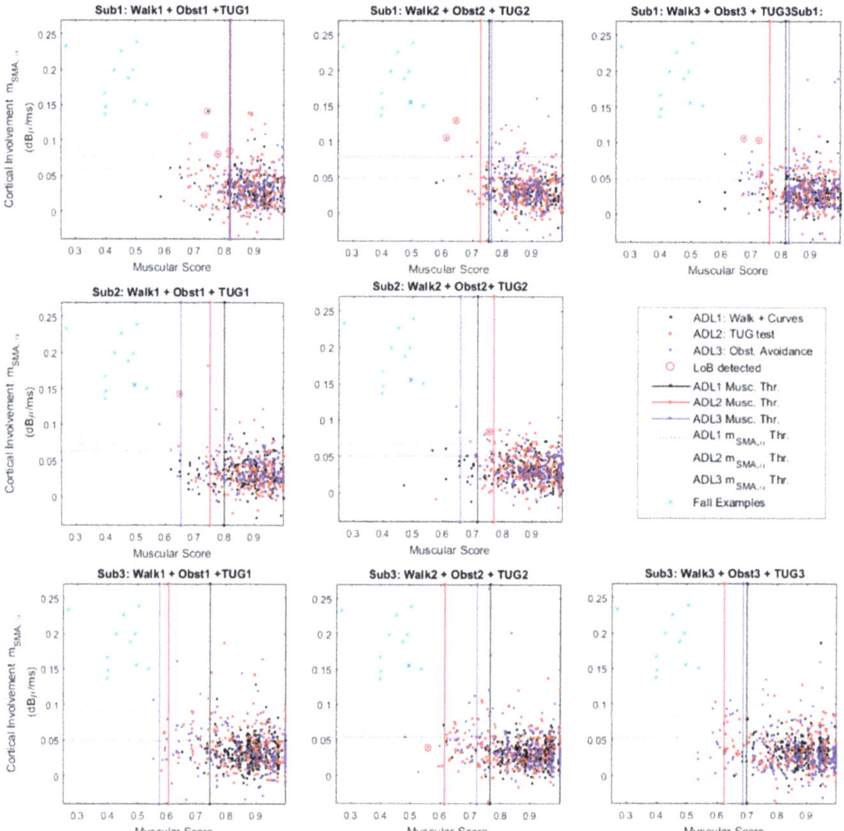

Figure 5. Cortico-muscular involvement planes for the 3 analyzed subjects. The figure merges contractions from three different Activities of Daily Life (ADL): curves (black), timed up and go (TUG) test (red) and obstacle avoidance (blue). The panels show the false alarms (red circles) and a comparison group (blue crosses) that represents the typical values occurring during a fall-related master trigger (MT) contraction.

The figure shows three panels per subject, except for Sub. 2 that did not perform the obstacle avoidance during the third test trial. Each single panel shows a 2D plane in which the x-axis reports the muscular score (MS in the following) as defined in Section 2.2.3, while the y-axis refers to the $\hat{m}_{SMA,\alpha}$ values, obtained according to Equation (5). Each point on the plane has coordinates {MS(i), $\hat{m}_{SMA,\alpha}(i)$} where "i" is the i-th MT contraction that led to features extraction. Each single point identifies two features that contributes to the final classification of the specific contraction.

The panel also shows 3 thresholds that are constant along the x-axis (solid red, blue and black lines). These thresholds are those that operate on the muscular score (i.e., MS), acting as shown in Figure 5 by considering the median on an observation window of 15 consequent contractions.

This means that the system slowly adapts to the worst thresholds among the evaluated ones.

For example, considering The Sub.1–Test: 2, the worst MS-related threshold is linked to the ADL3: obstacle avoidance (Figure 5). It means that during its steady functioning, the system will recognize as a real dangerous situation only MS below the obstacle avoidance related threshold (i.e., ADL3 Musc. Thr.—Figure 5).

The panels also provide y-axis constant thresholds that refers to virtual upper limits for $\hat{m}_{SMA,\alpha}$. These thresholds act in a similar manner of the previously presented ones. Considering for example data from Sub.3–Trial 2 in Figure 5, this means that during its steady functioning the system will recognize a dangerous situation only for $\hat{m}_{SMA,\alpha}$ above the obstacle avoidance threshold (i.e., ADL3 $\hat{m}_{SMA,\alpha}$ Thr.—Figure 5). Ultimately, real dangerous situations for both muscular and cortical involvements lie in the top-left rectangles delimited by the leftmost MS-related threshold and the highest $\hat{m}_{SMA,\alpha}$-related threshold. On these panels, some contractions wrongly recognized as losses of balance highlighted via red circles. In this context, we must consider it as incorrect classification, reducing the specificity as per Equation (5).

Finally, for the sake of comparison, Figure 5 reports as blue crosses some slip-related coordinates {MS (i), $\hat{m}_{SMA,\alpha}(\mathrm{i})$}. These coordinates have been extracted from the walking versus slip dataset (Section 3.1) considering two points per each analyzed subject. This shows how the two groups ADL1,2,3 and slips could be easily divided in clusters.

To give a complete overview of the system robustness against ADL, Table 3 summarizes the experimental results in terms of specificity from each protocol carried out (ADL 1, 2 and 3 Sp. (%)), as well the single test (Tasks 1, 2, 3 Sp. (%)) and subject-related (Sub. Sp. (%)) specificities. Analyzing data in Table 3 and in Figure 5 it is possible to state that, overall, the system showed a specificity of 98.91 ± 0.44% in steady walking steps' recognition (see Table 2), 99.62 ± 0.66% in sudden curves successfully detection, 98.95 ± 1.27% of correct recognition in contractions related to TUG tests. Finally, during the balance-challenging obstacle avoidance protocol the specificity reached 98.43 ± 0.88%.

Table 3. ADL-related specificity extraction.

Curves	**Task 1 Sp. (%)**	**Task 2 Sp. (%)**	**Task 3 Sp. (%)**	**Subject Sp. (%)**	**ADL1 Sp. (%)**
Sub. 1	96.55 (28/29)	100.00 (23/23)	100.00 (26/26)	98.85 ± 1.99	99.62 ± 0.66
Sub. 2	100.00 (34/34)	100.00 (36/36)	100.00 (35/35)	100.00	
Sub. 3	100.00 (51/51)	100.00 (48/48)	100.00 (49/49)	100.00	
TUG	**Task 1 Sp. (%)**	**Task 2 Sp. (%)**	**Task 3 Sp. (%)**	**Subject Sp. (%)**	**ADL2 Sp. (%)**
Sub. 1	97.50 (39/40)	98.21 (55/56)	96.00 (62/64)	97.52 ± 0.67	98.95 ± 1.27
Sub. 2	97.91 (47/48)	100.00 (56/56)	100.00 (48/48)	99.30 ± 1.20	
Sub. 3	100.00 (64/64)	100.00 (64/64)	100.00 (72/72)	100.00	
Obst. Avoidance	**Task 1 Sp. (%)**	**Task 2 Sp. (%)**	**Task 3 Sp. (%)**	**Subject Sp. (%)**	**ADL3 Sp. (%)**
Sub. 1	95.24 (40/42)	98.61 (71/72)	98.44 (63/64)	97.48 ± 1.89	98.43 ± 0.88
Sub. 2	100.00 (30/30)	97.22 (35/36)	-	98.61 ± 1.96	
Sub. 3	100.00 (30/30)	97.61 (41/42)	100.00 (36/36)	99.21 ± 1.37	

In order to provide a comparison with the state-of-the-art solutions, already presented in the introduction, Table 4 summarizes some specific features for each analyzed work. In particular, Table 4 focuses on recognized classes, the system performance in terms of accuracy and efficiency and the applicability tabs. Data from Table 4 show how the proposed system ensures very competitive specificity value (i.e., 98.9%).

Table 4. Comparison of proposed architecture's performance.

Reference	Recognized Classes		System Performance			Applicability		
	ADL	Near Falls	Se (%)	Sp. (%)	DT (ms)	OL	Clin.	FD
[11]	Walking, Standing, correct chair transition, lying, picking up objects, ascending and descending stairs	Slip, trip, incorrect chair transfer, misstep (*recovery*)	80.0–96.0	90.8 – 99.2	offline	✗	✔	✗
[12]	Walking, Standing	Slips (*recovery*)	88.5	92.9	680.00	✗	✔	✗
[13]	Walking, Standing	Slips (*recovery*)	92.7	98.0	351.00	✔	✔	✔
[14]	Walking, Standing, correct chair transition, lying, picking up objects, ascending and descending stairs	Slip, trip, incorrect chair transfer, misstep (*recovery*)	95.2	97.6	469.00	✔	✔	✔
[15]	Walking, Standing	Slip (*recovery*)	97.6	98.0	403.00	✗	✔	✔
This Work	Walking, Standing, Sudden curves, Chair transitions (TUG), Obstacle avoidance	Slips (*recovery*)	**93.3**	**98.9** [1]	**370.62**	✔	✔	✔

[1] The Sp(%) value has been evaluated as the average between the four specificity values (i.e., steady walking, curves, TUG, obstacle).

3.3. Acquisition Equipment Features

Once the algorithm robustness against ADL recognition is verified, the system's applicability to ordinary life imposes another constraint: wearability. The chosen equipment should address the wearability constraints, which according to [36] can be summarized briefly in three macro-categories: (i) encumbrance (ii) biomechanical effects and (iii) comfort.

Considering the former constraint, the physical dimensions of the wearable will be paramount. These dimensions include the size, weight and the distribution of the weight of the wearable on the body.

Secondly, the functional placement of the sensor nodes may affect the posture and musculoskeletal loading of the wearer. Finally, the sensors' node placement must avoid discomfort, favoring regular movements (e.g., walking or sitting) and non-biased postures.

To continuously analyze and characterize the subjects' cortical and muscular dynamics in several different ordinary life scenarios, the here-proposed experimental setup consists of a 32-channel wireless EEG headset (g.Nautilus Research) by g.Tec [28] and 10 wireless surface EMG nodes (Cometa Wave Plus) by Cometa Systems srl [29]. Table 5 provides information about the acquisition equipment comprising the set-up. For each device, the table reports the number of monitored nodes or channels, equipment features such as the size and weight, as well as the electrode characteristics and device parameters: wireless transmission range and protocol, resolution and sampling frequency. Table 5 demonstrates how the equipment choice ensures a fully wireless and low-encumbrance solution, validating the applicability in an indoor monitoring scenario. Despite this, the use of gel-based or pre-gelled electrodes could be considered uncomfortable for long-time acquisition. In this respect, the system can be considered reliable for 4 hours' acquisition, before the need to refill the gel to ensure the right input impedance to the amplifier.

Table 5. EEG/EMG acquisition devices features.

Signal	Num.	Equipment Features	Electrode		Transmission Range	Resolution (Sampling FrEquation)
			Size (mm)	Type		
EEG	13 channels	**EEG Headset:** **Back Head Station:** 70 × 55 × 30 mm Weight: 145 g **Headset:** Full-scalp elastic cap Weight: 12 g *Wireless* 10 h continuous acquisition @ 500Hz	16 × 10 × 5	Active Gel based Sintered Ag/AgCl probe	**Modulo RF:** XVV-MEGA22M00 (IEEE 802.15.4 WPAN @ 2.4GHz) **Indoor Range:** 17 m with 2.3–2.9dBm	24 bit (@500 Hz)
EMG	10 nodes	**EMG Single Node:** 33 × 23 × 19 mm Weight: 12 g Wireless 12 h continuous acquisition @ 2048 Hz	18 × 12 × 5	Active Pre-Gelled Sintered Ag/AgCl holder ring	**Private protocol:** Y9SMPTX 2.402–2.48 GHz **Indoor Range:** 15 m (+3dBm)	16 bit (@2048 Hz ↓ 500 Hz)

4. Discussion

The detection of near falls is an emerging area of research that is contextually growing with the development of an increasing number of miniaturized and power-efficient wearable devices [7]. Supported by accumulated evidence on fall detection [7–23], the clinical utility of this investigation involves the unobtrusive and continuous monitoring of activities of daily life in populations at high risk of falling. This kind of strategy (i.e., near fall detection) can be useful to identify issues to be further addressed to prevent falls, associated injuries or simply improve the efficiency of already existing pre-impact fall-detection architectures.

In this study, we have proposed a novel wearable architecture that exploits electrophysiological signals from brain and lower limb muscles to discriminate a near-fall scenario (i.e., unexpected slippages) from an activities of daily life. The proposed system realizes a turnkey solution, which can adapt its function to the user neuromuscular rhythms, without any long and fatiguing learning stage.

Results in Section 3.1 showed how the proposed architecture demands about 370.62 ± 60.85 ms to carry out a binary classification (i.e., ADL vs. near fall). As stated in the same section, the overall computation chain of muscular and cortical units requires, on average, 22 ms to be completed.

The remaining time, i.e., ~350 ms, with its high variability (see Table 2), is related to the muscle that has been selected as a master trigger. In fact, it should be reminded that the system starts working from the contraction onset of the gastrocnemius (right or left independently).

The times related to this physiological process remain hard to determine with certainty. In this respect, the response times of the gastrocnemius constitute unavoidable delays in recognizing losses of balance and largely determine the efficiency of the system. Further investigation should be conducted in order to find another muscle bundle that can: (i) uniquely define a gait phase, (ii) activate itself faster in a perturbation context, (iii) ensure repeatability during the contraction timing when the near-fall scenario occurs.

The detection times achieved are competitive with respect to the state-of-the-art solutions, highlighting the system applicability in contexts of postural recovery strategies implementation [17].

Concerning the results in Section 3.2 that analyze the system robustness against the ADL, an interesting evaluation should be undertaken into the losses of balance detected below the worst thresholds, such as Sub1-Test1 and Sub2-Test2 in Figure 5. In these cases, offline checks verified that the threshold was slowly adapting to the final value, causing transitional "false alarm" (wrongly loss of balance detection). Within this, the threshold adapting procedure should be improved and speeded up, while keeping high sensitivity and specificity.

Another noteworthy case is that shown in Sub3- Test 2 (Figure 5). In that case, it seems that a loss of balance is detected below the cortical thresholds. It is important to remember that these panels show only the $\hat{m}_{SMA,\alpha}$ values, nevertheless, we must consider that the analyzed problem is hyper-dimensional, because we should take into account other four bands of interest and the remaining 3 cortical groups.

Moreover, in Figure 5 it is notable that the Sub. 1 experienced a high number of false alarms with respect to the following two subjects. This result could be related to the protocol improvement asked on-going to the last two participants. This improvement mainly concerns the sit-down and stand-up movements during the TUG. In fact, since the rejection algorithm rASR has not been optimized to reject the muscular artifacts from strong contractions of the deltoids, the EEG acquisitions were spoiled by unpredicted artifacts. In this respect, further investigations are still ongoing aiming to extend the range of applicability of the implemented rASR algorithm.

5. Conclusions

In this paper, we proposed and validated a novel architecture for the losses of balance recognition. The proposed system, optimized for unexpected slippages, addressed some still open challenges related to the daily life applicability of this kind of system. Design and verification constraints mainly concern the need for high specificity and system robustness against ADL. In this respect, the proposed algorithm has been tested on five different tasks: sudden curves, chair transfers via the timed up and go test, balance-challenging obstacle avoidance and, of course slip-induced loss of balance. To ensure the ordinary life suitability, the proposed architecture has been fully based on wearable and wireless acquisition devices. Specifically, the architecture exploits electrophysiological measurements from 10 EMG electrodes and 13 EEG channels. The collected data are analyzed by the muscular unit, hosted by a STM32L4 microcontroller, and the cortical unit, which is implemented on a central computation unit via Simulink modeling. The first realizes a binary ON/OFF pattern from muscular activity (10 EMGs) and triggers the cortical unit that evaluates the contraction-related cortical involvements in terms of EEG responsiveness.

This parameter is evaluated as the variation behavior in the EEG PSD, considering five bands of interest. The neuromuscular features from both the computation units are sent to a clinical evidence based logical network. It embeds a set of automatically adaptive thresholds, which follow the user rhythms. Experimental validation on 9 healthy subjects showed that the system could react in a time compliant with fall-detection architectures constraints (i.e., 370.62 ± 60.85 ms). It also ensures a fall detection sensitivity of the $93.33 \pm 5.16\%$. During the ADL tests the system showed a specificity of $98.91 \pm 0.44\%$ in steady walking steps' recognition, $99.61 \pm 0.66\%$ in successful sudden curves detection, and $98.95 \pm 1.27\%$ of correct recognition in contractions related to TUG tests. Finally, during the balance-challenging obstacle avoidance protocol the specificity reached the $98.42 \pm 0.90\%$.

These preliminary results show promising accuracy values that, jointly with the system wearability (wireless acquisition devices), make the system potentially suitable for daily life application. Moreover, the achieved detection time (i.e., ~371 ms) is conservatively below 550 ms, which is considered as the maximum intervention limit for the implementation of countermeasures aimed at restoring the balance of the subject [17]. It ensures the system applicability to improving a fall-detection strategy.

Some drawbacks that need future larger and higher quality studies concern the acquisition devices and the generalization of the implemented method. In fact, the use of wireless sensors (EEG/EMG) theoretically ensures the system's wearability. Nevertheless, future perspectives concern the study of more comfortable solutions able to provide the same electrophysiological patterns, e.g., by using textile-based sensors arrays. The second weak point under investigation is the muscle to be selected as the master trigger to provide a quasi-deterministic delay, improving the system efficiency as a logical consequence.

Author Contributions: Conceptualization, G.M. and D.D.V.; methodology, G.M. and D.D.V.; software, G.M; validation, G.M.; formal analysis, G.M.; investigation, G.M. and D.D.V.; resources, D.D.V.; data curation G.M.; writing—original draft preparation, G.M. and D.D.V.; writing—review and editing, G.M and D.D.V.; supervision, D.D.V.; project administration, D.D.V. All authors have read and agreed to the published version of the manuscript.

Funding: This work was supported by the project AMICO (Assistenza Medicale In COntextual awareness, AMICO_Project_ARS01_00900) by National Programs (PON) of the Italian Ministry of Education, Universities and Research (MIUR): Decree n.267.

Acknowledgments: The authors would like to thank Vito Monaco, Federica Aprigliano, Silvestro Micera and the staff of Locomotion Biomechanics Lab at The BioRobotics Institute of the Scuola Superiore Sant'Anna (56025 Pontedera, Pisa, Italy), for their valuable contribution and assistance in data collection, for making available equipment for the experimental validation, and for the time/resources spent in the experimental investigation.

Conflicts of Interest: The authors declare no conflict of interest. The funders had no role in the design of the study; in the collection, analyses, or interpretation of data; in the writing of the manuscript; or in the decision to publish the results.

References

1. Pelicioni, P.H.S.; Menant, J.C.; Latt, M.D.; Lord, S.R. Falls in Parkinson's Disease Subtypes: Risk Factors, Locations and Circumstances. *Int. J. Environ. Res. Public Health* **2019**, *16*, 2216. [CrossRef] [PubMed]
2. Bloem, B.R.; Hausdorff, J.M.; Visser, J.E.; Giladi, N. Falls and freezing of gait in Parkinson's disease: A review of two interconnected, episodic phenomena. *Mov. Disord.* **2004**, *19*, 871–884. [CrossRef] [PubMed]
3. Bloem, B.R.; Grimbergen, Y.A.M.; Cramer, M.; Willemsen, M.; Zwinderman, A.H. Prospective assessment of falls in Parkinson's disease. *J. Neurol.* **2001**, *248*, 950–958. [CrossRef] [PubMed]
4. Paul, S.S.; Sherrington, C.; Canning, C.G.; Fung, V.S.C.; Close, J.C.T.; Lord, S.R. The relative contribution of physical and cognitive fall risk factors in people with Parkinson's disease: A large prospective cohort study. *Neurorehabilit. Neural Repair* **2014**, *28*, 282–290. [CrossRef] [PubMed]
5. Allen, N.E.; Schwarzel, A.K.; Canning, C.G. Recurrent falls in Parkinson's disease: A systematic review. *Parkinsons Dis.* **2013**, *2013*, 906274. [CrossRef] [PubMed]
6. Stack, E.; Ashburn, A. Fall events described by people with Parkinson's disease: implications for clinical interviewing and the research agenda. *Physiother. Res. Int.* **1999**, *4*, 190–200. [CrossRef]
7. Ivan, P.; Okubo, Y.; Sturnieks, D.; Lord, S.R.; Brodie, M.A. Detection of near falls using wearable devices: A systematic review. *J. Geriatr. Phys. Ther.* **2019**, *42*, 48–56.
8. Bianchi, F.; Redmond, S.J.; Narayanan, M.R.; Cerutti, S.; Lovell, N.H. Barometric pressure and triaxial accelerometry-based falls event detection. *IEEE Trans. Neural Syst. Rehabil. Eng.* **2010**, *18*, 619–627. [CrossRef]
9. Casilari, E.; Luque, R.; Moron, M.J. Analysis of android device-based solutions for fall detection. *Sensors* **2015**, *15*, 17827–17894. [CrossRef]
10. Srygley, J.M.; Herman, T.; Giladi, N.; Hausdorff, J.M. Self-report of missteps in older adults: A valid proxy of fall risk? *Arch. Phys. Med. Rehabil.* **2009**, *90*, 786–792. [CrossRef]
11. Aziz, O.; Park, E.J.; Mori, G.; Robinovitch, S.N. Distinguishing near-falls from daily activities with wearable accelerometers and gyroscopes using Support Vector Machines. In Proceedings of the 2012 Annual International Conference of the IEEE Engineering in Medicine and Biology Society, San Diego, CA, USA, 28 August–1 September 2012; pp. 5837–5840. [CrossRef]
12. Hu, X.; Qu, X. An individual-specific fall detection model based on the statistical process control chart. *Saf. Sci.* **2014**, *64*, 13–21. [CrossRef]
13. Martelli, D.; Artoni, F.; Monaco, V.; Sabatini, A.M.; Micera, S. Pre-impact fall detection: Optimal sensor positioning based on a machine learning paradigm. *PLoS ONE* **2014**, *9*, e92037. [CrossRef] [PubMed]
14. Lee, J.K.; Robinovitch, S.N.; Park, E.J. Inertial sensing-based pre-impact detection of falls involving near-fall scenarios. *IEEE Trans. Neural Syst. Rehabil. Eng.* **2015**, *23*, 258–266. [CrossRef] [PubMed]
15. Monaco, V.; Tropea, P.; Aprigliano, F.; Martelli, D.; Parri, A.; Cortese, M.; Molino-Lova, R.; Vitiello, N.; Micera, S. An ecologically controlled exoskeleton can improve balance recovery after slippage. *Sci. Rep.* **2017**, *7*, 46721. [CrossRef]
16. Hu, X.; Qu, X. Pre-impact fall detection. *Biomed. Eng. Online* **2016**, *15*, 61. [CrossRef]
17. Lajoie, Y.; Gallagher, S.P. Predicting falls within the elderly community: Comparison of postural sway reaction time the Berg balance scale and the activities-specific balance confidence (ABC) scale for comparing fallers and non-fallers. *Arch. Gerontol. Geriatr.* **2004**, *38*, 11–26. [CrossRef]
18. Winter, D.A. *Biomechanics and Motor Control of Human Movement*; John Wiley & Sons: Hoboken, NJ, USA, 2009; p. 370.

19. Varghese, J.P.; McIlroy, R.E.; Barnett-Cowan, M. Perturbation-evoked potentials: Significance and application in balance control research. *Neurosci. Biobehav. Rev.* **2017**, *83*, 267–280. [CrossRef]
20. Cavanagh, J.F.; Frank, M.J. Frontal theta as a mechanism for cognitive control. *Trends Cogn. Sci.* **2014**, *18*, 414–421. [CrossRef]
21. Klimesch, W.; Fellinger, R.; Freunberger, R. Alpha oscillations and early stages of visual encoding. *Front Psychol.* **2011**, *2*, 118. [CrossRef]
22. Engel, A.K.; Fries, P. Beta-band oscillations—signalling the status quo? *Curr. Opin. Neurobiol.* **2010**, *20*, 156–165. [CrossRef]
23. Neuper, C.; Pfurtscheller, G. Event-related dynamics of cortical rhythms: frequency-specific features and functional correlates. *Int. J. Psychophysiol.* **2011**, *43*, 41–58. [CrossRef]
24. De Venuto, D.; Annese, V.F.; de Tommaso, M.; Vecchio, E.; Vincentelli, A.L.S. Combining EEG and EMG Signals in a Wireless System for Preventing Fall in Neurodegenerative Diseases. In *Biosystems & Biorobotics;* Andò, B., Siciliano, P., Marletta, V., Monteriù, A., Eds.; Springer: Cham, Switzerland, 2015.
25. Berger, W.; Dietz, V.; Quintern, J. Corrective reactions to stumbling in man: Neuronal co-ordination of bilateral leg muscle activity during gait. *J. Physiol.* **1984**, *357*, 109–125. [CrossRef] [PubMed]
26. Mezzina, G.; Aprigliano, F.; Micera, S.; Monaco, V.; De Venuto, D. EEG/EMG based Architecture for the Early Detection of Slip-induced Lack of Balance. In Proceedings of the 2019 IEEE 8th International Workshop on Advances in Sensors and Interfaces (IWASI), Otranto, Italy, 13–14 June 2019; pp. 9–14. [CrossRef]
27. De Venuto, D.; Ohletz, M.J.; Ricco, B. Automatic repositioning technique for digital cell based window comparators and implementation within mixed-signal DfT schemes. In Proceedings of the Fourth International Symposium on Quality Electronic Design 2003, San Jose, CA, USA, 24–26 March 2003; pp. 431–437. [CrossRef]
28. g.Nautilus Research Headset by g.Tec. Application note. Available online: http://www.gtec.at/Products/Hardware-andAccessories/g.Nautilus-Specs-Features (accessed on 31 October 2019).
29. EMG Wave Plus by Cometa srl. Application Note. Available online: https://www.cometasystems.com/products/wave-plus-wireless-emg (accessed on 31 October 2019).
30. De Venuto, D.; Tio Castro, D.; Ponomarev, Y.; Stikvoort, E. 0.8μW 12-bit SAR ADC sensors interface for RFID applications. *Microelectron. J.* **2010**, *41*, 746–751. [CrossRef]
31. De Luca, C.J.; Gilmore, L.D.; Kuznetsov, M.; Roy, S.H. Filtering the surface EMG signal: Movement artifact and baseline noise contamination. *J. Biomech.* **2010**, *43*, 1573–1579. [CrossRef]
32. Luciani, L.B.; Genovese, V.; Monaco, V.; Odetti, L.; Cattin, E.; Micera, S. Design and Evaluation of a new mechatronic platform for assessment and prevention of fall risks. *J. Neuroeng. Rehabil.* **2012**, *9*, 51. [CrossRef]
33. De Venuto, D.; Ohletz, M.J.; Ricco, B. Testing of analogue circuits via (standard) digital gates. In Proceedings of the International Symposium on Quality Electronic Design, San Jose, CA, USA, 18–21 March 2002; pp. 112–119. [CrossRef]
34. Blum, S.; Jacobsen, N.S.J.; Bleichner, M.G.; Debener, S. A Riemannian modification of Artifact Subspace Reconstruction for EEG artifact handling. *Front. Human Neurosci.* **2019**, *13*, 141. [CrossRef]
35. De Venuto, D.; Ohletz, M.J.; Riccò, B. Digital Window Comparator DfT Scheme for Mixed-Signal ICs. *J. Electron. Test.* **2002**, *18*, 121. [CrossRef]
36. Knight, J.F.; Deen-Williams, D.; Arvanitis, T.N.; Baber, C.; Sotiriou, S.; Anastopoulou, S.; Gargalakos, M. Assessing the Wearability of Wearable Computers. In Proceedings of the 2006 10th IEEE International Symposium on Wearable Computers, Montreux, Switzerland, 11–14 October 2006; pp. 75–82. [CrossRef]

MDPI
St. Alban-Anlage 66
4052 Basel
Switzerland
Tel. +41 61 683 77 34
Fax +41 61 302 89 18
www.mdpi.com

Sensors Editorial Office
E-mail: sensors@mdpi.com
www.mdpi.com/journal/sensors

www.ingramcontent.com/pod-product-compliance
Lightning Source LLC
LaVergne TN
LVHW070150170726
843515LV00007B/1884